The Ecology of Fungal Entomopathogens

Helen Roy · Fernando Vega · Dave Chandler ·
Mark Goettel · Judith Pell · Eric Wajnberg
Editors

The Ecology of Fungal Entomopathogens

Previously published in *BioControl*, Volume 55, Issue 1, 2010

Springer

Editors
Helen E. Roy
CEH Wallingford
Biological Records Centre
Crowmarsh Gifford
Wallingford, Oxon
United Kingdom OX 10 8BB
hele@ceh.as.uk

Fernando E. Vega
USDA - ARS
Plant Sciences Institute &
Invasive Insect Biocontrol
Beltsville MD 20705
Bldg. 011A, BARC-West
USA

Dave Chandler
Warwick HRI, University of Warwick
Wellesboune
CV35 9EF Warwick
United Kingdom

Mark S. Goettel
Agriculture & Agri-Food Canada
Lethbridge AB T1J 4B1
Canada

Judith Pell
Rothamsted Research
AL5 2JQ Harpenden
United Kingdom

Eric Wajnberg
Institut National de la Recherche
Agronomique (INRA)
400 route des Chappes
06903 Sophia Antipolis CX
France
wajnberg@sophia.inra.fr

ISBN: 978-90-481-3965-1

Springer Dordrecht Heidelberg London New York

Library of Congress Control Number: 2010920281

Cover illustration: Scatophaga – Photograph by Jørgen Eilenberg

Printed on acid-free paper

Springer is part of Springer Science+Business Media (www.springer.com)

Contents

BioControl (2010) 55:1–6
DOI 10.1007/s10526-009-9244-7

Deep space and hidden depths: understanding the evolution and ecology of fungal entomopathogens

Helen E. Roy · Eoin L. Brodie · Dave Chandler · Mark S. Goettel · Judith K. Pell · Eric Wajnberg · Fernando E. Vega

Received: 22 September 2009 / Accepted: 15 October 2009 / Published online: 17 November 2009

Abstract Entomopathogens are important natural enemies of many insect and mite species and as such have been recognised as providing an important ecosystem service. Indeed, fungal entomopathogens have been widely investigated as biological control agents of pest insects in attempts to improve the sustainability of crop protection. However, even though our understanding of the ecology of fungal entomopathogens has vastly increased since the early 1800s, we still require in-depth ecological research that can expand our scientific horizons in a manner that facilitates widespread adoption of these organisms as efficient biological control agents. Fungal entomopathogens have evolved some intricate interactions with arthropods, plants and other microorganisms. The full importance and complexity of these relationships is only just becoming apparent. It is important to shift our thinking from conventional biological control, to an understanding of an as yet unknown "deep space". The use of molecular techniques and phylogenetic analyses have helped us move in this direction, and have provided important insights on fungal relationships. Nevertheless, new techniques such as the PhyloChip and pyrosequencing might help us see beyond the familiar fields, into areas that could help us forge a new understanding of the ecology of fungal entomopathogens.

H. E. Roy (✉)
NERC Centre for Ecology & Hydrology, Wallingford, Oxfordshire OX10 8BB, UK
e-mail: Hele@ceh.ac.uk

E. L. Brodie
Ecology Department, Earth Sciences Division, Lawrence Berkeley National Laboratory, Berkeley, CA 94720, USA

D. Chandler
Warwick HRI, University of Warwick, Wellesbourne, Warwick CV35 9EF, UK

M. S. Goettel
Agriculture and Agri-Food Canada, Lethbridge Research Centre, 5403-1 Avenue South, P.O. Box 3000, Lethbridge, AB T1J 4B1, Canada

J. K. Pell
Department of Plant and Invertebrate Ecology, Rothamsted Research, Harpenden, Hertfordshire AL5 2JQ, UK

E. Wajnberg
INRA, 400 Route des Chappes, BP 167, 06903 Sophia Antipolis Cedex, France

F. E. Vega
Sustainable Perennial Crops Laboratory, United States Department of Agriculture, Agricultural Research Service, Building 001, BARC-West Beltsville, MD 20705, USA

Keywords Ecology · Evolution · Entomopathogenic fungi · Biological control · Tri-trophic interactions · Modelling · Rhizosphere · Endophytes · Behavioural ecology · Molecular tools

Fungi have a profound impact on global ecosystems. They modify our habitats and are essential for many ecosystem functions
Blackwell et al. (2006).

It has been estimated that the Kingdom Fungi consists of 1.5 million species (Hawksworth 2001; Mueller and Schmit 2007; Schmit and Mueller 2007), with approximately 110,000 described species (Kirk et al. 2008). Of these, 700 species in 90 genera are recognized as insect pathogens (Roberts and Humber 1981), and approximately 170 pest control products have been developed based on at least 12 species of fungal entomopathogens (de Faria and Wraight 2007). Undoubtedly, fungal entomopathogens are important natural enemies of many insect and mite species and as such, provide an important ecosystem service contributing to pest control with minimal detectable negative effects on the environment (Vestergaard et al. 2003). However, the small subset of fungi developed as biological control agents have had limited success. Our ability to employ them effectively and reliably for pest control in the field has not matched up to expectations (Vestergaard et al. 2003; Chandler et al. 2008; Vega et al. 2009). In part, this may be because of variable and unpredictable levels of efficacy compared to chemical pesticides (Waage 1997; Vega et al. 2009), but we also lack some basic understanding of their ecology and evolution (Vega et al. 2009).

The importance of basic knowledge, theory and predictive ability in the use of biological control agents has been recognised for some time (Gurr et al. 2000). However, the dearth of basic information on fungal entomopathogens is pronounced even though these organisms have historically dominated the field of microbial control (Lord 2005). Vega et al. (2009) have proposed the need for "a new paradigm for fungal entomopathogens that should refocus our efforts and hopefully lead to exciting new findings." In this special issue of BioControl we report on some of the latest research, innovations and ideas relating to fungal entomopathogens within an ecological context.

One of the most significant challenges facing insect pathologists is to understand the evolutionary history and relationships amongst fungal entomopathogens. Intricate interactions with arthropods, plants and other microorganisms are evident, but the full importance and complexity of these relationships is just becoming apparent. The advent of new molecular tools over the last few decades has dramatically improved the resolution of fungal systematics and there have been huge advances in this field (Blackwell et al. 2006; Hibbett et al. 2007; Humber 2008; Blackwell 2009; Enkerli and Widmer 2009). The acquisition of a phylogeny enables us to examine evolutionary relationships and better understand and predict ecological interactions (Blackwell 2009). Molecular tools will provide methods for examining the host-pathogen dynamics in complex environments (Enkerli and Widmer 2009). Enkerli and Widmer (2009) comprehensively review the tools available within the context of population ecology studies.

Meyling and Hajek (2009) provide an excellent background to ecological interactions relevant to fungal entomopathogens from a community ecology perspective. An ecological context is important for increasing our empirical understanding of host-parasite interactions and improving the efficacy of these microbes as biological control agents. Fungal entomopathogens often exist as patches in a spatially heterogeneous matrix (Rodríguez and Torres-Sorando 2001) and metapopulation dynamics could be particularly pertinent to describing these spatially distinct populations that are connected by dispersal. Meyling and Hajek (2009) describe how insects and their fungal pathogens could be used as model species for exploring metapopulation theory using experimental and predictive models.

In recent years there have been intriguing advances in our appreciation of the role of fungal entomopathogens beyond their applied role as biological control agents of insects. Pathogens have traditionally been neglected in life history studies and often considered as having negligible impact (Hawkins et al. 1997). Hesketh et al. (2009) review the role of fungal entomopathogens as natural enemies of insects in semi-natural habitats, describing the theoretical host-pathogen models available to examine their role in population regulation. The need to consider the complexity, and particularly the heterogeneity, of semi-natural habitats within the context of theoretical models and as a framework for empirical studies is highlighted. However, Hesketh et al. (2009) acknowledge that fundamental gaps in our understanding of fungal entomopathogens from an ecological perspective, coupled with a lack of empirical data to test theoretical predictions, is impeding progress.

Ecological understanding has never been more vital than in this period of unprecedented environmental change:

> Changes in biodiversity due to human activities were more rapid in the past 50 years than at any time in human history, and the drivers of change that cause biodiversity loss and lead to changes in ecosystem services are either steady, show no evidence of declining over time, or are increasing in intensity (Millennium Ecosystem Assessment 2005).

Many studies on the effects of the major drivers of biodiversity loss (habitat destruction, invasive species, exploitation, climate change and pollution) involve just one trophic level and often just one species. Fungal entomopathogens provide an additional trophic level that should be included in such studies, particularly in relation to climate change and habitat destruction (Roy and Cottrell 2008; Roy et al. 2009). Cory and Ericsson (2009) review the literature on tri-trophic interactions encompassing fungal entomopathogens. The promising roles of plant volatiles and plant surface chemistry on ecological interactions between host insects and their pathogenic fungi are described. Although intriguing concepts such as the "bodyguard hypothesis" have been examined and demonstrated for natural enemies such as parasitoids and predators, there is a lack of empirical evidence for this in fungal entomopathogens. This is likely because there has simply been limited research in this field. Cory and Ericsson (2009) assess the relevance of plant-mediated effects on fungal entomopathogens and urge researchers to focus work on the considerable gaps in knowledge concerning fungal entomopathogens and tritrophic interactions.

Behavioural ecology will be critical in the exploration of tritrophic interactions. Baverstock et al. (2009) provide a review of fungal entomopathogens and insect behaviour. The behavioural response of an insect to a fungal pathogen will not only have a direct effect on the efficacy of the fungus as a biological control agent but also provide us with a model system for understanding interactions within guilds. Simple laboratory bioassays can provide a measure of insect mortality in the presence of a pathogen but experiments designed to include elements of spatial complexity are critical to improving accuracy of predictions. The papers reviewed by Baverstock et al. (2009) demonstrate this and reveal manipulations of host behaviour induced by fungi and countermeasures employed by the host (Roy et al. 2006). The often complex interactions between fungus and host are being unravelled through eloquent research and the importance of these often subtle behavioural modifications in determining the success or failure of biological control cannot be underplayed.

The opportunities and challenges provided by the soil environment, and specifically the rhizosphere, have long been recognised (Vega et al. 2009) but only now are the subtleties slowly being revealed (Bruck 2009). There is no doubt that the ecology of fungal entomopathogens in the rhizosphere is a neglected area of research within insect pathology. A better understanding of their ecology in the rhizosphere will not only help in the development of successful microbial control strategies against root-feeding insect pests, but is also certain to reveal intriguing insights into the subterranean "hidden depths" of fungal entomopathogens.

Ownley et al. (2009) review the ecology and evolution of fungal entomopathogens as antagonists of plant pathogens. Simultaneous biological control of both insect pests and plant pathogens has been reported for the hypocrealean fungal entomopathogens, *Beauveria bassiana* and *Lecanicillium* spp. and accumulating evidence shows that *Beauveria* spp. can colonize a wide array of plant species endophytically. Furthermore, traits that are important for insect pathogenicity are also involved in pathogenicity to phytopathogens.

From 1845 to 1916, Elie Metchnikoff assessed an insect disease of wheatchafers later identified as the hypocrealean fungus *Metarhizium anisopliae* (Lord 2005). These early studies inspired many to focus their research on assessing the potential of fungal entomopathogens as microbial control agents. A series of papers in this special issue of BioControl explore advances in their use for biological control of pest insects. Jackson et al. (2009) eloquently describe the importance of linking ecology with formulation and production of fungal entomopathogens for biological control. The commercial drivers of formulation (maximising yield, storage stability and ease of application) are often in conflict with ecological considerations. However, efficacy can be improved dramatically by considering ecological factors such as the importance of environmental conditions on the host-pathogen interaction (Jackson et al. 2009).

Biological control strategies include classical, inundative augmentation and conservation approaches. Hajek and Delalibera (2009) examine the use of fungal entomopathogens in classical biological control and conclude that they have been used more frequently than other types of pathogens and provide a sustainable avenue for controlling arthropod pests, especially the increasing numbers of invasive species. Inundative biological control strategies rely on the released organism exerting control without subsequent transmission and reproduction in a similar way to a synthetic pesticide; the chemical paradigm. Jaronski (2009) aptly demonstrates the drawbacks of taking this approach in isolation with fungal entomopathogens. In most cases, effective application of sufficient inoculum to rapidly reduce pest numbers to below economic threshold levels is financially and logistically prohibitive. Biotic, abiotic and economic realities certainly restrict such an approach in most field situations although there have been some notable successes in controlling pest insects in glasshouses. Through a better understanding of the ecology of fungal entomopathogens and the dynamics of the pest, crop and environment, it may be possible to employ inundative application of fungi within ecologically based integrated pest management systems. However, it will be imperative that such strategies encompass the complex and multifaceted interactions that the released organism must contend with. The review on conservation biological control by Pell et al. (2009) explores the novel ways in which fungal entomopathogens can be enhanced in the environment. Understanding the dynamics of fungal entomopathogens at the field and landscape scale is imperative for implementing conservation biological strategies. There have been a number of eloquent studies demonstrating the potential of such an approach and these are comprehensively reviewed by Pell et al. (2009).

The realm of ecology is vast and deciphering insect-fungal pathogen interactions within an ecological context will take us on voyages beyond our imagination. New and innovative methods will provide the inspiration to explore the hidden depths and deep space of these interactions. The PhyloChip microarray hybridization technique might point at what the future holds for mycological research. At present, the PhyloChip allows for the identification of bacterial and archaeal organisms using 16S rRNA-targeted oligonucleotide microarrays (Brodie et al. 2007; DeSantis et al. 2007). The method takes advantage of the variation in the 16S rRNA gene to capture the broad range of microbial diversity that may be present in a given sample, without the need for microbial cultivation. This high-throughput technique makes it possible to identify overall microbial diversity, and combined with dissection of specific insect tissues (e.g., foregut, midgut, hindgut), determine microbial communities in these tissues. A version is currently being developed for the analysis of fungal community diversity. Similarly, sequencing technologies such as 454-pyrosequencing now permit large numbers of shorter sequences (pyrotags) to be obtained from a large number of samples by employing sequence barcoding techniques (Hamady et al. 2008). These approaches allow deeper profiling of complex microbial communities from the deep-sea (Sogin et al. 2006) to the gut microbiota of humans and 59 other mammals (Ley et al. 2008). Greif and Currah (2007) have shown that fungal entomopathogens are common components of the surface mycota of arthropods, and that they are not necessarily restricted to diseased insects. Once a microarray technique similar to the PhyloChip or pyrotag sequencing has been developed for fungal entomopathogens, what would their uses reveal in insects? Will fungal entomopathogens be found to be common inhabitants of the cuticle of uninfected insects? Could they also be common internal inhabitants of uninfected insects? Furthermore, using microarray techniques for sampling fungal entomopathogens as plant endophytes might reveal that they are much more common and globally distributed than is presently thought. Would the same situation occur in the rhizosphere? If the answer to any of these questions were positive, what would this imply for our understanding of fungal entomopathogens?

There might be a "deep space" that will only be revealed when we start to decipher the myriad fungal inhabitants in insects and plants, which at present remain in "hidden depths". The importance of these interactions has been superbly described by Berenbaum and Eisner (2008):

> There is no limit to what remains to be discovered in that interactive zone between macroorganism and microbe, where so many biological mutualisms and antagonisms play out. Microbes blanket the planet, and in their infinite variety they must be involved in infinite interactions. Deciphering these could lead to a vast increase in ecological knowledge, as well as to the isolation of natural products of unforeseen function.

Let the adventure begin!

Acknowledgments The Rockefeller Foundation Bellagio Study and Conference Center in Italy hosted the June 2008 meeting *Entomopathogenic fungi in sustainable agriculture: use against insects and beyond* (organised by F. E. Vega and M. S. Goettel). This meeting was the inspiration for this special edition and we express our sincere gratitude to the staff at The Rockefeller Foundation and at the Bellagio Study and Conference Centre. HER is supported by the Natural Environment Research Council. JKP is supported by the Department for Environment, Food and Rural Affairs of the United Kingdom (Defra) and the Biotechnology and Biological Sciences Research Council of the United Kingdom (BBSRC). Rothamsted Research is an Institute of BBSRC.

References

Baverstock J, Roy H, Pell JK (2009) Entomopathogenic fungi and insect behaviour: from unsuspecting hosts to targeted vectors. BioControl. doi: 10.1007/s10526-009-9238-5 (this SI)

Berenbaum M, Eisner T (2008) Bugs' bugs. Science 322:52–53

Blackwell M (2009) Fungal evolution and taxonomy. BioControl. doi:10.1007/s10526-009-9243-8 (this SI)

Blackwell M, Hibbett DS, Taylor JW, Spatafora JW (2006) Research coordination networks: a phylogeny of the kingdom Fungi (Deep Hypha). Mycologia 98:829–837

Brodie EL, DeSantis TZ, Moberg Parker JP, Zubietta IX, Piceno YM, Andersen GL (2007) Urban aerosols harbor diverse and dynamic bacterial populations. Proc Natl Acad Sci USA 104:299–304

Bruck D (2009) Fungal entomopathogens in the rhizosphere. BioControl. doi:10.1007/s10526-009-9236-7 (this SI)

Chandler D, Davidson G, Grant WP, Greaves J, Tatchell GM (2008) Microbial biopesticides for integrated crop management: an assessment of environmental and regulatory sustainability. Trends in Food Science & Technology 19:275–283

Cory JS, Ericsson JD (2009) Fungal entomopathogens in a tritrophic context. BioControl. doi:10.1007/s10526-009-9247-4 (this SI)

de Faria MR, Wraight SP (2007) Mycoinsecticides and mycoacaricides: a comprehensive list with worldwide coverage and international classification of formulation types. Biol Control 43:237–256

DeSantis TZ, Brodie EL, Moberg JP, Zubieta IX, Piceno YM, Andersen GL (2007) High-density universal 16S rRNA microarray analysis reveals broader diversity than typical clone library when sampling the environment. Microb Ecol 53:371–383

Enkerli J, Widmer F (2009) Molecular ecology of fungal entomopathogens: the molecular genetic tools and their application. BioControl. doi:10.1007/s10526-009-9251-8 (this SI)

Greif MD, Currah RS (2007) Patterns in the occurrence of saprophytic fungi carried by arthropods caught in traps baited with rotted wood and dung. Mycologia 99:7–19

Gurr GM, Barlow ND, Memmott J, Wratten SD, Greathead DJ (2000) A history of methodological, theoretical and empirical approaches to biological control. In: Gurr G, Wratten SD (eds) Biological control: measures of success. Kluwer Academic Publishers, Dordrecht, The Netherlands, pp 3–37

Hajek AE, Delalibera I (2009) Fungal pathogens as classical biological control agents against invasive arthropods. BioControl. doi:10.1007/s10526-009-9253-6 (this SI)

Hamady M, Walker JJ, Harris JK, Gold NJ, Knight R (2008) Error-correcting barcoded primers for pyrosequencing hundreds of samples in multiplex. Nat Methods 5: 235–237

Hawkins BA, Cornell HV, Hochberg ME (1997) Predators, parasitoids and pathogens as mortality agents in phytophagous insect populations. Ecology 78:2145–2152

Hawksworth DL (2001) The magnitude of fungal diversity: the 1.5 million species estimate revisited. Mycol Res 105:1422–1432

Hesketh H, Roy HE, Eilenberg J, Pell JK, Hails RS (2009) Modelling complexity: entomopathogenic fungi in semi-natural populations of insects. BioControl. doi: 10.1007/s10526-009-9249-2 (this SI)

Hibbett DS, Binder M, Bischoff JF, Blackwell M, Cannon PF, Eriksson OE, Huhndorf S, James T, Kirk PM, Lücking R, Thorsten Lumbsch H, Lutzoni F, Brandon Matheny P, McLaughlin DJ, Powell MJ, Redhead S, Schoch CL, Spatafora JW, Stalpers JA, Vilgalys R, Aime MC, Aptroot A, Bauer R, Begerow D, Benny GL, Castlebury LA, Crous PW, Dai YC, Gams W, Geiser DM, Griffith GW, Gueidan C, Hawksworth DL, Hestmark G, Hosaka K, Humber RA, Hyde KD, Ironside JE, Kõljalg U, Kurtzman CP, Larsson K-H, Lichtwardt R, Longcore J, Miądlikowska J, Miller A, Moncalvo JM, Mozley-Standridge S, Oberwinkler F, Parmasto E, Reeb V, Rogers JD, Roux C, Ryvarden L, Sampaio JP, Schüßler A, Sugiyama J, Thorn RG, Tibell L, Untereiner WA, Walker C, Wang Z, Weir A, Weiss M, White MM, Winka K, Yao YJ, Zhang N (2007) A higher-level phylogenetic classification of the Fungi. Mycol Res 111:509–547

Humber RA (2008) Evolution of entomopathogenicity in fungi. J Invertebr Pathol 98:262–266

Jackson M, Dunlap CA, Jaronski S (2009) Ecological considerations in producing and formulating fungal entomopathogens for use in insect biocontrol. BioControl. doi:10.1007/s10526-009-9240-y (this SI)

Jaronski S (2009) Ecological factors in the inundative use of entomopathogenic fungi. BioControl. doi:10.1007/s10526-009-9248-3 (this SI)
Kirk PM, Cannon PF, Minter DW, Stalpers JA (2008) Dictionary of the fungi, 10th edn. CAB International, Wallingford, UK
Ley RE, Hamady M, Lozupone C, Turnbaugh PJ, Ramey RR, Bircher JS, Schlegel ML, Tucker TA, Schrenzel MD, Knight R, Gordon JI (2008) Evolution of mammals and their gut microbes. Science 320:1647–1651
Lord J (2005) From Metchnikoff to Monsanto and beyond: the path of microbial control. J Invertebr Pathol 89:19–29
Meyling NV, Hajek AE (2009) Principles from community and metapopulation ecology and their application to fungal entomopathogens. BioControl. doi:10.1007/s10526-009-9246-5 (this SI)
Millennium Ecosystem Assessment (2005) Ecosystems and human well-being: biodiversity synthesis. World Resources Institute, Washington, DC
Mueller GM, Schmit JP (2007) Fungal biodiversity: what do we know? What can we predict? Biodivers Conserv 16:1–5
Ownley B, Gwinn KD, Vega FE (2009) Endophytic fungal entomopathogens with activity against plant pathogens: ecology and evolution. BioControl. doi:10.1007/s10526-009-9241-x (this SI)
Pell JK, Hannam J, Steinkraus DS (2009) Conservation biological control using entomopathogenic fungi. BioControl. doi:10.1007/s10526-009-9245-6 (this SI)
Roberts DW, Humber RA (1981) Entomogenous fungi. In: Cole GT, Kendrick B (eds) Biology of conidial fungi. Academic Press, New York, pp 201–236
Rodríguez DJ, Torres-Sorando L (2001) Models of infectious diseases in spatially heterogeneous environments. Bull Math Biol 63:547–571
Roy HE, Cottrell T (2008) Forgotten natural enemies: interactions between coccinellids and insect-parasitic fungi. European Journal of Entomology 105:391–398
Roy HE, Steinkraus D, Eilenberg E, Pell JK, Hajek A (2006) Bizarre interactions and endgames: entomopathogenic fungi and their arthropod hosts. Ann Rev Entomol 51:331–357
Roy HE, Hails RS, Hesketh H, Roy DB, Pell JK (2009) Beyond biological control: non-pest insects and their pathogens in a changing world. Insect Conservation and Biodiversity 2:65–72
Schmit JP, Mueller GM (2007) An estimate of the lower limit of global fungal diversity. Biodivers Conserv 16:99–111
Sogin ML, Morrison HG, Huber JA, Mark Welch D, Huse SM, Neal PR, Arrieta JM, Herndl GJ (2006) Microbial diversity in the deep sea and the underexplored "rare biosphere". Proc Natl Acad Sci USA 103:12115–12120
Vega FE, Goettel MS, Blackwell M, Chandler D, Jackson MA, Keller S, Koike M, Maniania NK, Monzón A, Ownley BH, Pell JK, Rangel DEN, Roy HE (2009) Fungal entomopathogens: new insights on their ecology. Fungal Ecol 2:149–159
Vestergaard S, Cherry A, Keller S, Goettel M (2003) Safety of hyphomycete fungi as microbial control agents. In: Hokkanen HMT, Hajek AE (eds) Environmental impacts of microbial insecticides. Kluwer Academic Publishers, Dordrecht, pp 35–62
Waage JK (1997) Biopesticides at the crossroads: IPM products or chemical clones? In Evans HF (Chair) Microbial insecticides: novelty or necessity? British Crop Protection Council Symposium Proceedings No. 68, Farnham, UK, pp 11–20

Author Biographies

Helen E. Roy leads zoological research in the Biological Records Centre at the NERC Centre for Ecology & Hydrology (UK). The focus of her research is insect community interactions with particular emphasis on the effects of environmental change. She is an associate editor of BioControl.

Eoin L. Brodie develops and applies culture independent approaches to analyze microbial communities and conducts research on climate change impacts on their structure and function.

Dave Chandler is an insect pathologist at the University of Warwick, UK. He has studied entomopathogenic fungi for just over 20 years. He has particular interests in entomopathogenic fungi as biocontrol agents of horticultural crops, fungal physiology and ecology, and the pathogens of honeybees.

Mark S. Goettel is an insect pathologist at the Lethbridge Research Centre of Agriculture & Agri-Food Canada, specializing in the development of entomopathogenic fungi as microbial control agents of insects. In addition to this research, he has been extensively involved in the review and revision of the regulations for registration of microbial control agents and has addressed regulatory and safety issues at the international level. He is currently President of the Society for Invertebrate Pathology and has been Editor-in-Chief of Biocontrol Science & Technology since 2000.

Judith K. Pell heads the Insect Pathology Group in the Department for Plant and Invertebrate Ecology at Rothamsted Research, UK. She leads research on the ecology of entomopathogenic fungi, to elucidate their role in population regulation and community structure and to inform biological control strategies. Specifically: intraguild interactions; the relationships between guild diversity, habitat diversity and ecosystem function; pathogen-induced host behavioural change.

Eric Wajnberg is a population biologist specialised in behavioural ecology, statistical modelling and population genetics. He is also an expert in biological control, with more than 20 years experience of working with insect parasitoids. He has been the Editor in Chief of BioControl since 2006.

Fernando E. Vega is an entomologist with the United States Department of Agriculture, Agricultural Research Service, in Beltsville, Maryland, USA. He conducts research on biological methods to control the coffee berry borer, the most important insect pest of coffee throughout the world. He is co-editor, with Meredith Blackwell, of *Insect-Fungal Associations: Ecology and Evolution*, published by Oxford University Press in 2005, and serves as an Editorial Board Member for *Fungal Ecology*.

BioControl (2010) 55:7–16
DOI 10.1007/s10526-009-9243-8

Fungal evolution and taxonomy

Meredith Blackwell

Received: 30 September 2009 / Accepted: 15 October 2009 / Published online: 5 November 2009

Abstract Fungi and insects are closely associated in many terrestrial and some aquatic habitats. In addition to the pathogenic associations, many more interactions involve fungal spore dispersal. Recent advances in the study of insect-associated fungi have come from phylogenic analyses with increased taxon sampling and additional DNA loci. In addition to providing stable phylogenies, some molecular studies have begun to unravel problems of dating of evolutionary events, convergent evolution and host switching. These studies also enlighten our understanding of fungal ecology and the development of organismal interactions. Mycologists continue to rely heavily, however, on identified specimens based on morphology to incorporate more of the estimated 1.5 million species of fungi in phylogenetic studies.

Keywords Insect fungi · Fungal phylogeny · Hypocreales

Introduction

In the eighteenth century Otto von Munchhausen, a contemporary of Linnaeus, determined the nature of fungi: his observations lead him to believe that fungi were the dwellings of animals (Findlay 1982). Mycologists now understand the organismal nature of an estimated 1.5 million species of fungi, and they also have learned much more about the associations between fungi and animals, especially insects, occurring in habitats they share. We have confirmed that some fungi are indeed the houses and sustenance of animals. We also have found that fungi sometimes live within insects and other arthropods. Insects also are essential to carry fungi from depleted substrate to a new home (Figs. 1, 2).

There are many interactions between fungi and insects ranging from transient to obligate associations, some of which kill insects, but a large number that benefit either the insect or the fungus or in which the benefit is reciprocal. Among basidiomycetes there are classic examples of farming interactions in which Old World termites cultivate a monophyletic group of fungi and New World leaf-cutting ants cultivate two distinct cultivar groups (Currie et al. 2003; Munkacsi et al. 2004; Little and Currie 2008). Other basidiomycetes (e.g., species of *Septobasidium*) parasitize scale insects, although most of the scales in the colony are protected from insect parasites within chambers of the fungal thallus (Henk and Vilgalys 2007). Many insects are adapted for living their entire lives within the fruiting bodies of basidiomycetes, where they ingest the tissue and reproduce, leaving only to find a fresh fungus when the old substrate is depleted. In addition various basidiomycetes are adapted for dispersal by insects.

Handling Editor: Helen Roy.

M. Blackwell (✉)
Department of Biological Sciences, Louisiana State University, Baton Rouge, LA 70803, USA
e-mail: mblackwell@lsu.edu

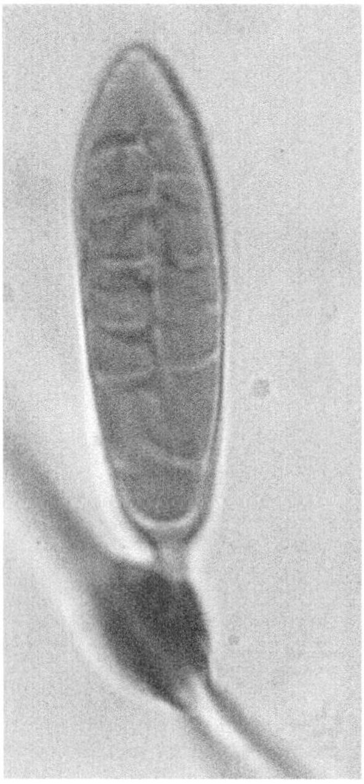

Fig. 1 The capilliconidium of *Basidiobolus ranarum*, once placed in an unclassified genus, *Amphoromorpha*, is attached to an insect seta by a darkened attachment region. Although *B. ranarum* also has more obvious forcibly discharged spores, insect dispersal is important in the life cycle of the fungus, as evidenced by the development of the fungus from the excrement of many insect-eating amphibians. Stained with glycerol acid fuchsin. Source: Weir and Blackwell (2005)

Fig. 2 Stinkhorns such as *Mutinus* sp. are adapted for insect dispersal. Their fetid odors attract flies and other insects to the slimy slurry of spores. Many spores adhere to the insect body and later are deposited in habitats favorable for germination. Photo: Nhu Nguyen

Fig. 3 Insects may be hosts for small poorly known fungi such as *Muiogone medusae*, probably an asexual ascomycete. Source: Weir and Blackwell (2005)

Ascomycete associations with arthropods include numerous dispersal and fertilization interactions. There also are farming interactions between bark beetles and ascomycetes in several orders. Some ascomycetes parasitize insects and other arthropods (Figs. 3, 4), and some of the virulent pathogens are discussed throughout this special issue. Ascomycetes, especially yeasts, are found in the insect gut, sometimes as endosymbionts in special compartments, and the yeasts may detoxify plant materials or provide enzymes to attack plant cell walls that are intractable to the insects (Vega and Dowd 2005). This is a powerful interaction that allows insects to move into habitats they otherwise could not utilize at so little genetic expense.

In the past, mycologists relied heavily on morphological characters to suggest relationships among fungi, although in the case of insect-associated fungi, morphology often has been the result of convergent evolution. The use of DNA characters has helped to improve and stabilize our understanding of fungal relationships and to develop a phylogenetic classification to the level of order (Hibbett et al. 2007). This work has progressed from phylogenetic trees based on analyses of a partial gene to multiple genes to entire genomes. The new classification (Hibbett et al. 2007) is anticipated to remain stable because it is based on a multiple locus phylogeny and increased taxon sampling rather than subjective opinions based on few, often plastic, morphological characters. Figure 5 provides a skeletal outline of major taxonomic groups in the new classification. This is an ongoing process, and new lineages and taxa have already been added to the classification because fungi

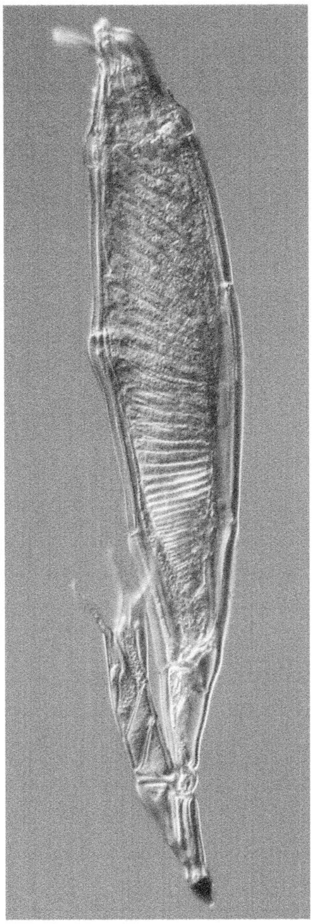

Fig. 4 *Prolixandromyces triandrus* (Laboulbeniales) is one member of a speciose group of ascomycetes that are obligate biotrophic ectoparasites of arthropods. The lack of a germ tube and determinate growth of the thallus are derived characters that set the group apart from all other fungi. One species originally was described as a parasitic worm of nycteribid bat flies. Photo: Alex Weir

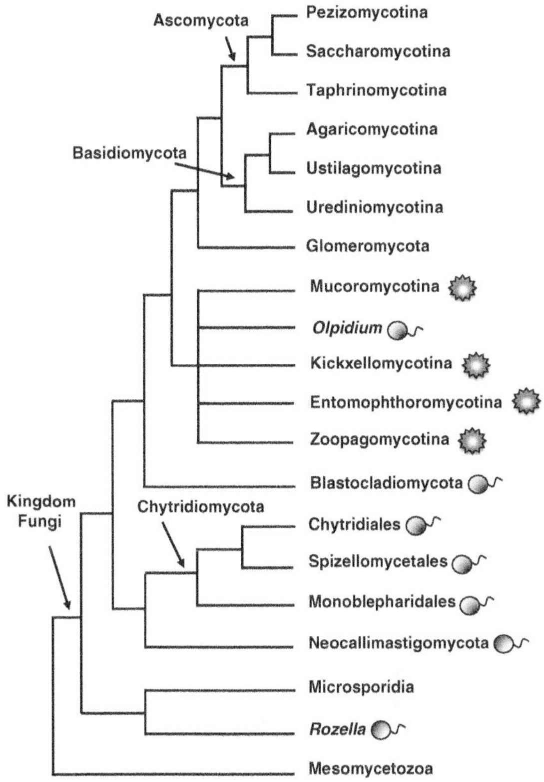

Fig. 5 The tree presents the major groups of fungi in the new classification (Hibbett et al. 2007; White et al. 2006). The best-known fungal parasites of insects are classified in Pezizomycotina, sometimes called filamentous ascomycetes (see Fig. 6 and text for more detail). Note the greater diversity of zoosporic (indicated by flagellate cell diagram) and zygosporic (indicated by zygospore diagram) fungal groups than previously appreciated (e.g., Alexopoulos et al. 1996). The flagellum appears to have been lost on more than one occasion

remain under-sampled (Blackwell et al. 2006; Hibbett et al. 2007). For example, recently discovered fungal diversity includes a relatively large ascomycete lineage, Soil Clone Group 1, that has been identified only from environmental samples. This clade is known from widely spaced localities including several sites in northern Europe and North America (Schadt et al. 2003; Vandenkoornhuyse et al. 2002). It is now possible to detect fungi we have never seen as well as those that are not culturable.

Molecular techniques have revolutionized the study of insect-associated fungi. For example, asexual morphs of fungi, many of which are insect pathogens, previously were placed as form genera in groups such as deuteromycota or fungi imperfecti, based solely on artificial morphological characteristics. We now can place the asexual fungi among their nearest sexual relatives, and terms including deuteromycota have been abandoned completely (Blackwell et al. 2006; Hibbett et al. 2007). Just as exciting is the identification of some insect fungi that only recently have been determined for the first time (e.g., attine ant associates and Laboulbeniales). Identification of the species involved in the attine associations, in particular, has renewed interest in the system and has allowed for population studies, determination of the approximate age of the interactions, and new evolutionary perspectives. The new phylogenetic information on many fungi has revolutionized our understanding of the associations between fungi and other organisms. Now we not only recognize individual fungi in all their guises, but we can track common evolutionary histories of all the participants in the associations.

Past fungal-insect associations

In addition to fungal fossil evidence, molecular clock divergence rates provide estimates of the timing of historical interactions between insects and fungi. Because the fungal fossil record is scanty, the DNA estimates will give earlier dates than fossils, and these dates are dependent on the calibration points used in estimating ages of certain fungal lineages. Based on molecules, fungi are estimated to have at least a billion year history on earth (Matheny et al. 2009; Taylor and Berbee 2006). Fossils provide evidence only of more recent fungal activities because fungi are small and often ephemeral. It is possible, however, to discover fossil fungi, especially when the hosts and specialized substrates are targeted, and these serve as the essential reference points for calibration.

There is evidence that well-developed fungal communities were present in the Devonian (416–359 million years ago (Ma)) with assemblages comprised of several phyla already associated with vascular plants in the Rhynie Chert (400 Ma) of Scotland. The Rhynie fungi have been recognized as chytrids, Glomeromycota and Ascomycota. Fungal diversity grew rapidly with the increase in terrestrial environments. The rise of insects began somewhat later than that of fungi with the appearance of the first wingless insects in the Devonian. The Carboniferous Period (359–299 Ma) witnessed the diversification of plants and winged insects as well as fungi, and these primarily flourished on the land. The first mushroom fossil was relatively late in the fossil record, appearing only about 90 Ma in the Cretaceous (145.5–65.5 Ma). Winged insect fossils were found in Devonian deposits, and some groups such as cockroaches, associated with certain fungi today, were present in late Paleozoic (before 251 Ma) and into the early Mesozoic (after 251 Ma), when fossils of most modern insect orders appeared. Some of these insects, including roaches, termites, dung beetles and wood wasps are closely associated with fungi today (Blackwell 2000).

There is little early fossil evidence of fungus-arthropod associations, but several such associations have been reported from the Carboniferous. White and Taylor (1989) reported an eccrinid trichomycete (eccrinids are now considered to be members of the Mesomycetozoa rather than fungi) in Late Carboniferous associated with what was assumed to be the gut lining of an arthropod. Trace fossils of millipedes, common hosts of eccrinids today, were first reported in Late Ordovician (488–444 Ma). This report, however, has not been confirmed. Many ephemeral fungi and associated insects have been preserved in amber. A coral fungus with a sand fly larva (Diptera: Psychodidae) preserved in early Cretaceous amber (ca. 100 Ma) suggests that the flies may have been feeding on the fungus (Poinar et al. 2006). Cretaceous amber also yielded an *Ophiocordyceps*-like anamorph associated with scale insects, corresponding well with a hypothesized Jurassic origin of *Cordyceps*-like fungi based on DNA divergence rates (Sung et al. 2008; Nikoh and Fukatsu 2000). Other insect-associated fungi, including a laboulbeniomycete on a dipteran, have been discovered by targeting amber-preserved insects that are known hosts of extant fungi (Rossi et al. 2005). The laboulbeniomycete is from Baltic amber (55–35 Ma) that was later redeposited (22 Ma). Younger Dominican Republic amber has yielded a number of insect pathogens such as Entomophthorales on a termite, *Beauveria* on a worker ant and a *Mucor*-like fungus on a walking stick (Poinar 1992).

In cases where there are good calibration points as in fossils of fungus-eating insects, phylogenetic studies can provide information on the history of insect-fungus associations. For example, many extant beetles are closely associated with fungi for habitats and nutritional resources. A molecular study aimed at discerning the history of lifestyle preferences of Coleoptera suggested that beetles evolved about 285 Ma (Hunt et al. 2007) with at least 15 independent origins of fungus feeding (e.g., certain clades of Staphylinidae, Scarabaeoidea, Leiodidae). A major shift to fungal feeding by speciose cucujoid beetles is estimated to have occurred about 236 Ma and represents a relatively derived condition with some reversals (e.g., Chrysomeloidea, Curculionoidea; Hunt et al. 2007). One might suspect that fossil mushrooms would have damage from browsing insects, but this is not the case for the few early mushroom fossils known (D. Hibbett, personal communication 2009).

Phylogeny and phylogenetic classification

The simple morphology and heterotrophic nutrition of fungi was used for many years as the basic criteria for their identification. These traits, however, did not distinguish fungi from other groups of organisms with a similar ecology. It has taken several hundred years to refine a classification of fungi by searching out characters from life cycles, biochemical pathways and ultrastructural anatomy. Characters such as site of meiosis in the life cycle, flagellation, cell wall carbohydrate, mitochondrial structure and pathway of lysine synthesis were used to define a monophyletic kingdom Fungi (Alexopoulos et al. 1996). Although these characters served to separate fungi from unrelated organisms [e.g., slime molds (Myxomycetes) and water molds (Oomycota)], these "all or none" characters did not allow mycologists to group the organisms based on their similarities. Eventually, the use of rDNA overcame this deficit and brought the advantage of large numbers of characters that could be analyzed by phylogenetic methods to answer questions about evolutionary pathways. The non-photosynthetic Oomycetes, such as *Lagenidium giganteum*, important in attempts to control aquatic insect larvae (Kerwin and Petersen 1997), are now grouped as straminipiles with brown algae and other photosynthetic protists that contain chlorophylls *a* and *c*. Other organisms previously considered to be fungi also have been excluded from the kingdom on the basis of DNA analyses, and these include two of the groups of arthropod-associated "trichomycetes" in the orders Amoebidiales and Eccrinales that now are placed in the Mesomycetozoa, a group basal to fungi.

Improved molecular techniques and analysis methods and a dynamic community of mycologists came together in an effort to improve taxon sampling and acquire trees based on multiple alleles [see articles in Mycologia 98(6)]. The phylogenetic studies were the basis of a phylogenetic classification (Fig. 6) that has been widely accepted and to which more taxa are being added (Hibbett et al. 2007). Many mycologists worked to achieve the partially resolved tree upon which a phylogenetic classification to the level of order could be established. The tree is based on the best data available, often multiple DNA loci [see Mycologia 98(6) and http://aftol.org/] and sometimes whole genomes (Robbertse et al. 2006). Work on structural and biochemical characters continues so

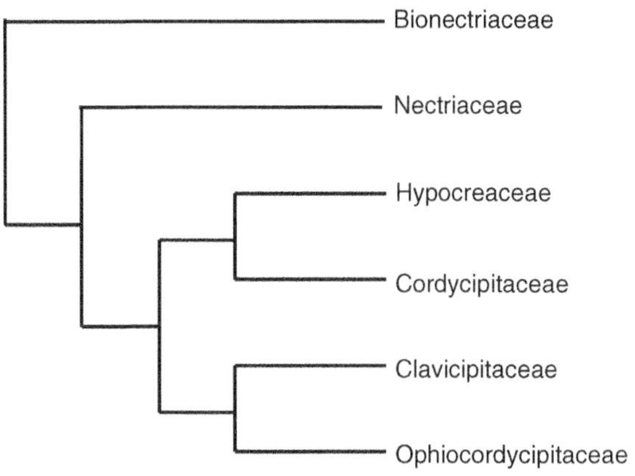

Fig. 6 Tree showing the relationships of families classified in Hypocreales (after Sung et al. 2007; http://cordyceps.us/). The order contains many insect pathogens and has been the subject of studies of interkingdom host switching (Spatafora et al. 2007)

that these characters can be integrated into other phylogenetic information (http://aftol.org/, especially http://aftol.umn.edu/). A number of characters are known to be useful at certain taxonomic levels, and include flagellar apparatus in flagellated fungi, septal pore plugs of Agaricomycotina, and type of membrane sterol in certain zygomycetes and basidiomycetes. A short overview of the major higher taxa recognized using molecular characters follows. See Humber (2008) for a more detailed discussion of the phylogenetic placement of insect-associated fungi.

Basal fungi

Current fungal trees (White et al. 2006; Hibbett et al. 2007) show a greater diversity of early diverging fungi than was previously known, especially when the derived nonflagellated anaerobic microsporidian parasites are included. Early in the use of small subunit ribosomal RNA gene (SSU rDNA) sequences for phylogenetic analyses, Microsporidia were considered to be basal pre-mitochondrial eukaryotes. More recently, however, based on protein-coding genes, these vertebrate and insect parasites appear to be among basal fungal groups or just basal to fungi in phylogenetic trees. The phylogenetic position of microsporidians, near or within Fungi, is supported by the traits that indicate the derived condition of the group. If microsporidians are included in Fungi, they will stand with Entomophthorales as one of the few basal groups of fungi that have widespread associations with arthropods. In the past, zoosporic fungi known as chytrids were considered members of a single phylum, and it was assumed that flagellation was lost on only one occasion. Based on analyses with additional genes and increased taxon sampling, the flagellated phylum Blastocladiomycota does not form a monophyletic group with other flagellated phyla (Chytridiomycota and Neocallimastigomycota), and flagellation appears to have been lost on more than one occasion. The Blastocladiomycota contains some parasites of aquatic insects. These fungi differ from all other true fungi because meiosis is sporic, resulting in an alternation of generations (diplobiontic life cycle) between a diploid sporothallus and a haploid gametothallus. The *Coelomomyces* relies on two different aquatic arthropod hosts to complete its alternating life cycle. This discovery helped to explain why it had been so difficult to reinfect mosquitoes in lab experiments (Whisler et al. 1974). In addition, the phylogenetic position of specialized, flagellated, intracellular parasites in two genera (*Rozella* and *Olpidium*) are not well resolved, and they lie outside other flagellated clades in current trees (Fig. 5).

Zygosporic fungi are not resolved as a monophyletic group, but can be placed informally in three or more clades (White et al. 2006; Hibbett et al. 2007). "Zygomycota I" contains a core group of mucoralean fungi. A related group, Mortierellales, has species sometimes associated with insects. "Zygomycota II," contains a monophyletic group, the DKH clade consisting of Dimargaritales, Harpellales, Kickxellales and Zoopagales. The species in the DHK clade have septate hyphae characterized by distinctive septal pore plugs. These fungi are often parasitic or predaceous on invertebrate animals including insects and in some cases other fungi. Members of Harpellales are well known as gut inhabitants of arthropods. "Zygomycota III" consists of Entomophthorales. *Basidiobolus* (Fig. 1), a traditional member of the order is not included and the position of the genus is still not clear (Fig. 5). Many of these species are insect pathogens (*Entomophaga* and *Entomophthora*), some with strict specificity (*Massospora* and *Strongwellsia*). *Basidiobolus* has dispersal interactions with insects, and infections of mammals may result (Blackwell and Malloch 1989). The insect pathogenic aspect of Entomophthorales was discussed by Humber (2008).

Glomeromycota

Members of the Glomeromycota are obligate arbuscular mycorrhizal (AM) fungi that are widespread associates of the roots of many plants. AM fungi were once considered to be zygomycetes, although they do not produce zygospores. One small group of species placed in Endogenales once were considered close relatives of AM fungi, but are classified in a clade with Mucorales and other zygosporic fungi (Fig. 5).

Ascomycota

Among the Ascomycota, many previously proposed evolutionary senarios have not been supported. The phylum is divided into three subphyla (Taphrinomycotina, Saccharomycotina and Pezizomycotina), and recent phylogenetic analyses have revealed several surprising finds (Fig. 5). For example, discomycetes (apothecial ascomycetes) were assumed to be highly derived forms, but phylogenetic studies using DNA characters indicate that these ascomycetes are basal members of the large group of mainly filamentous apothecial ascomycetes that we now call Pezizomycotina. It is of interest that species of *Neolecta* in one of the basal ascomycetes group, Taphrinomycotina, possess apothecial ascomata. Not only apothecia but also other ascomata are evolutionary labile and do not define monophyletic groups. Although there are no well-known associations between members of the Taphrinomycotina and insects, many members of Saccharomycotina and Pezizomycotina are insect associates (Suh et al. 2004; Humber 2008). Insects are important dispersers of plant pathogens, especially tree diseases caused by members of Pezizomycotina. It is within the Pezizomycotina that the most important insect pathogens are classified. These include members of the Hypocreales that have interactions not only with arthropods, but plants and other fungi as well. Recent work on the order revealed that the well-known insect parasite, *Cordyceps*, is not monophyletic, and species have been placed in three separate families (Fig. 6; Table 1). The phenomenon of host-switching in the Hypocreales is discussed below (See host switching). Other insect-associated members of Pezizomycotina include the bee parasites in the genus *Ascosphaera* (Eurotiales) and *Podonectria* (Tubeufiaceae) on scale insects. The Laboulbeniomycetes (Fig. 4) are all

Table 1 Classification of arthropod parasites previously placed in Clavicipitaceae

Clavicipitaceae: *Conoideocrella, Hypocrella, Metacordyceps, Moelleriella, Orbiocrella, Regiocrella, Samuelsia, Shimizuomyces, Villosiclava* (Major anamorphs: *Metarhizium, Paecilomyces* s.l. *Pochonia*)
Cordycipitaceae: *Ascopolyporus, Cordyceps, Hyperdermium, Torrubiella* (Major anamorphs: *Akanthomyces, Beauveria, Isaria, Lecanicillium, Simplicillium*)
Ophiocordycipitaceae: *Cordyceps* s.l., *Elaphocordyceps, Ophiocordyceps* (Major anamorphs: *Haptocillium, Hirsutella, Hymenostilbe*)

These ascomycetes and their anamorphs are now classified in three lineages reflected in changes of family level taxa: Cordycipitaceae, Clavicipitceae and Ophiocordycipitaceae. About 160 additional taxa originally classified as species of *Cordyceps* sensu Kobayasi and Mains require additional study in order to place them in the new phylogenetic classification and to determine their anamorphs. Several anamorph genera are probably polyphyletic (Sung et al. 2007; Spatafora et al. 2007; http://cordyceps.us/)

associated with insects and other arthropods, most as biotrophic parasites. No other group of ascomycetes except the Hypocreales, however, has so many associations with arthropods (Fig. 6; Table 1).

Basidiomycota

Basidiomycetes are classified in three subphyla, Pucciniomycotina, Ustilaginomycotina and Agaricomycotina (Fig. 5). The basal clade, Pucciniomycotina, contains the rust fungi, important plant pathogens, some of which have insect associations, especially fertilization and dispersal by chrysomelid beetles, flies and butterflies. Also included in Pucciniomycotina are species of Septobasidiales, parasites of scale insects. Smut fungi, Ustilaginomycotina, are plant pathogens. Members of Agaricomycotina have many associations with insects, including providing habitat for insects and other invertebrate animals. Several lineages of fungi of this subphylum are cultivated by ants and termites, and many members are dispersed by insects, including wood decaying fungi injected into wood by siricid wood wasps.

Applications of molecular methods to the study of insect-associated fungi

Distinguishing convergent evolution

In the past, mycologists were aware of the difficulties in detecting relationships among certain insect-associated fungi. For example, insect-associated fungi possess a suite of morphological characters involving spore-producing structures. The so-called ophiostomatoid fungi (e.g., *Ophiostoma*) are ascomycetes with long necked perithecia, evanescent asci, and adhesive spores collected in droplets at the perithecial tip, traits that promote ascospore dispersal by insects (Blackwell et al. 1993). The very characters used for defining and grouping ophiostomatoid taxa have proven deceptive, and DNA sequences were required to sort them into their independent lineages. For example, all species of several genera once were considered congeneric with the species placed in *Ceratocystis*. What is remarkable is that all of the genera now are placed in distinct orders: Ophiostomatales (*Ophiostoma*), Microascales (*Ceratocystis, Sphaeronaemella*), Laboulbeniales (*Pyxidiophora*), and *Kathistes*, in a separate unnamed order (Blackwell 1994; Blackwell et al. 2003). In addition to convergence among ophiostomatoid sexual states, there are many examples of convergence of coniodigenous cells and conidia. These include species of *Chalara*, the anamorph of insect-associated fungi in *Ceratocystis* that also is an anamorph of at least five orders of ascomycetes (Nag Raj and Kendrick 1993). *Ophiostoma* and *Ceratocystis* have similar asexual relatives, all placed in *Ambrosiella*, implying that traits of both asexual and sexual states are being selected upon for insect associations (Cassar and Blackwell 1996).

Another example of a presumed convergent character among insect-associated ascomycetes is the hat shaped (galeate) ascospore. This trait is found among a number of clades of insect-associated Saccharomycotina (especially previous members of the genus *Pichia*) and several clades of Pezizomycotina (Eurotiales, Ophiostomatales). Hat-shaped ascospores were once the basis for a taxonomic revision including yeasts and the galeate-spored Pezizomycotina in a common family (Redhead and Malloch 1977).

There also are examples of what appear to be rapid divergence. Obligate arthropod parasites (Laboulbeniomycetes, Fig. 4) previously have been placed in four different fungal phyla as well as in floridean red algae. In addition, certain species also were considered to be insect setae or even parasitic worms (Blackwell 1994). Some insect-associated Basidiomycota such as *Septobasidium* are morphologically distinct from near relatives, and molecular characters were required to place these organisms among their rust fungus relatives.

Host switching

Molecular techniques provide opportunities to trace changes in nutritional modes of fungi. Current patterns of fungal-insect parasitism may be explained by the "related host hypothesis" reported for certain attine ant-associated fungi in which cospeciation patterns are detected by congruence of species level phylogenetic trees of interacting ants and fungi (Little and Currie 2008). Far more often, however, fungi with close arthropod associations display a pattern of host switching, so that closely related fungi are not necessarily associated with closely related insects and vice versa (Nikoh and Fukatsu 2000). The related host hypothesis has been used to explain some of the "interkingdom host shifts" evident among clades of the Hypocreales (Spatafora et al. 2007; Sung et al. 2007; http://cordyceps.us/). These studies provide an excellent understanding of widespread host shifts and were cited as well-designed studies to show such changes and to make the corresponding nomenclatural changes that so often lag behind the phylogenetic work (Spatafora et al. 2007; Sung et al. 2007; http://cordyceps.us/; Vega et al. 2009). One less well-known case involves not only host switching, but also a dramatic change in life histories. A small group of endosymbionts of plant hoppers arose from pathogenic members of Ophiocordycipitaceae to evolve into an obligate association for both fungi and insects (Suh et al. 2001).

The host habitat hypothesis (Nikoh and Fukatsu 2000) was proposed to explain the associations of distantly related hosts of Hypocreales, but other examples are found among other fungi and their hosts. The "host habitat hypothesis" also may explain distributions of obligate biotrophic parasites (Laboulbeniales, Septobasidiales, Pucciniomycotina), pathogens (Entomophthorales, Blastocladiomycota, Eurotiales) and perhaps even commensals (Harpellales and Asellariales). Other examples provide insight into the host habitat hypothesis. *Laboulbenia ecitonis* is a species that parasitizes unrelated hosts that are inhabitants of legionary ant nests. The hosts, including histerid and staphylinid beetles, two species of mites and the ants themselves, are relatively confined in a common habitat (Benjamin 1965). Removal of all hosts except a mite species could appear as a rapid host shift to an unrelated host, especially if the nearest fungal relative were determined to be restricted to related ant hosts.

Some species of Laboulbeniales have broad host distributions, while strict specificity is assumed for others. De Kesel's (1996) experimental study provided insight into how host isolation and subsequent fungal specialization might occur after removal of some hosts. Assume that a generalist fungus is associated with a number of arthropod hosts; subsequently most of the potential host taxa disappear from the habitat, and the fungal parasite becomes geographically isolated on the single remaining host. Isolation followed by divergence of the fungus could lead to specialization on that particular host (Suh et al. 2005). For example, a single carabid species, the only one available, was the usual host for a laboulbenialean fungus (De Kesel 1996). The fungus, however, was able to infect some, but not all carabids. Beetles outside of Carabidae, however, never served as hosts. This study also indicates that there is sometimes a host genetic component in infection, and the absence of infection in some potential hosts may indicate that divergence and host switching are in progress.

Future considerations

Progress in evolutionary understanding and phylogenetic classification continues to be made as more taxa are sampled and more genes and genomes become available for analysis. For some time there will be a great need for biologists who know the biology and ecology of the organisms, and can collect and identify them for molecular studies. Obtaining correctly identified fungi is of paramount importance to understanding the evolutionary relationships among fungi, which will help us understand their evolutionary history.

Acknowledgments Funding from the National Science Foundation (NSF-0732671), Assembling the Fungal Tree of Life: Resolving the evolutionary history of the Fungi is gratefully acknowledged.

References

Alexopoulos CJ, Mims CW, Blackwell M (1996) Introductory mycology. Wiley, New York

Benjamin RK (1965) Study in specificity: minute fungi parasitize living arthropods. Nat Hist 74:42–49

Blackwell M (1994) Minute mycological mysteries: the influence of arthropods on the lives of fungi. Mycologia 86:1–17

Blackwell M (2000) Evolution: terrestrial life—fungal from the start? Science 289:1884–1885

Blackwell M, Malloch D (1989) Similarity of *Amphoromorpha* and secondary capilliconidia of *Basidiobolus*. Mycologia 81:735–741

Blackwell M, Spatafora JW, Malloch D, Taylor JW (1993) Consideration of higher taxonomic relationships involving Pyxidiophora. In: Wingfield MJ, Seifert KA, Webber JF (eds) *Ceratocystis* and *Ophiostoma*: taxonomy, ecology and pathology. American Phytopathological Society, St. Paul, pp 105–108

Blackwell M, Henk D, Jones KG (2003) Extreme morphological divergence: phylogenetic position of a termite ectoparasite. Mycologia 95:987–992

Blackwell M, Hibbett DS, Taylor JW, Spatafora JW (2006) Research coordination networks: a phylogeny for kingdom fungi (Deep Hypha). Mycologia 98:829–837

Cassar SC, Blackwell M (1996) Non-monophyly of ambrosia fungi in *Ambrosiella*. Mycologia 88:596–601

Currie CR, Wong B, Stuart AE, Schultz TR, Rehner SA, Mueller UG, Sung G-H, Spatafora JW, Straus NA (2003) Ancient tripartite coevolution in the attine ant-microbe symbiosis. Science 299:386–388

De Kesel A (1996) Host specificity and habitat preference of *Laboulbenia slackensis*. Mycologia 88:565–573

Findlay WPK (1982) Fungi: folklore, fiction, and fact. Mad River Press, Eureka

Henk DA, Vilgalys R (2007) Molecular phylogeny suggests a single origin of insect symbiosis in the Pucciniomycetes with support for some relationships within the genus *Septobasidium*. Am J Bot 94:1515–1526

Hibbett DS, Binder M, Bischoff JF, Blackwell M, Cannon PF, Eriksson OE, Huhndorf S, James T, Kirk PM, Lücking R, Thorsten Lumbsch H, Lutzoni F, Brandon Matheny P, McLaughlin DJ, Powell MJ, Redhead S, Schoch CL, Spatafora JW, Stalpers JA, Vilgalys R, Aime MC, Aptroot A, Bauer R, Begerow D, Benny GL, Castlebury LA, Crous PW, Dai YC, Gams W, Geiser DM, Griffith GW, Gueidan C, Hawksworth DL, Hestmark G, Hosaka K, Humber RA, Hyde KD, Ironside JE, Kõljalg U, Kurtzman CP, Larsson K-H, Lichtwardt R, Longcore J, Miądlikowska J, Miller A, Moncalvo JM, Mozley-Standridge S, Oberwinkler F, Parmasto E, Reeb V, Rogers JD, Roux C, Ryvarden L, Sampaio JP, Schüßler A, Sugiyama J, Thorn RG, Tibell L, Untereiner WA, Walker C, Wang Z, Weir A, Weiss M, White MM, Winka K, Yao YJ, Zhang N (2007) A higher-level phylogenetic classification of the Fungi. Mycol Res 111:509–547

Humber RA (2008) Evolution of entomopathogenicity in fungi. J Invertebr Pathol 98:262–266

Hunt T, Bergsten J, Levkanicova Z, Papadopoulou A, John O, Wild R, Hammond PM, Ahrens D, Balke M, Caterino MS, Gómez-Zurita J, Ribera I, Barraclough TG, Bocakova M, Bocak L, Vogler AP (2007) A comprehensive phylogeny of beetles reveals the evolutionary origins of a superradiation. Science 318:1913–1916

Kerwin JL, Petersen EE (1997) Fungi: Oomycetes and Chytridiomycetes. In: Lacey LA (ed) Manual of techniques in insect pathology. Academic Press, New York, pp 251–268

Little AEF, Currie CR (2008) Black yeast symbionts compromise the efficiency of antibiotic defenses in fungus-growing ants. Ecology 89:1216–1222

Matheny PB, Aime MC, Bougher NL, Buyck B, Desjardin DE, Horak E, Kropp BR, Jean Lodge D, Soytong K, Trappe JM, Hibbett DS (2009) Out of the palaeotropics? Historical biogeography and diversification of the cosmopolitan ectomycorrhizal mushroom family Inocybaceae. J Biogeogr 36:577–592

Munkacsi AB, Pan JJ, Villesen P, Mueller UG, Blackwell M, McLaughlin DJ (2004) Convergent coevolution in the domestication of coral mushrooms by fungus-growing ants. Proc R Soc Lond B Biol Sci 271:1777–1782

Nag Raj TR, Kendrick WB (1993) The anamorph as genetic determinant in the holomorph: the *Chalara* connection in the ascomycetes, with special reference to the ophiostomatoid fungi. In: Wingfield MJ, Seifert KA, Webber JF (eds) *Ceratocystis* and *Ophiostoma*: taxonomy, ecology and pathology. American Phytopathological Society, St. Paul, pp 61–70

Nikoh N, Fukatsu T (2000) Interkingdom host jumping underground: phylogenetic analysis of entomopathogenic fungi of the genus *Cordyceps*. Mol Biol Evol 17:629–638

Poinar GO Jr (1992) Life in amber. Stanford University Press, California

Poinar GO Jr, Jacobson RL, Eisenberger CL (2006) Early Cretaceous phlebotomine sand fly larvae (Diptera: Psychodidae). Proc Entomol Soc Washington 108:785–792

Redhead SA, Malloch DA (1977) The Endomycetaceae: new concepts, new taxa. Canad J Bot 55:1701–1711

Robbertse B, Reeves JB, Schoch CL, Spatafora JW (2006) A phylogenomic analysis of the Ascomycota. Fungal Genet Biol 43:715–725

Rossi W, Kotrba M, Triebel D (2005) A new species of *Stigmatomyces* from Baltic amber, the first fossil record of Laboulbeniomycetes. Mycol Res 109:271–274

Schadt CW, Martin AP, Lipson DA, Schmidt SK (2003) Seasonal dynamics of previously unknown fungal lineages in tundra soils. Science 301:1359–1361

Spatafora JW, Sung GH, Sung JM, Hywel-Jones NL, White JF Jr (2007) Phylogenetic evidence for an animal pathogen origin of ergot and the grass endophytes. Mol Ecol 16:1701–1711

Suh S-O, Noda H, Blackwell M (2001) Insect symbiosis: derivation of yeast-like endosymbionts within an entomopathogenic filamentous lineage. Mol Biol Evol 18:995–1000

Suh S-O, McHugh JV, Blackwell M (2004) Expansion of the *Candida tanzawaensis* yeast clade: 16 novel *Candida* species from basidiocarp-feeding beetles. Int J Syst Evol Microbiol 54:2409–2429

Suh S-O, McHugh JV, Pollock D, Blackwell M (2005) The beetle gut: a hyperdiverse source of novel yeasts. Mycol Res 109:261–265

Sung G-H, Hywel-Jones NL, Sung J-M, Luangsa-ard JJ, Shrestha B, Spatafora JW (2007) Phylogenetic classification of *Cordyceps* and the clavicipitaceous fungi. Stud Mycol 57:5–59

Sung GH, Poinar GO Jr, Spatafora JW (2008) The oldest fossil evidence of animal parasitism by fungi supports a Cretaceous diversification of fungal-arthropod symbioses. Mol Phylogenet Evol 49:495–502

Taylor JW, Berbee ML (2006) Dating divergences in the fungal tree of life: review and new analyses. Mycologia 98:838–849

Vandenkoornhuyse P, Baldauf SL, Leyval C, Straczek J, Young JPW (2002) Extensive fungal diversity in plant roots. Science 295:2051

Vega FE, Dowd PF (2005) The role of yeasts as insect endosymbionts. In: Vega FE, Blackwell M (eds) Insect-fungal associations: ecology and evolution. Oxford University Press, New York, pp 211–243

Vega FE, Goettel MS, Blackwell M, Chandler D, Jackson MA, Keller S, Koike M, Maniania NK, Monzón A, Ownley B, Pell JK, Rangel DEN, Roy HE (2009) Fungal entomopathogens: new insights on their ecology. Fungal Ecol 2:149–159

Weir A, Blackwell M (2005) Phylogeny of arthropod ectoparasitic ascomycetes. In: Vega FE, Blackwell M (eds) Insect-fungal associations: ecology and evolution. Oxford, UK, pp 119–145

Whisler HC, Zebold SL, Shemanchuk JA (1974) Alternate host for mosquito parasite *Coelomomyces*. Nature 251:715–716

White JF Jr, Taylor TN (1989) A trichomycete-like fossil from the Triassic of Antarctica. Mycologia 81:643–646

White MM, James TY, O'Donnell K, Cafaro MJ, Tanabe Y, Sugiyama J (2006) Phylogeny of the Zygomycota based on nuclear ribosomal sequence data. Mycologia 98:872–884

Author Biography

Meredith Blackwell is a fungal biologist, who is interested in arthropod associated fungi. Her current research deals with biodiversity of insect gut fungi. She has been involved in several community-wide projects, including the Research Coordination Network: A phylogeny for kingdom Fungi (Deep Hypha) and Assembling the Fungal Tree of Life II (AFToL II).

BioControl (2010) 55:17–37
DOI 10.1007/s10526-009-9251-8

Molecular ecology of fungal entomopathogens: molecular genetic tools and their applications in population and fate studies

Jürg Enkerli · Franco Widmer

Received: 2 October 2009 / Accepted: 22 October 2009 / Published online: 14 November 2009

Abstract The power of molecular genetic techniques to address ecological research questions has opened a distinct interdisciplinary research area collectively referred to as molecular ecology. Molecular ecology combines aspects of diverse research fields like population and evolutionary genetics, as well as biodiversity, conservation biology, behavioural ecology, or species-habitat interactions. Molecular techniques detect specific DNA sequence characteristics that are used as genetic markers to discriminate individuals or taxonomic groups, for instance in analyses of population and community structures, for elucidation of phylogenetic relationships, or for the characterization and monitoring of specific strains in the environment. Here, we summarize the PCR-based molecular techniques used in molecular ecological research on fungal entomopathogens and discuss novel techniques that may have relevance to the studies of entomopathogenic fungi in the future. We discuss the flow chart of the molecular ecology approaches and we highlight some of the critical steps involved. There are still many unresolved questions in the understanding of the ecology of fungal entomopathogens. These include population characteristics and relations of genotypes and habitats as well as host-pathogen interactions. Molecular tools can provide substantial support for ecological research and offer insight into this far inaccessible systems. Application of molecular ecology approaches will stimulate and accelerate new research in the field of entomophathogen ecology.

Keywords Cultivation-dependent analysis · Cultivation-independent analysis · Genotyping · Monitoring · Genetic diversity

Handling Editor: Helen Roy.

J. Enkerli (✉) · F. Widmer
Molecular Ecology, Agroscope Reckenholz-Tänikon Research Station ART, Reckenholzstrasse 191, 8046 Zurich, Switzerland
e-mail: juerg.enkerli@art.admin.ch

F. Widmer
e-mail: franco.widmer@art.admin.ch

Introduction

Molecular ecology is a large interdisciplinary area of research, which comprises diverse fields including population and evolutionary genetics, biodiversity, conservation biology, behavioural ecology, and species-habitat interactions (Beebee and Rowe 2008). Molecular ecologists use molecular genetic techniques to address specific ecological problems and questions. During the past two decades a tremendous variety of molecular tools has been developed and these have a considerable relevance to molecular ecology. However, it is important to realize that the use of molecular genetic techniques represents one aspect of molecular ecology and that this research area relies on a combination of various sciences like biology, ecology, biochemistry and molecular genetics. In this paper we will first introduce the basic steps used in molecular ecological analyses. Furthermore, we will discuss critical issues related to such analyses and we will present the most important PCR-related techniques and tools used in molecular ecology. However, we will not include hybridization-based techniques like micro array or fluorescent in situ hybridization (FISH). For such techniques see Amann et al. (1995, 2001) and Sessitsch et al. (2006). We will use studies as examples to illustrate implementation of the techniques in investigations on molecular ecology of fungal entomopathogens or to indicate the potential for their use. There are a number of papers in this special issue, which highlight the relevance of molecular techniques to studying the ecology of fungal entomopathogens (Blackwell 2009; Hajek and Delalibera 2009; Hesketh et al. 2009; Meyling and Hajek 2009; Ownley et al. 2009).

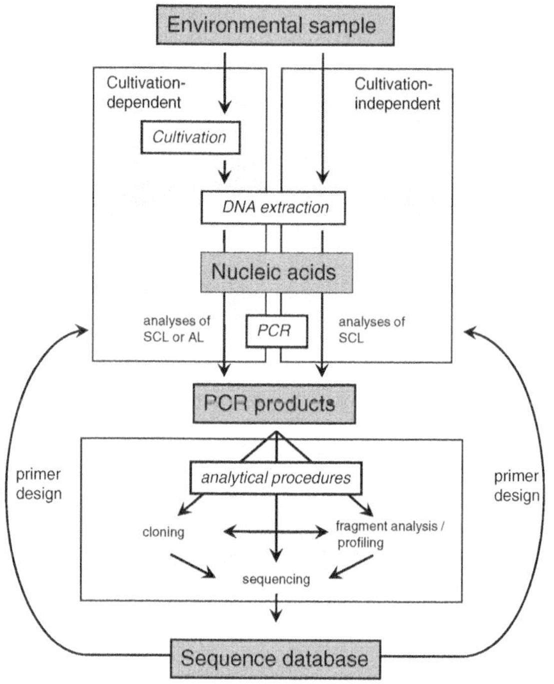

Fig. 1 Flow chart illustrating the principles for analyses of environmental samples using PCR-based molecular techniques. Cultivation-dependent and cultivation-independent analyses are highlighted in light grey boxes and analytical procedures are indicated with a hatched box. Details are described in the text. SCL sequence-characterized loci, AL anonymous loci

The principle of molecular ecological analyses

The development of the polymerase chain reaction (PCR) represents a milestone in the history of biological research (Mullis et al. 1986, 1994). This facilitated the development of a range of new approaches for genetic analysis. PCR allows specific amplification of DNA regions or fragments from a DNA sample and relies on the use of short oligonucleotide primers complementary to the target regions flanking the amplicon. Therefore, PCR represents a core technology in molecular ecological analyses. In Fig. 1 the analytical steps used in PCR-based molecular ecological analyses are illustrated in a flow chart, which is also referred to as the "full cycle approach" (Amann et al. 1995). The process starts with the collection of environmental samples, which may consist of any type of biological material, representing individual organisms like a plant, an animal, or an axenic culture of microorganism, or complex samples of mixtures of different organisms like mycorrhiza colonized plant root or soil samples. Accordingly, analyses are performed cultivation-dependent, where the target organism is available in pure culture prior to analyses or cultivation-independent, where investigations are performed directly on environmental samples such as soil without isolation and cultivation of the target organism. In the first step, nucleic acids are extracted from the samples and subjected to PCR. There are two types of target loci that are amplified by PCR (Table 1). The first target type represents sequence-characterized loci (SCL), which are defined DNA regions that are amplified from the genome of an organism, e.g. the small subunit ribosomal RNA (SSU rRNA) gene. The second target type represents anonymous loci (AL), which usually occur as multiple randomly distributed target regions in genomes and are simultaneously amplified using defined PCR primers. The complexity of the sample and the specific research question dictates which target type to select. In cultivation-dependent analysis, where the sample consists of DNA exclusively from the organism under investigation, both target types can be amplified. However, in cultivation-independent analysis it is only feasible to amplify SCL. DNA samples used in cultivation-independent analyses represent mixtures of organisms (metagenomic DNA), and individual genomes

Table 1 PCR-based molecular ecology techniques used in cultivation-dependent and cultivation-independent analyses

Target type	Analytical procedure	Applied in	
		Cultivation-dependent analyses	Cultivation-independent analyses
SCL	PCR-LP	+	+[a]
	PCR–RFLP	+	+[b]
	PCR-SSCP	+	+
	PCR-DGGE	+	+
	PCR-TGGE	+	+
	Sequencing	+	+
AL	RAPD	+	−
	UP-PCR	+	−
	Rep-PCR	+	−
	AFLP	+	−

SCL sequence-characterized loci, AL anonymous loci, LP length polymorphism, RFLP restriction fragment length polymorphism, SSCP single-strand conformation polymorphism, DGGE denaturing gradient gel electrophoresis, TGGE temperature gradient gel electrophoresis, RAPD random amplified polymorphic DNA, UP universally primed, Rep Repetitive element, AFLP amplified fragment length polymorphism

[a] RISA, ribosomal intergenic spacer analysis

[b] RFLP, ARDRA, amplified ribosomal DNA restriction analysis or T-RFLP terminal restriction fragment length polymorphism

can not be targeted by analysis of AL. PCR products generated during cultivation-dependent as well as independent analyses are subsequently assessed using different analytical procedures, such as electrophoretic sizing, or sequencing. This allows detection of specific sequence characteristics such as length variability or nucleotide polymorphisms, which serve as genetic markers to discriminate individuals. Such data represent the fundamental information necessary in molecular ecological studies, for instance to identify individuals, to analyze population structures of a species, or to describe community structures in complex samples. Sequence data obtained can be deposited in public sequence databases like GenBank of the National Center for Biotechnology Information (NCBI, Bethesda, MD, USA) or the Ribosomal Database Project (RDP, Michigan State University).

PCR amplifications are performed either with random, specific, or universal primers. Random primers are short oligonucleotides of usually ten nucleotides of arbitrary sequence. Such primers allow amplification of randomly selected DNA regions (AL), which is utilized in Randomly Amplified Polymorphic DNA (RAPD) analyses (see below). Specific primers are designed using specific sequence characteristics to restrict PCR amplification to a narrow range of organisms like a species. For fungal entomopathogens specific PCR primers have been designed to amplify target loci for instance at the subspecies (Castrillo et al. 2003; Destefano et al. 2004) or species level (Entz et al. 2005; Castrillo et al. 2007; Fournier et al. 2008; Guzman-Franco et al. 2008). Specificity and sensitivity for amplification of a target locus can be further increased by use of a nested PCR approach. This approach consists of a two-step PCR where the target locus is amplified with one primer pair followed by a second PCR with a primer pair that amplifies an internal region of the first PCR product. Such an approach has been applied for specific amplification of the internal transcribed spacer (ITS) region of the rRNA gene cluster from isolates of the *Entomophthora muscae* (Entomophthoromycotina: Entomophthorales) species complex (Thomsen and Jensen 2002). Universal primers are designed in conserved regions of the locus of interest to allow the amplification of a wide range of organism across phylogenetically related groups. For fungi, universal primers are available that allow amplification of loci for instance from different phyla (Borneman and Hartin 2000; Lynch and Thorn 2006) or the fungal kingdom (White et al. 1990; Zhou et al. 2000). Universal primers can be used to amplify and analyze a target region from an organism for which the locus has not been sequenced before. Cloning and sequencing of such products can provide sequence information that may be used to design new primers, which are specific for the target organism (Fig. 1). As fungal taxonomy is still in progress and phylogenetic information is constantly growing, it is important to continuously re-evaluate primer specificities and amplification ranges of fungal primers (Rehner and Buckley 2005; Bischoff et al. 2009).

Exploration of molecular data and their interpretation in ecological contexts requires thorough statistical assessment. A number of statistical software packages have been developed, e.g. PAUP (Swofford 2002); PHYLIP (Felsenstein 2009); The Software R (R Development Core Team 2008); Statistica, StatSoft, Tulsa, OK, USA; Canoco: Microcomputer Power, Ithaca, NY, USA that allow to perform a wide range of detailed assessments that are important for instance to perform phylogenetic inferences based on DNA sequence data (Rehner and Buckley 2005; Hartmann and Widmer 2006; Bischoff et al. 2009; Blackwell 2009) as well as descriptive and explorative analyses of genetic profiling-based data on genetic diversities or community structures (Hughes et al. 2001; Rees et al. 2004; Hughes and Hellmann 2005; Hartmann et al. 2006; Ramette 2007; Hartmann and Widmer 2008; Schwarzenbach et al. 2009). Statistical assessment of data can be very demanding and may depend on the amount and type of data available and the specific research question addressed. Therefore, it is important to consider statistical analyses required when planning an experiment.

Critical issues in PCR-based molecular ecological analyses

DNA extraction

Reliable isolation of high quality DNA is an important issue in molecular ecological analyses. Numerous protocols for extraction of DNA have been developed and a variety of them have been applied, for instance, to extract DNA from insects (Entz et al. 2005; Fournier et al. 2008; Guzman-Franco et al. 2008; Agboton et al. 2009), plants (Jensen and Eilenberg 2001; Destefano et al. 2004; Fournier et al. 2008), plant surfaces (Castrillo et al. 2003; Castrillo et al. 2008), or soil samples (Entz et al. 2005; Schwarzenbach et al. 2007b; Castrillo et al. 2008) used in cultivation-independent detection and analysis of fungal entomopathogens in environmental samples. The basic steps used for DNA extraction include suspension of the sample in a buffer and subsequent cell lysis performed either chemically by using phenol, detergents (SDS, CTAB), and/or lysing enzymes (e.g. proteinase K, lysozyme) or physically, for instance using sonication, freezing-thawing, bead-beating, grinding procedures or a combination of any of those (Sambrook and Russell 2001; Kowalchuk et al. 2004). Subsequently, raw DNA extracts are purified applying phenol/chloroform extractions, column purification and/or precipitation, e.g., with potassium acetate, ethanol, isopropanol, and/or polyethylene glycol (Widmer et al. 1996; Sambrook and Russell 2001; Kowalchuk et al. 2004). While the protocols for DNA extraction from tissue samples like insects and plants are comparable in their efficiency and quality, extraction of high quality DNA from soil samples remains a challenge because of the highly complex composition of soil (Martin-Laurent et al. 2001; Feinstein et al. 2009). Protocols have been specifically adapted for extraction of soil DNA (Akkermans et al. 1995; Bürgmann et al. 2001; Kowalchuk et al. 2004) and various commercial kits are available. However, yield and DNA quality vary considerably among techniques (Lloyd-Jones and Hunter 2001; Kabir et al. 2003; Roh et al. 2006; Whitehouse and Hottel 2007). Protocols optimized for a particular soil type may not be efficient for extraction of soil DNA from an other soil type due to differences in the chemical or physical composition of the soil sample (Frostegård et al. 1999; Kabir et al. 2003). Furthermore, efficient extraction requires successful cell lysis, which is dependent on the stages of the organisms present in the soil, i.e., single cells, mycelium or spore/spore type (Frostegård et al. 1999). Lysis conditions may have to be harsher for fungi that are present as resting spores (Castrillo et al. 2007) than for those that are present as mycelium. However, it was shown that successive extractions from the same soil sample provides more complete extraction of soil DNA and may partially compensate for variation in extraction efficiencies among extraction protocols and/or samples of different soil types (Bürgmann et al. 2001; Feinstein et al. 2009).

PCR inhibition

Another critical issue is the purity of the extracted DNA. Very often PCR inhibiting factors are co-extracted from environmental samples (Wilson 1997; Poussier et al. 2002) particularly when extracting DNA from soil samples (Watson and Blackwell 2000). Such factors include for instance humic compounds, like humic (Tebbe and Vahjen 1993), fluvic- and tannic acids (Kreader 1996) or other similar substances (Watson and Blackwell 2000). Accordingly, soil types that are particularly rich in such compounds may pose extra problems. For example, Bürgmann et al. (2001) have shown that DNA extracts from strongly acidic forest soils inhibited PCR.

Along with the development and optimization of DNA extraction protocols, different approaches have been followed to eliminate the problem of PCR inhibition: (1) Avoidance of co-extraction of PCR inhibiting factors by improving DNA extraction protocols for instance by increasing salt concentrations in the lysis buffer (LaMontagne et al. 2002); (2) Removal of PCR inhibiting factors from extracted DNA using clean up procedures like polyvinylpolypyrrolidone spin columns (Widmer et al. 1996; Poussier et al. 2002), Sephadex G-200 spin columns (Kuske et al. 1998; Miller et al. 1999), Sepharose resins (Jackson et al. 1997; Miller et al. 1999), or DNA precipitation with isopropanol (Zhou et al. 1996; LaMontagne et al. 2002) or polyethylene glycol 8000 (Widmer et al. 1996; Arbeli and Fuentes 2007); (3) Decreasing inhibition of PCR by adding proteins like bovine serum albumin (BSA) (Romanowski et al. 1993; Fournier et al. 2008), phage T4 gene 32 protein (Kreader 1996; Poussier et al. 2002), or skim milk (Arbeli and Fuentes 2007) to scavenge inhibitors and to protect DNA polymerases; (4) Dilution of DNA extracts to lower the concentration of the inhibiting factors in the PCR (Miller et al. 1999; Arbeli and Fuentes 2007; Schwarzenbach et al. 2007a).

It is difficult to predict whether PCR inhibiting factors are present in a particular sample or not. Therefore, testing for PCR inhibition is an important step in data validation, particularly when quantifying target organisms by use of quantitative PCR. PCR inhibition has been assessed by adding serial dilutions of potential inhibitors to PCR (Kreader 1996) or by spiking soil DNA with defined amounts of template DNA (Bürgmann et al. 2001; Castrillo et al. 2007; Fournier et al. 2008). For example Bürgmann et al. (2001) and Fournier et al. (2008) added defined numbers of plasmid DNA copies to soil samples, performed end-point PCR and quantified product yields by gel electrophoresis while Castrillo et al. (2007) spiked soil samples with different quantities of genomic DNA containing the target region and performed quantitative PCR. Unfortunately, neither of these approaches allow for quantification of PCR inhibition. Recently, a more general approach has been described for quantification of such effects (Schneider et al. 2009). In this approach, known amounts of recombinant DNA template are spiked into serial dilutions of soil DNA and quantified by real-time PCR followed by statistical analyses.

PCR bias and formation of artefacts

Although PCR represents a powerful tool in ecological research, it does have limitations (Kanagawa 2003; Anderson and Cairney 2004). Most of these limitations are relevant when performing cultivation-independent analyses on complex environmental samples such as soil. The term PCR bias for instance refers to the fact that homologous genes with different sequences such as the rRNA genes may PCR amply at different rates even though the PCR primers perfectly match them (Kanagawa 2003; Anderson and Cairney 2004). This phenomenon has been explained by sequence-specific secondary structure formation of the single-stranded PCR template or by higher rehybridization rates of abundant templates. Classical PCR artefacts include chimera-formation, which may be the result of partial single stranded templates in a PCR (Qiu et al. 2001; Kanagawa 2003). Such templates may be produced if severely sheared DNA is used as template or if DNA synthesis time during PCR is too short. Another typical PCR artefact is the spontaneous introduction of point mutations when using DNA polymerases that synthesize DNA with low fidelity (Cariello et al. 1991; Qiu et al. 2001). While this is less relevant for community profiling approaches, such as Denaturing Gradient Gel Electrophoresis (DGGE) or Ribosomal Internal Spacer Analysis (RISA) (see below), it may become critical if PCR products are sequenced and used for phylogenetic studies. In such cases proof-reading DNA polymerases should be used, which synthesize DNA with high fidelity. Finally, primer annealing artefacts are also particularly relevant if PCR is performed on highly complex samples, such as metagenomic soil DNA extracts (Kanagawa 2003; Anderson and Cairney 2004; Lynch and Thorn 2006). PCR primers may anneal to targets even though mismatches occur. Design of primers that clearly distinguish targets with several mismatches and performing hot-start PCR reduce the risk to amplify non-specific targets. The problems related to the PCR bias and PCR artefacts are gaining increasing attention and have been addressed in several publications (Kanagawa 2003; Frey et al. 2006; Hartmann et al. 2007; Hartmann and Widmer 2008). For molecular ecological analyses it is important to know them and to be aware that unknown biases still may exist. This will help to reduce the impact they may have on the data produced.

Cultivation-dependent analyses

Analysis of sequence-characterized loci

Amplification of SCL can be used as Sequence-Characterized Amplified Region (SCAR) markers. For this application, specific primers are used that selectively amplify the marker from a target organism, such as a fungal species. Presence or absence of an amplification product indicates presence or absence of the target organism. Such a qualitative detection can be used for instance to identify a cultivated organism at a species or isolate level that may be difficult to be discriminated by other means (Tymon et al. 2004; Agboton et al. 2009). Primers used for cultivation-dependent analysis of SCAR markers often are applicable also in cultivation-independent detection and vice versa (see below). PCR amplification products of SCL can be subjected to a large number down stream analytical procedures for more detailed discrimination and are described in greater details in the following paragraphs.

PCR-length polymorphism

Products obtained from amplification of SCL can be assessed for length polymorphism by comparing obtained product sizes using gel electrophoresis. For instance PCR-LP (PCR-length polymorphism) of the ITS region of entomophthoralean fungi allowed discrimination of different species (Nielsen et al. 2001; Hajek et al. 2003; Tymon et al. 2004) or isolates (Rohel et al. 1997). Similarly, Neuvéglise et al. (1997) have detected length variation in the LSU rRNA gene among isolates of *Beauveria brongniartii* (Ascomycota: Hypocreales). However, PCR-LP of such loci is rather limited in its resolution and often used as an initial step in analyses of SCL.

Another type of SCL that is analyzed for length polymorphism is simple sequence repeat (SSR) or microsatellite markers. SSR markers currently represent the most popular genetic marker used to infer population structure, genetic variation, and relatedness (Avis 2004). SSR loci consist of tandem repeats of 1–6 nucleotides, and they are dispersed throughout the genome of most organisms (Goldstein and Schlötterer 1999) including microorganism (Field and Wills 1996). Alleles of a given locus may vary in the repeat numbers resulting in length polymorphism of the SSR alleles. SSR loci are individually amplified by PCR using pairs of PCR primers specific to the unique DNA sequences flanking the SSR. Allele sizes are subsequently determined using electrophoresis techniques, such as capillary electrophoresis. SSR markers are highly polymorphic and they are potentially independently segregating, which are important criteria when discriminating closely related organisms and performing population genetic analyses (Avis 2004). SSR markers have been isolated and characterized for various entomopathogenic species, i.e., *Ascosphaeara apis* (Ascomycota: Ascosphaerales) (Rehner and Evans 2009), *B. brongniartii* (Enkerli et al. 2001), *Beauveria bassiana* (Ascomycota: Hypocreales) (Rehner and Buckley 2003), *Metarhizium anisopliae* (Ascomycota: Hypocreales) (Enkerli et al. 2005; Oulevey et al. 2009) and *Paecilomyces fumosoroseus* (= *Isaria fumosorosea*) (Ascomycota: Hypocreales) (Dalleau-Clouet et al. 2005). They have been used to elucidate genetic diversity and population structures (Enkerli et al. 2001; Gauthier et al. 2007; Velásquez et al. 2007), to investigate insect-host associations (Dalleau-Clouet et al. 2005; Leland et al. 2005), to identify and characterize strains with potential for use in biological control (Leland et al. 2005; McGuire et al. 2005), and to monitor isolates released for biological control purposes (Enkerli et al. 2004; Wang et al. 2004). SSR analysis has been used to investigate the persistence of *B. brongniartii* strains that were applied for biological control of the European maybeetle, *Melolontha melolontha* (Coleoptera: Scarabaeidae), at seven different grassland sites in Switzerland, and that were still present up to 14 years after application (Enkerli et al. 2004). At some sites the applied strain as well as indigenous *B. brongniartii* strains were detected, while at other sites only the applied strain was present. The results of this study suggested that *B. brongniartii* strains can establish and coexist with indigenous populations in the same habitat and provide a long term biological control of *M. melolontha*.

PCR-restriction fragment length polymorphism

PCR–RFLP (PCR-restriction fragment length polymorphism) has been the most widely applied procedure to analyze products amplified from SCL over the past two decades. PCR–RFLP is based on amplification of a specific SCL from different target organisms followed by digestion with restriction endonucleases. Subsequently, restriction products are separated by gel electrophoresis and analyzed for fragment length polymorphisms. PCR–RFLP has been used intensively for fungal entomopathogens to assess genotype variability at the genus- or species-level (Tymon et al. 2004), to investigate population structures (Coates et al. 2002a) and host-pathogen associations (Neuvéglise et al. 1994). The ITS region of the rRNA gene cluster has been the main focus for this type of analyses and it has been applied for example to *B. bassiana* (Coates et al. 2002a; Aquino de Muro et al. 2005; Velásquez et al. 2007), *B. brongniartii* (Neuvéglise et al. 1994; Wada et al. 2003), *P. fumosoroseus* (=*I. fumosorosea*) (Fargues et al. 2002), *E. muscae* (Jensen et al. 2001; Thomsen and Jensen 2002), *Pandora neoaphidis* (Entomophthoromycotina: Entomophthorales) (Rohel et al. 1997; Francis et al. 2004; Tymon et al. 2004), *Zoophthora radicans* (Entomophthoromycotina: Entomophthorales) (Guzman-Franco et al. 2008), and *Conidiobolus* spp. (Tymon et al. 2004). Furthermore, PCR–RFLP analysis has been applied to the large subunit (LSU) rRNA gene to assess genetic variability and relationship among *B. brongniartii* isolates (Neuvéglise et al. 1997) and within and/or among the entomophthoralean genera *Entomophthora*, *Eryniopsis* and *Entomophaga* (Jensen and Eilenberg 2001; Hajek et al. 2003). Outside of the rRNA gene complex the PCR–RFLP approach has been applied to the pathogenicity related Pr1 protease gene and three chitinase genes to investigate strain relatedness and population structure in *B. bassiana* (Wang et al. 2003b) and *M. anisopliae* (Leal et al. 1997; Enkerli et al. 2009). Pr1 PCR–RFLP analysis in combination with other genotyping tools has been applied to demonstrate that the genetic structure of *M. anisopliae* is habitat dependent, i.e., isolates originating from agricultural or forested habitats belong to separate genetic groups (Bidochka et al. 2001).

PCR-single strand conformation polymorphism, PCR-denaturing gradient gel electrophoresis, and PCR-temperature gradient gel electrophoresis

SSCP (PCR-single strand conformation polymorphism), DGGE (PCR-denaturing gradient gel electrophoresis), and TGGE (PCR-temperature gradient gel electrophoresis) are used to detect genetic differences in PCR products obtained from SCL. SSCP relies on differences in secondary structure of single stranded DNA assessed by gel electrophoresis under non-denaturing conditions (Schwieger and Tebbe 1998), whereas DGGE and TGGE detect differences in DNA double-strand stability assessed by gel electrophoresis through a denaturant or temperature gradient (Muyzer and Smalla 1998). SSCP analysis have allowed discrimination of isolates of *Lecanicillium lecanii* (Ascomycota: Hypocreales) (Sugimoto et al. 2003), *B. bassiana* (Hegedus and Khachatourians 1996) or *Nomuraea riley* (Ascomycota: Hypocreales) (Devi et al. 2007) by targeting loci coding for the mitochondrial small and large subunit rRNA, β-tubulin, or histon 4. DGGE has been used to discriminate *B. bassiana* isolates based on amplification products obtained from the ITS region (Pantou et al. 2003). TGGE has not been applied to entomopathogenic fungi, however it has a similar potential to SSCP or DGGE for identification of genotypes as demonstrated by studies on different yeast (Hernan-Gomez et al. 2000; Manzano et al. 2005) or bacterial species (Wagner-Dobler et al. 2000; Tominaga 2006). These three analytical procedures are technically demanding and they have not been used to analyze large numbers of isolates, which may be necessary when investigating and comparing genetic diversity of populations. Furthermore, they represent techniques that are preferentially applied for cultivation-independent community structure analyses (see below).

Sequencing

Sequencing of PCR amplified SCL has become a very powerful procedure during the past decade. Sequencing reactions are routinely performed by use of commercial kits followed by automated analysis on specifically designed sequencing equipment such as capillary electrophoretic analyzers. Sequencing of SCL like the SSU (Nagahama et al. 1995; Jensen et al. 1998; Bidochka et al. 1999; Nikoh and Fukatsu 2000; Coates et al. 2002b), the LSU (Rakotonirainy et al. 1994; Pantou et al. 2003; Wang et al. 2003a), the ITS (Neuvéglise et al. 1994; Bidochka et al. 1999; Zare et al. 1999; Fargues et al. 2002; Liu et al. 2002; Aquino de Muro et al. 2005; Bidochka et al. 2005; Glare et al. 2008), and the intergenic spacer region (IGS) (Pantou et al. 2003) of the rRNA gene cluster, elongation factor 1-α (EF 1-α) (Rehner and Buckley 2005; Glare et al. 2008; Meyling et al. 2009), the largest (RPB1) and second largest (RPB2) subunit of RNA polymerase II (Bischoff et al. 2009), β-tubulin (Bischoff et al. 2009), or genes of the mitochondrial genome (Nikoh and Fukatsu 2000; Ghikas et al. 2006; Kouvelis et al. 2008a, b; Sosa-Gomez et al. 2009) have been used to assess genetic variation among species or isolates, to investigate host-pathogen associations, to elucidate community composition or to infer phylogenetic relation among various entomopathogenic taxa.

With the tremendous advances made during the past decade sequencing has become routine, and it is now possible to perform comprehensive analyses that are based on multiple genes. For instance, in population genetics multi-locus sequence typing (MLST) approaches are used for species definition and recognition or assessment of population structures (Taylor and Fisher 2003). Similarly, phylogenetic studies are increasingly based on multi-gene analyses (see also Blackwell 2009) as demonstrated by the recently published phylogeny of the genus *Beauveria* (Rehner and Buckley 2005), the *M. anisopliae* lineage (Bischoff et al. 2009) or the six-gene based general phylogeny of fungi (James et al. 2006). However, besides its importance in analyses of SCL, sequencing is also important when isolating and characterizing new loci. Availability of sequence information allows to investigate gene structure and function (Fang et al. 2005; Wang and St Leger 2007), and it provides the base for the development of universal and/or specific primers as indicated in Fig. 1 (Tymon et al. 2004; Fournier et al. 2008; Agboton et al. 2009; Enkerli et al. 2009). Such primers may be used for the development of tools that allow identification, detection, or quantification of the target organism in complex environmental samples (see cultivation-independent analyses).

Single Nucleotide Polymorphism (SNP) markers are a type of SCL, which are related to the MLST approach mentioned above (Morin et al. 2004; Kim and Misra 2007). SNP signatures represent single base pair positions with different sequence signatures (alleles) in individuals of a population (Brookes 1999). SNPs have been detected in genomes of many taxa (Kim and Misra 2007) including fungi (e.g., Bain et al. 2007; Kristensen et al. 2007; Xu et al. 2007; Lambreghts et al. 2009; Munoz et al. 2009). They have been widely applied in biomedical fields (Kim and Misra 2007) and have a great potential for use in ecology, evolution, and conservation biology (Morin et al. 2004). The discovery and development of SNP markers relies on the availability of high-quality sequence information from target loci of representative individuals. Such sequences may be collected from public data bases or they may have to be generated by sequencing the target locus in defined individuals, which involves a substantial sequencing effort (Morin et al. 2004). SNP alleles are detected with allele specific reactions, which are based on primer extension, hybridization, ligation, or enzymatic cleavage (Kim and Misra 2007). Among fungal entomopathogens a SNP-based genotyping assay has only been developed for the entomophthoralean species *P. neoaphidis* (Fournier et al. 2009). Methods to discriminate genotypes of this fungus have been limited and the development of the SNP typing assay provides a powerful tool for investigating the ecology of this fungus.

Analysis of anonymous loci

The greatest advantages of analyses based on AL compared to SCL, is the fact that no sequence information of the target organism is required. Analyses can be performed on DNA of any organism as long as it has been isolated from pure culture. PCR amplification products of AL are analyzed by electrophoretic techniques like gel or capillary electrophoresis. Resulting banding patterns or profiles are compared assuming that fragments of the same size represent identical loci. However, fragments of identical size but different sequence (size homoplasy) cannot be distinguished, which could lead to misinterpretations (Rieseberg 1996; Vekemans et al. 2002) and reflects one of the main disadvantages of analysis of AL. Nevertheless, single discriminating bands identified with this methodology can be isolated, characterized, and converted into SCL-markers such as SSR (Groppe et al. 1995) or SCAR-markers (Brugmans et al. 2003; Castrillo et al. 2003; Sudisha et al. 2009), which are preferred for diagnostic purposes like detecting or monitoring specific strains in the environment (see analysis of SCL).

Randomly amplified polymorphic DNA

The most common method used in the past to analyse AL is RAPD (randomly amplified polymorphic DNA) (Williams et al. 1990), also referred to as Arbitrary Amplified PCR (Welsh and McClelland 1990). RAPD-PCR is performed with random primers. Due to the short sequence of the primers there are numerous primer binding sites throughout the genome. Fragments are amplified if two primers bind to the template DNA in suitable distance (up to 2000 bp) and opposite orientation. RAPD-PCR produces numerous fragments amplified from different anonymous regions of the genome. Resulting products are separated by gel electrophoresis, which provides RAPD-banding profiles that allow to analyze presence or absence of bands. This profiling technique has been widely employed to study intraspecific variation within species like *Entomophaga grylli* (Entomophthoromycotina: Entomophthorales) (Bidochka et al. 1995), *E. muscae* (Jensen et al. 2001), *P. neoaphidis* (Rohel et al. 1997; Nielsen et al. 2001; Tymon and Pell 2005), *Zoophthora phytonomi* (Entomophthoromycotina: Entomophthorales) (Hajek et al. 1996), *Z. radicans* (Hodge et al. 1995), *B. bassiana* (Maurer et al. 1997; Fernandes et al. 2006), *B. brongniartii* (Cravanzola et al. 1997; Piatti et al. 1998), *Hirsutella thompsonii* (Ascomycota: Hypocreales)(Mozes-Koch et al. 1995; Aghajanzadeh et al. 2007), *L. lecanii* (Mor et al. 1996), *M. anisopliae* (Fegan et al. 1993; Leal et al. 1994; Velásquez et al. 2007), *N. rileyi* (Boucias et al. 2000; Vargas et al. 2003), and *Paecilomyces farinosus* (=*Isaria farinosus*) (Ascomycota: Hypocreales)(Chew et al. 1998). Studies have focused for instance on associations of fungal genotypes with specific hosts (Hodge et al. 1995; Bridge et al. 1997; Maurer et al. 1997; Jensen et al. 2001) or on correlations between RAPD profiles and geographical origin of a species (Leal et al. 1994; Hajek et al. 1996; Boucias et al. 2000; Nielsen et al. 2001). Furthermore, RAPD has been used to investigate the fate of released *Z. radicans* isolates (Hodge et al. 1995) or to trace the origin of a possibly introduced strain of *Z. phytonomi* (Hajek et al. 1996). Even though the RAPD technique has been used extensively, it has the disadvantages of low reproducibility among laboratories (Perez et al. 1998). Therefore, other markers like SSR markers are often preferred. However, RAPD analysis is a fast and simple method, which may be of use as a first step to assess genetic variability among isolates.

Universally primed PCR

Universally primed PCR (UP-PCR) is a technique which is closely related to RAPD (Bulat and Mironenko 1990), and relies on the use of single primers. However, UP-PCR primers are 15–20 bp in length, which allows the use of PCR conditions that are more stringent than the conditions used in RAPD analysis. As a result UP-PCR amplifications are more specific and reproducible than RAPD analyses (Bulat and Mironenko 1990). This technique provides a simple and fast way to assess genetic variability among fungal isolates. UP-PCR has been used to discriminate isolates of *E. muscae* (Jensen et al. 2001) or *L. lecanii* (Mitina et al. 2007), or to investigate the genetic diversity of *B. bassiana* isolates collected from the phylloplane of different hedgerow plants (Meyling and Eilenberg 2006) .

Repetitive element PCR

Various other methods used to analyze AL have collectively been termed repetitive element PCR [rep-PCR (Repetitive element PCR), Versalovic et al. 1994]. The common principle of these methods is based on repetitive DNA elements that are distributed across the genome as annealing sites for specific primers. Fragments between two repeated DNA elements, i.e., inter fragments, are amplified if two repeated DNA elements are close enough and provide matching primer binding sites in opposite directions. Several methods have been developed, which are adapted to different types of repetitive elements. They include Enterobacterial Repetitive Intergenic Consensus-PCR (ERIC-PCR) (Versalovic et al. 1994), BOX-PCR (Versalovic et al. 1994), Inter Simple Sequence Repeats (ISSR-PCR) (Zietkiewicz et al. 1994), and Inter Retrotransposon Amplified Polymorphism PCR (IRAP-PCR) (George et al. 1998). Rep-PCR approaches are fast, and simple to perform. Like UP-PCR they generally are more reproducible and reliable than RAPD analyses because primers are longer and therefore allow use of more stringent conditions during PCR. ISSR-PCR was used for instanced to elucidate genetic diversity and population structure of *B. bassiana* (Aquino de Muro et al. 2005; Wang et al. 2005b; Estrada et al. 2007) and *E. muscae* (Lihme et al. 2009). Furthermore, a combination of ISSR, ERIC, and RAPD was applied to investigate genetic variability among *P. neoaphidis* isolates and to confirm the monophyletic descent of this species (Tymon and Pell 2005).

Amplified restriction fragment length polymorphism

The AFLP (amplified restriction fragment length polymorphism) method consists of three elements: the specific restriction of genomic DNA, the ligation of adapters, and the subsequent amplification of the fragments by PCR (Vos et al. 1995). Genomic DNA is typically digested with two restriction enzymes and synthetic oligonucleotide adapters are ligated to the cohesive ends of the restriction fragments. Subsequently, the restriction fragments of unknown sequence are amplified with PCR primers corresponding to the restriction site and adapter sequence. For most organisms, the complexity of resulting fragments has to be reduced to allow appropriate resolution (50–100 fragments) of the PCR products, for example by capillary electrophoresis. This is done by performing a second selective PCR using primers with 1–3 additional bases at the 3′ end. For example, in AFLP analyses of *B. bassiana* (Aquino de Muro et al. 2005), *Entomophaga maimaiga* (Entomophthoromycotina: Entomophthoralses)(Nielsen et al. 2005) or *H. thompsonii* (Tigano et al. 2006), 2–3 additional bases were used in the selective PCR. AFLP allows for efficient detection of polymorphisms and due to the use of long and specific primers it provides more robust and reproducible amplification than RAPD, UP-PCR or rep-PCR (Meudt and Clarke 2007). However, AFLP is more laborious and requires more technical expertise. During the past decade AFLP has become a widely applied method to investigate population structure and diversity of animals, plants and fungi and it has been applied to *B. bassiana* (Aquino de Muro et al. 2003; Aquino de Muro et al. 2005), *E. maimaiga* (Nielsen et al. 2005), *H. thompsonii* (Tigano et al. 2006), *M. anisopliae* (Inglis et al. 2008), and *N. rileyi* (Boucias et al. 2000; Devi et al. 2007). The use of AFLP in combination with ISSR-PCR revealed a possible correlation between intra-specific groupings and geographical origin of *B. bassiana* isolates (Aquino de Muro et al. 2005). Furthermore, this technique allowed unravelling the origin of the founder population of *E. maimaiga*, a pathogen of the gypsy moth, *Lymantria dispar* (Lepidoptera: Lymantriidae), in the USA, which was introduced in the USA in 1910 but reported in the field only since 1989 (Nielsen et al. 2005).

Cultivation-independent analyses

Detection of entomopathogenic species or single strains

Cultivation-independent approaches for detection of species and/or strains in environmental samples like insect cadavers or soil samples are often more efficient than cultivation-based approaches as they allow to circumvent time consuming cultivation steps (Schwarzenbach et al. 2007b). Furthermore, they allow investigation of species that are difficult to isolate and/or cultivate, or are morphologically difficult to identify (Fournier et al. 2008; Guzman-Franco et al. 2008). However, as targeted organisms are not cultivated with this approach they are not directly amenable for subsequent physiological or phylogenetic investigations. Defined primer-specificity is critical in cultivation-independent detection of a specific organism. The specificity of detection depends on specific sequence signatures of the targets sequence and therefore requires sufficient and reliable sequence information. The more sequences of closely related organisms or taxa one has available the more specific primers may be designed. Therefore, specificity of a primer has to be continuously reconsidered with the growing number of sequences available in public data bases.

Similar to analyses of SCL in cultivation-dependent analyses, specific primers are used to amplify a locus from the target organism. PCR products are analyzed by gel-electrophoresis and assessed for presence or absence of the specific product, which reflects presence or absence of the target organism in the sample. Species-specific detection tools for cultivation-independent analyses have been developed for *E. maimaiga* (Castrillo et al. 2007), *M. anisopliae* var. *acridum* (Entz et al. 2005), *Neozygites tanaijoae* (Entomophthoromycotina: Entomophthorales) (Agboton et al. 2009), *Pandora blunckii* (Entomophthoromycotina: Entomophthoralses)(Guzman-Franco et al. 2008), *Pandora kondoiensis* (Entomophthoromycotina: Entomophthorales) (Tymon et al. 2004), *P. neoaphidis* (Tymon et al. 2004; Fournier et al. 2008; Fournier et al. 2009), *Z. radicans* (Guzman-Franco et al. 2008) targeting the ITS/SSU region, and *B. brongniartii* (Schwarzenbach et al. 2007b) targeting a SSR. Detection at the strain-level has been developed for instance for *B. bassiana* strain IMI391510 (Bell et al. 2009), *M. anisopliae* var. *acridum* strain IMI330189 (Bell et al. 2009), *M. anisopliae* var. *anisopliae* strains E_9, B/Vi and C (Destefano et al. 2004) targeting the ITS region and *B. bassiana* strains GHA (Castrillo et al. 2003; Castrillo et al. 2008) and F-263 (Takatsuka 2007) targeting a SCAR. Cultivation-independent detection has for example been applied to investigate overwintering strategies of *P. neoaphidis* (Fournier et al. 2008). Using species-specific amplification of a fragment of the rRNA gene cluster of *P. neoaphidis* it was possible to detect *P. neoaphidis* DNA during winter and spring in topsoil samples collected from a nettle field harbouring infected aphids in fall. Furthermore, in a study on the *E. muscae* species complex, specific cultivation-independent amplification was combined with RFLP analysis (Thomsen and Jensen 2002). A nested PCR technique with *Entomophthora*-specific primers was used to specifically amplify the ITS II region from resting spore-bearing fly cadavers. Subsequent RFLP analysis on the obtained PCR products allowed identification of the different sub-groups of the *E. muscae* complex, which is not possible based on the morphology of isolated resting spores. In addition, the results of this study confirmed the previously observed correlation between *E. muscae* sub-groups and fly host species, which was based on cultivation-dependent analyses (Jensen et al. 2001).

In various cases, the tools have been adapted for quantitative detection of a species or a strain by use of real time PCR. Such approaches have been developed at the species-level for *B. brongniartii* (Schwarzenbach et al. 2009), *E. maimaiga* (Castrillo et al. 2007), and *P. neoaphidis* (Enkerli, unpublished) and at the strain-level for *B. bassiana* strains GHA (Castrillo et al. 2008) and IMI391510 (Bell et al. 2009), and *M. anisopliae* var. *acridum* strain IMI330189 (Bell et al. 2009). Cultivation-independent quantitative detection tools offer new ways for ecological studies on these fungi and for monitoring applied biological control agents (BCA). For example, resting spores of *E. maimaiga* were successfully quantified in different soil types by use of this PCR application (Castrillo et al. 2007). *E. maimaiga* can not be cultivated from soil, thus the quantitative PCR applications will allow monitoring spore titres in situ, which will contribute to the understanding of the life cycle and ecology of this species.

Analysis of community structures

Cultivation-independent analysis of community structures allows to determine presence and relative abundance of specific genotypes and comparison of structures of bacterial and/or fungal communities in complex samples (Kirk et al. 2004). For this approach specific marker genes are amplified from complex DNA samples and subsequently resolved by use of various analytical procedures. Resulting profiles represent relative images of the community structure present in the sample and may suffer from PCR biases. Therefore, this approach does not allow for quantitative assessments. For such purposes quantitative PCR approaches have to be applied (see above). Analyses rely on the use of universal primers that define the target group according to their amplification range. The SSU rRNA gene and the ITS region of the rRNA gene cluster have been the main target loci for analyses of fungal community structures (Anderson and Cairney 2004), however other loci, e.g., EF1-α, have been targeted as well (Yergeau et al. 2005). A large number of different primers has been designed during the past 20 years that are used for community structure analyses of different taxonomic groups, i.e., on different phylogenetic levels from genus up to the fungal kingdom (Smit et al. 1999; Borneman and Hartin 2000; Anderson and Cairney 2004; de Souza et al. 2004; Green et al. 2004; Yergeau et al. 2005; Oliveira et al. 2009). Various genetic profiling procedures are available to resolve the targeted community structures. They include LP (Suzuki et al. 1998) and RISA (Fisher and Triplett 1999; Ranjard et al. 2001), which are used to reveal length polymorphisms in the amplified marker gene fragments. PCR–RFLP- and Amplified Ribosomal DNA Restriction Analysis (ARDRA, Widmer et al. 2001; Tun et al. 2002; Hunt et al. 2004), or terminal RFLP (T-RFLP, Liu et al. 1997; Lord et al. 2002; Schwarzenbach et al. 2007a), which are applied to distinguish sequences based on variations in the location of restriction enzyme recognition sites. DGGE or TGGE, which rely on differences in DNA duplex stability (Muyzer and Smalla 1998; van Elsas et al. 2000) or SSCP, which detects differences in secondary structure of single-stranded DNA (Schwieger and Tebbe 1998).

There is an increasing number of studies where these techniques have successfully been applied for analyzing and comparing complex microbial communities. For example, T-RFLP has been applied to investigate fungal diversity in agricultural land that was turned into fallow fields (Klamer and Hedlund 2004). DGGE has been used to analyze the effects of different cultivation factors on plant pathogenic *Fusarium* and arbuscular mycorrhizal fungi communities in asparagus fields (Yergeau et al. 2006) or to investigate differences in ascomycetes rhizosphere communities in different crops (Viebahn et al. 2005). Furthermore, TGGE and SSCP analyses have been performed to asses and compare structures of soil fungal communities in different forest soils (He et al. 2005). So far cultivation-independent community analyses have not been performed specifically on fungal entomopathogens. However, in a recent study RISA has been applied to assess potential effects of the *B. brongniartii* BCA on soil fungal communities in microcosms (Schwarzenbach et al. 2009). RISA has revealed that application of this BCA has only marginally affected soil fungal communities, while strong and significant effects have been caused by dying *M. melolontha* larvae killed by an insecticide or the BCA. This study has demonstrated the use of such approaches for effect assessment of fungal BCAs.

Profiles of amplified marker regions do not provide direct information on the identity of the genotypes detected. One way to improve taxonomic information content of community structure analysis is to isolate and sequence specific bands from gel electrophoretic analyses, such as DGGE (Muyzer and Smalla 1998), T-RFLP (Widmer et al. 2006), or SSCP (Schwieger and Tebbe 1998). Sequences may subsequently be identified by similarity searches in public data bases like GenBank or RDP. This approach has been used to describe and compare ascomycete taxa present in the rhizosphere of wheat in monoculture and wheat in rotation with potato (Viebahn et al. 2005) or to identify mycorrhizal species present in the rhizoshpere of maize plants (Oliveira et al. 2009). Obtained sequence information then may allow for designing primers for a strain- or species-specific PCR detection (see above) (Pesaro and Widmer 2006; Widmer et al. 2006).

An approach that has been pursued in various studies is to shotgun clone and sequence entire PCR amplification products of marker regions, and to subsequently identify obtained sequences by performing similarity searches in public databases as described above. This approach has been applied to identify and compare fungal communities, e.g., in plant roots (Vandenkoornhuyse et al. 2002), plant rhizospheres (Smit et al. 1999), and in various types of soils including agricultural (Lynch and Thorn 2006; Midgley et al. 2007), grassland (Midgley et al. 2007), tundra (Schadt et al. 2003), and forest (He et al. 2005; O'Brien et al. 2005) soils. In these studies analyses have been performed on different taxonomic levels using fungus (Smit et al. 1999; Schadt et al. 2003) or basidiomycete-specific primers (Lynch and Thorn 2006; Midgley et al. 2007). Even though fungal entomopathogens have not been specifically targeted in any of these studies, entomopathogenic genera have been detected within the fungal communities, i.e. *Paecilomyces* spp. (= *Isaria spp.*) (Smit et al. 1999) and *Entomophthora* spp. (Lynch and Thorn 2006). Interestingly, in the latter case, the genus *Entomophthora,* which belongs to the Entomophthoramycotina (Hibbett et al. 2007) has been detected with a primer pair designed to specifically amplify Basidiomycetes. This indicates that amplification ranges of primers are not strict, i.e., a primer pair designed to amplify the bulk of a phylogenetically related group may cross amplify individuals from non-target groups. Furthermore, whether a specific group of organisms can be detected by such a global approach largely depends on its abundance. It is necessary to screen large numbers of clones to reach saturated resolution and to detect also organisms of low abundance. Large scale or next generation sequencing approaches like pyrosequencing provide the capacity to generate thousands of sequences, which may allow to obtain maximal resolution and species representation (Ronaghi et al.

1998; Margulies et al. 2005). This technology allows for instance shotgun sequencing of single genomes (Margulies et al. 2005), massive parallel sequencing of PCR amplified target regions like rRNA genes or ITS regions from metagenomic samples (Christen 2008; Petrosino et al. 2009), and whole genome shotgun sequencing of metagenomic DNA samples (Tringe and Rubin 2005; Petrosino et al. 2009), where partial genomes of diverse organisms are sequenced simultaneously. Such a metagenomic sequencing approach has for instance been applied to survey microorganisms associated with honey bee colony collapse disorder (CCD, Cox-Foster et al. 2007). Metagenomic pyrosequencing and subsequent nucleotide sequence comparisons using data base searches revealed the presence of viruses, parasites, metazoan, bacteria, and fungi, including the entomopathogenic species *Pandora delphacis* in CCD-positive bee samples. Among the identified microorganisms only Israeli acute paralysis virus of bee was strongly correlated with colony collapse disorder.

Conclusions

A large body of information and data has been acquired over the past years on various ecological aspects of fungal entomopathogens. The use of molecular techniques has increasingly influenced ecological research on fungal entomopathogens and in combination with other disciplines has contributed to progress made during the past decade. Knowledge on the life cycle of various entomopathogenic fungal species has been improved, their natural occurrence and dispersal mechanisms has been investigated and new qualities as endophytes, rhizosphere colonizers, plant growth promoters, or antagonists of plant diseases have been discovered (Meyling and Eilenberg 2007; Vega et al. 2009; Bruck 2009; Ownley et al. 2009).

However, there are still many open questions remaining, particularly regarding ecological aspects like population characteristics and relation of genetic structure and habitat type as well as host associations. Fungal entomopathogens perform important ecosystem functions by controlling insect population levels and this function constitutes an essential aspect in the process of self regulation in the environmental network. However, many details and links of this function are still not fully understood and need further investigation. Improvement of this basic knowledge will also help to further explore the potential of these fungi in biological control and to develop integrated control strategies.

Molecular ecological approaches applying tools like SSR or SNP high resolution genetic markers may allow better description of population structures and help to understand global as well as local migration and dissemination patterns. Application of cultivation-independent detection and quantification techniques may improve efficiency of how these organisms can be monitored and may help to explore so far unknown stages of their life cycle. Furthermore, the use of cultivation-independent analyses of community structures using profiling and/or large scale sequencing approaches offer ways to investigate and understand how fungal entomopathogens interact with environmental factors.

Genome sequencing has become readily accessible and complete or near complete DNA sequences are available for many microorganisms. Whole genome sequences on the one hand may provide the basis for a more profound understanding of functions of a microorganism and on the other hand offer the potential for developing defined genetic modifications of specific genetic traits and specific molecular diagnostics. However, currently there is only the genome sequence for one fungal entomopathogen *Ascospharera apis* (Qin et al. 2006) available. Therefore, there is definite need to obtain more sequences of entire genomes of fungal enotomopathogens to allow for comparative genomics of this functionally important group of fungi. New and comprehensive genome sequence information will provide a more profound approach towards gene function, which will allow to further improve existing or develop new DNA microarray technologies (Freimoser et al. 2005; Wang et al. 2005a) and develop new post-genomic approaches. Research on fungal entomopathogen ecology will benefit from molecular ecology tools and applications also in the future.

References

Agboton BV, Delalibera I, Hanna R, von Tiedemann A (2009) Molecular detection and differentiation of Brazilian and African isolates of the entomopathogen *Neozygites tanajoae* (Entomophthorales: Neozygitaceae) with PCR using specific primers. Biocontrol Sci Technol 19:67–79

Aghajanzadeh S, Prasad DT, Mallik B (2007) Genetic diversity in *Hirsutella thompsonii* isolates based on random amplified polymorphic DNA analysis. BioControl 52: 375–383

Akkermans ADL, Van Elsas JD, DeBruijn FJ (1995) Molecular microbial ecology manual. Kluwer, Dordrecht

Amann RI, Ludwig W, Schleifer KH (1995) Phylogenetic identification and in situ detection of individual microbial-cells without cultivation. Microbiol Rev 59:143–169

Amann R, Fuchs BM, Behrens S (2001) The identification of microorganisms by fluorescence in situ hybridisation. Curr Opin Biotechnol 12:231–236

Anderson IC, Cairney JWG (2004) Diversity and ecology of soil fungal communities: increased understanding through the application of molecular techniques. Environ Microbiol 6:769–779

Aquino de Muro M, Mehta S, Moore D (2003) The use of amplified fragment length polymorphism for molecular analysis of *Beauveria bassiana* isolates from Kenya and other countries, and their correlation with host and geographical origin. FEMS Microbiol Lett 229:249–257

Aquino de Muro M, Elliott S, Moore D, Parker BL, Skinner M, Reid W, El Bouhssini M (2005) Molecular characterisation of *Beauveria bassiana* isolates obtained from overwintering sites of Sunn Pests (*Eurygaster* and *Aelia* species). Mycol Res 109:294–306

Arbeli Z, Fuentes CL (2007) Improved purification and PCR amplification of DNA from environmental samples. FEMS Microbiol Lett 272:269–275

Avis JC (2004) Molecular markers, natural history, and evolution. Sinauer Associates, Sunderland

Bain JM, Tavanti A, Davidson AD, Jacobsen MD, Shaw D, Gow NAR, Odds FC (2007) Multilocus sequence typing of the pathogenic fungus *Aspergillus fumigatus*. J Clin Microbiol 45:1469–1477

Beebee T, Rowe G (2008) An introduction to molecular ecology. Oxford University Press, Oxford

Bell AS, Blanford S, Jenkins N, Thomas MB, Read AF (2009) Real-time quantitative PCR for analysis of candidate fungal biopesticides against malaria: technique validation and first applications. J Invertebr Pathol 100:160–168

Bidochka MJ, Walsh SRA, Ramos ME, St Leger RJ, Silver JC, Roberts DW (1995) Pathotypes in the *Entomophaga grylli* species complex of grasshopper pathogens differentiated with random amplification of polymorphic DNA and cloned-DNA probes. Appl Environ Microbiol 61: 556–560

Bidochka MJ, St Leger RJ, Stuart A, Gowanlock K (1999) Nuclear rDNA phylogeny in the fungal genus *Verticillium* and its relationship to insect and plant virulence, extracellular proteases and carbohydrases. Microbiology 145: 955–963

Bidochka MJ, Kamp AM, Lavender TM, Dekoning J, De Croos JN (2001) Habitat association in two genetic groups of the insect-pathogenic fungus *Metarhizium anisopliae*: uncovering cryptic species? Appl Environ Microbiol 67: 1335–1342

Bidochka MJ, Small CLN, Spironello M (2005) Recombination within sympatric cryptic species of the insect pathogenic fungus *Metarhizium anisopliae*. Environ Microbiol 7:1361–1368

Bischoff JF, Rehner SA, Humber RA (2009) A multilocus phylogeny ofthe *Metarhizium anisopliae* lineage. Mycologia 101:512–530

Blackwell M (2009) Fungal evolution and taxonomy. BioControl. doi:10.1007/s10526-009-9243-8 (this SI)

Borneman J, Hartin RJ (2000) PCR primers that amplify fungal rRNA genes from environmental samples. Appl Environ Microbiol 66:4356–4360

Boucias DG, Tigano MS, Sosa-Gomez DR, Glare TR, Inglis PW (2000) Genotypic properties of the entomopathogenic fungus *Nomuraea rileyi*. Biol Control 19:124–138

Bridge PD, Prior C, Sagbohan J, Lomer CJ, Carey M, Buddie AG (1997) Molecular characterization of isolates of *Metarhizium* from locusts and grasshoppers. Biodivers Conserv 6:177–189

Brookes AJ (1999) The essence of SNPs. Gene 234:177–186

Bruck D (2009) Fungal entomopathogens in the rhizosphere. BioControl. doi:10.1007/s10526-009-9236-7 (this SI)

Brugmans B, van der Hulst RGM, Visser RGF, Lindhout P, van Eck HJ (2003) A new and versatile method for the successful conversion of AFLP markers into simple single locus markers. Nucleic Acids Res 31:e55

Bulat SA, Mironenko N (1990) Species identity of the phytopathogenic fungi *Pyrenophora teres* Drechsler and *P. graminea* Ito & Kuribayashi. Mikol Fitopatol 24:435–441

Bürgmann H, Pesaro M, Widmer F, Zeyer J (2001) A strategy for optimizing quality and quantity of DNA extracted from soil. J Microbiol Methods 45:7–20

Cariello NF, Thilly WG, Swenberg JA, Skopek TR (1991) Deletion mutagenesis during polymerase chain-reaction—dependence on DNA-polymerase. Gene 99:105–108

Castrillo LA, Vandenberg JD, Wraight SP (2003) Strain-specific detection of introduced *Beauveria bassiana* in agricultural fields by use of sequence-characterized amplified region markers. J Invertebr Pathol 82:75–83

Castrillo LA, Thomsen L, Juneja P, Hajek AE (2007) Detection and quantification of *Entomophaga maimaiga* resting spores in forest soil using real-time PCR. Mycol Res 111: 324–331

Castrillo L, Griggs MH, Vandenberg JD (2008) Quantitative detection of *Beauveria bassiana* GHA (Ascomycota: Hypocreales), a potential microbial control agent of the emerald ash borer, by use of real-time PCR. Biol Control 45:163–169

Chew JSK, Strongman DB, MacKay RM (1998) Comparisons of twenty isolates of the entomopathogen *Paecilomyces farinosus* by analysis of RAPD markers. Mycol Res 102: 1254–1258

Christen R (2008) Global sequencing: a review of current molecular data and new methods available to assess microbial diversity. Microbes Environ 23:253–268

Coates BS, Hellmich RL, Lewis LC (2002a) *Beauveria bassiana* haplotype determinatin based on nuclear rDNA internal transcribed spacer PCR-RFLP. Mycol Res 106:40–50

Coates BS, Hellmich RL, Lewis LC (2002b) Nuclear small subunit rRNA group I intron variation among *Beauveria* spp. provide tools tor strain identification and evidence of horizontal transfer. Curr Genet 41:414–424

Cox-Foster DL, Conlan S, Holmes EC, Palacios G, Evans JD, Moran NA, Quan PL, Briese T, Hornig M, Geiser DM, Martinson V, van Engelsdorp D, Kalkstein AL, Drysdale

A, Hui J, Zhai JH, Cui LW, Hutchison SK, Simons JF, Egholm M, Pettis JS, Lipkin WI (2007) A metagenomic survey of microbes in honey bee colony collapse disorder. Science 318:283–287

Cravanzola F, Piatti P, Bridge PD, Ozino OI (1997) Detection of genetic polymorphism by RAPD-PCR in strains of the entomopathogenic fungus *Beauveria brongniartii* isolated from the European cockchafer (*Melolontha* spp.). Lett Appl Microbiol 25:289–294

Dalleau-Clouet C, Gauthier N, Risterucci AM, Bon MC, Fargues J (2005) Isolation and characterization of microsatellite loci from the entomopathogenic hyphomycete, *Paecilomyces fumosoroseus*. Mol Ecol Notes 5:496–498

de Souza FA, Kowalchuk GA, Leeflang P, van Veen JA, Smit E (2004) PCR-denaturing gradient gel electrophoresis profiling of inter- and intraspecies 18S rRNA gene sequence heterogeneity is an accurate and sensitive method to assess species diversity of arbuscular mycorrhizal fungi of the genus Gigaspora. Appl Environ Microbiol 70:1413–1424

Destefano RHR, Destefano SAL, Messias CL (2004) Detection of *Metarhizium anisopliae* var. *anisopliae* within infected sugarcane borer *Diatraea saccharalis* (Lepidoptera, Pyralidae) using specific primers. Genet Mol Biol 27:245–252

Devi UK, Reineke A, Rao UCM, Reddy NRN, Khan APA (2007) AFLP and single-strand conformation polymorphism studies of recombination in the entomopathogenic fungus *Nomuraea rileyi*. Mycol Res 111:716–725

Enkerli J, Widmer F, Gessler C, Keller S (2001) Strain-specific microsatellite markers in the entomopathogenic fungus *Beauveria brongniartii*. Mycol Res 105:1079–1087

Enkerli J, Widmer F, Keller S (2004) Long-term field persistence of *Beauveria brongniartii* strains applied as biocontrol agents against European cockchafer larvae in Switzerland. Biol Control 29:115–123

Enkerli J, Kölliker R, Keller S, Widmer F (2005) Isolation and characterization of microsatellite markers from the entomopathogenic fungus *Metarhizium anisopliae*. Mol Ecol Notes 5:384–386

Enkerli J, Ghormade V, Oulevey C, Widmer F (2009) PCR-RFLP analysis of chitinase genes enables efficient genotyping of *Metarhizium anisopliae* var. *anisopliae*. J Invertebr Pathol 102:185–188

Entz SC, Johnson DL, Kawchuk LM (2005) Development of a PCR-based diagnostic assay for the specific detection of the entomopathogenic fungus *Metarhizium anisopliae* var. *acridum*. Mycol Res 109:1302–1312

Estrada ME, Camacho MV, Benito C (2007) The molecular diversity of different isolates of *Beauveria bassiana* (Bals.) Vuill. as assessed using intermicrosatellites (ISSRs). Cell Mol Biol Lett 12:240–252

Fang WG, Leng B, Xiao YH, Jin K, Ma JC, Fan YH, Feng J, Yang XY, Zhang YJ, Pei Y (2005) Cloning of *Beauveria bassiana* chitinase gene *Bbchit1* and its application to improve fungal strain virulence. Appl Environ Microbiol 71:363–370

Fargues J, Bon MC, Manguin S, Couteaudier Y (2002) Genetic variability among *Paecilomyces fumosoroseus* isolates from various geographical and host insect origins based on the rDNA-ITS regions. Mycol Res 106:1066–1074

Fegan M, Manners JM, Maclean DJ, Irwin JAG, Samuels KDZ, Holdom DG, Li DP (1993) Random amplified polymorphic DNA markers reveal a high degree of genetic diversity in the entomopathogenic fungus *Metarhizium anisopliae* var. *anisopliae*. J Gen Microbiol 139:2075–2081

Feinstein LM, Sul WJ, Blackwood CB (2009) Assessment of bias associated with incomplete extraction of microbial DNA from soil. Appl Environ Microbiol 75:5428–5433

Felsenstein J (2009) PHYLIP (Phylogeny Inference Package) version 3.69. Distributed by the author. Department of Genome Sciences, University of Washington, Seattle

Fernandes EKK, Costa GL, Moraes AML, Zahner V, Bittencourt V (2006) Study on morphology, pathogenicity, and genetic variability of *Beauveria bassiana* isolates obtained from *Boophilus microplus* tick. Parasitol Res 98:324–332

Field D, Wills C (1996) Long, polymorphic microsatellites in simple organisms. Proc R Soc Lond B Biol Sci 263: 209–215

Fisher MM, Triplett EW (1999) Automated approach for ribosomal intergenic spacer analysis of microbial diversity and its application to freshwater bacterial communities. Appl Environ Microbiol 65:4630–4636

Fournier A, Widmer F, Enkerli J (2009) Development of a single nucleotide polymorphism (SNP) assay for genotyping of *Pandora neoaphidis*. (submitted)

Fournier A, Enkerli J, Keller S, Widmer F (2008) A PCR-based tool for the cultivation-independent monitoring of *Pandora neoaphidis*. J Invertebr Pathol 99:49–56

Francis C, Anagnou-Veroniki N, Rouffaud MA, De Bievre C, Papierok B (2004) Morphometric and genetic variation among strains of two related species of *Erynia* (Zygomycota, Entomophthorales) isolated from aphids or Diptera in a limited geographical area Ingreece. J Mycol Med 14:171–180

Freimoser FM, Hu G, Leger RJ (2005) Variation in gene expression patterns as the insect pathogen *Metarhizium anisopliae* adapts to different host cuticles or nutrient deprivation in vitro. Microbiology 151:361–371

Frey JC, Angert ER, Pell AN (2006) Assessment of biases associated with profiling simple, model communities using terminal-restriction fragment length polymorphism-based analyses. J Microbiol Methods 67:9–19

Frostegård A, Courtois S, Ramisse V, Clerc S, Bernillon D, Le Gall F, Jeannin P, Nesme X, Simonet P (1999) Quantification of bias related to the extraciton of DNA directly form soils. Appl Environ Microbiol 65:5409–5420

Gauthier N, Dalleau-Clouet C, Fargues J (2007) Microsatellite variability in the entomopathogenic fungus *Paecilomyces fumosoroseus*: genetic diversity and population structure. Mycologia 99:693–704

George MLC, Nelson RJ, Zeigler RS, Leung H (1998) Rapid population analysis of *Magnaporthe grisea* by using rep-PCR and endogenous repetitive DNA sequences. Phytopathology 88:223–229

Ghikas DV, Kouvelis VN, Typas MA (2006) The complete mitochondrial genome of the entomopathogenic fungus *Metarhizium anisopliae* var. *anisopliae*: gene order and trn gene clusters reveal a common evolutionary course for all Sordariomycetes, while intergenic regions show variation. Arch Microbiol 185:393–401

Glare TR, Reay SD, Nelson TL, Moore R (2008) *Beauveria caledonica* is a naturally occurring pathogen of forest beetles. Mycol Res 112:352–360

Goldstein DB, Schlötterer C (1999) Microsatellites: evolution and applications. Oxford University Press, New York

Green SJ, Freeman S, Hadar Y, Minz D (2004) Molecular tools for isolate and community studies of Pyrenomycete fungi. Mycologia 96:439–451

Groppe K, Sanders I, Wiemken A, Boller T (1995) A microsatellite marker for studying the ecology and diversity of fungal endophytes (*Epichloe* spp.) in grasses. Appl Environ Microbiol 61:3943–3949

Guzman-Franco AW, Atkins SD, Alderson PG, Pell JK (2008) Development of species-specific diagnostic primers for *Zoophthora radicans* and *Pandora blunckii*; two co-occurring fungal pathogens of the diamondback moth, *Plutella xylostella*. Mycol Res 112:1227–1240

Hajek AE, Delalibera I (2009) Fungal pathogens as classical biological control agents against arthropods. BioControl. doi:10.1007/s10526-009-9253-6 (this SI)

Hajek AE, Hodge KT, Liebherr JK, Day WH, Vandenberg JD (1996) Use of RAPD analysis to trace the origin of the weevil pathogen *Zoophthora phytonomi* in North America. Mycol Res 100:349–355

Hajek AE, Jensen AB, Thomsen L, Hodge KT, Eilenberg J (2003) PCR-RFLP is used to investigate relations among species in the entomopathogenic genera *Eryniopsis* and *Entomophaga*. Mycologia 95:262–268

Hartmann M, Widmer F (2006) Community structure analyses are more sensititive to differences in soil bacterial communities than anonymous diversity indices. Appl Environ Microbiol 72:7804–7812

Hartmann M, Widmer F (2008) Reliability for detecting composition and changes of microbial communities by T-RFLP genetic profiling. FEMS Microbiol Ecol 63:249–260

Hartmann M, Fliessbach A, Oberholzer H-R, Widmer F (2006) Ranking the magnitude of crop and farming system effects on soil microbial biomass and genetic structure of bacterial communities. FEMS Microbiol Ecol 57:378–388

Hartmann M, Enkerli J, Widmer F (2007) Residual polymerase activity-induced bias in terminal restriction fragment length polymorphism analysis. Environ Microbiol 9: 555–559

He JZ, Xu ZH, Hughes J (2005) Analyses of soil fungal communities in adjacent natural forest and hoop pine plantation ecosystems of subtropical Australia using molecular approaches based on 18S rRNA genes. FEMS Microbiol Lett 247:91–100

Hegedus DD, Khachatourians GG (1996) Identification and differentiation of the entomopathogenic fungus *Beauveria bassiana* using polymerase chain reaction and singel-strand conformation polymorphism analysis. J Invertebr Pathol 67:289–299

Hernan-Gomez S, Espinosa JC, Ubeda JF (2000) Characterization of wine yeasts by temperature gradient gel electrophoresis (TGGE). FEMS Microbiol Lett 193:45–50

Hesketh H, Roy HE, Eilenberg J, Hails RS (2009) Challenges in modelling complexity of fungal entomopathogens in semi-natural populations of insects. BioControl. doi:10.1007/s10526-009-9249-2 (this SI)

Hibbett DS, Binder M, Bischoff JF, Blackwell M, Cannon PF, Eriksson OE, Huhndorf S, James T, Kirk PM, Lucking R, Lumbsch HT, Lutzoni F, Matheny PB, McLaughlin DJ, Powell MJ, Redhead S, Schoch CL, Spatafora JW, Stalpers JA, Vilgalys R, Aime MC, Aptroot A, Bauer R, Begerow D, Benny GL, Castlebury LA, Crous PW, Dai YC, Gams W, Geiser DM, Griffith GW, Gueidan C, Hawksworth DL, Hestmark G, Hosaka K, Humber RA, Hyde KD, Ironside JE, Koljalg U, Kurtzman CP, Larsson KH, Lichtwardt R, Longcore J, Miadlikowska J, Miller A, Moncalvo JM, Mozley-Standridge S, Oberwinkler F, Parmasto E, Reeb V, Rogers JD, Roux C, Ryvarden L, Sampaio JP, Schussler A, Sugiyama J, Thorn RG, Tibell L, Untereiner WA, Walker C, Wang Z, Weir A, Weiss M, White MM, Winka K, Yao YJ, Zhang N (2007) A higher-level phylogenetic classification of the Fungi. Mycol Res 111:509–547

Hodge KT, Sawyer AJ, Humber RA (1995) RAPD-PCR for identification of *Zoophthora radicans* isolates in biological control of the potato leafhopper. J Invertebr Pathol 65: 1–9

Hughes JB, Hellmann JJ (2005) The application of rarefaction techniques to molecular inventories of microbial diversity. Methods Enzymol 397:292–308

Hughes JB, Hellmann JJ, Ricketts TH, Bohannan BJM (2001) Counting the uncountable: statistical approaches to estimating microbial diversity. Appl Environ Microbiol 67: 4399–4406

Hunt J, Boddy L, Randerson PF, Rogers HJ (2004) An evaluation of 18S rDNA approaches for the study of fungal diversity in grassland soils. Microb Ecol 47:385–395

Inglis GD, Duke GM, Goettel MS, Kabaluk JT (2008) Genetic diversity of *Metarhizium anisopliae* var. *anisopliae* in southwestern British Columbia. J Invertebr Pathol 98: 101–113

Jackson CR, Harper JP, Willoughby D, Roden EE, Churchill PF (1997) A simple, efficient method for the separation of humic substances and DNA from environmental samples. Appl Environ Microbiol 63:4993–4995

James TY, Kauff F, Schoch CL, Matheny PB, Hofstetter V, Cox CJ, Celio G, Gueidan C, Fraker E, Miadlikowska J, Lumbsch HT, Rauhut A, Reeb V, Arnold AE, Amtoft A, Stajich JE, Hosaka K, Sung GH, Johnson D, O'Rourke B, Crockett M, Binder M, Curtis JM, Slot JC, Wang Z, Wilson AW, Schussler A, Longcore JE, O'Donnell K, Mozley-Standridge S, Porter D, Letcher PM, Powell MJ, Taylor JW, White MM, Griffith GW, Davies DR, Humber RA, Morton JB, Sugiyama J, Rossman AY, Rogers JD, Pfister DH, Hewitt D, Hansen K, Hambleton S, Shoemaker RA, Kohlmeyer J, Volkmann-Kohlmeyer B, Spotts RA, Serdani M, Crous PW, Hughes KW, Matsuura K, Langer E, Langer G, Untereiner WA, Lucking R, Budel B, Geiser DM, Aptroot A, Diederich P, Schmitt I, Schultz M, Yahr R, Hibbett DS, Lutzoni F, McLaughlin DJ, Spatafora JW, Vilgalys R (2006) Reconstructing the early evolution of fungi using a six-gene phylogeny. Nature 443:818–822

Jensen AB, Eilenberg J (2001) Genetic variation within the insect-pathogenic genus *Entomophthora*, focusing on the *E. muscae* complex, using PCR-RFLP of the ITS II and the LSU rDNA. Mycol Res 105:307–312

Jensen AB, Gargas A, Eilenberg J, Rosendahl S (1998) Relationships of the insect-pathogenic order entomophthorales (Zygomycota, Fungi) based on phylogenetic analyses of nuclear small subunit ribosomal DNA sequences (SSU rDNA). Fungal Genet Biol 24:325–334

Jensen AB, Thomsen L, Eilenberg J (2001) Intraspecific variation and host specificity of *Entomophthora muscae sensu stricto* isolates revealed by random amplified polymorphic DNA, universal primed PCR, PCR-restriction fragment length polymorphism, and conidial morphology. J Invertebr Pathol 78:251–259

Kabir S, Rajendran N, Amemiya T, Itoh T (2003) Quantitative measurement of fungal DNA extracted by three different methods using real-time polymerase chain reaction. J Biosci Bioeng 96:337–343

Kanagawa T (2003) Bias and artifacts in multitemplate polymerase chain reactions (PCR). J Biosci Bioeng 96:317–323

Kim S, Misra A (2007) SNP genotyping: technologies and biomedical applications. Annu Rev Biomed Eng 9:289–320

Kirk JL, Beaudette LA, Hart M, Moutoglis P, Khironomos JN, Lee H, Trevors JT (2004) Methods of studying soil microbial diversity. J Microbiol Methods 58:169–188

Klamer M, Hedlund K (2004) Fungal diversity in set-aide agricultural soil investigated using terminal-restriction fragment length polymorphism. Soil Biol Biochem 36: 983–988

Kouvelis VN, Ghikas DV, Edgington S, Typas MA, Moore D (2008a) Molecular characterization of isolates of *Beauveria bassiana* obtained from overwintering and summer populations of Sunn Pest (*Eurygaster integriceps*). Lett Appl Microbiol 46:414–420

Kouvelis VN, Sialakouma A, Typas MA (2008b) Mitochondrial gene sequences alone or combined with ITS region sequences provide firm molecular criteria for the classification of *Lecanicillium* species. Mycol Res 112:829–844

Kowalchuk GA, deBruijn FJ, Head IM, Akkermans ADL, Van Elsas JD (2004) Molecular microbial ecolgy manual. Kluwer, Dordrecht

Kreader CA (1996) Relief of amplification inhibition in PCR with bovine serum albumin or T4 gene 32 protein. Appl Environ Microbiol 62:1102–1106

Kristensen R, Berdal KG, Holst-Jensen A (2007) Simultaneous detection and identification of trichothecene- and moniliformin-producing *Fusarium* species based on multiplex SNP analysis. J Appl Microbiol 102:1071–1081

Kuske CR, Banton KL, Adorada DL, Stark PC, Hill KK, Jackson PJ (1998) Small-scale DNA sample preparation method for field PCR detection of microbial cells and spores in soil. Appl Environ Microbiol 64:2463–2472

Lambreghts R, Shi M, Belden WJ, de Caprio D, Park D, Henn MR, Galagan JE, Bastürkmen M, Birren BW, Sachs MS, Dunlap JC, Loros JJ (2009) A high-density single nucleotide polymorphism map for *Neurospora crassa*. Genetics 181:767–781

LaMontagne MG, Michel FC, Holden PA, Reddy CA (2002) Evaluation of extraction and purification methods for obtaining PCR-amplifiable DNA from compost for microbial community analysis. J Microbiol Methods 49: 255–264

Leal SCM, Bertioli DJ, Butt TM, Peberdy JF (1994) Characterization of isolates of the entomopathogenic fungus *Metarhizium anisopliae* by RAPD-PCR. Mycol Res 98:1077–1081

Leal SCM, Bertioli DJ, Butt TM, Carder JH, Burrows PR, Peberdy YF (1997) Amplification and restriction endonuclease digestion of the *Pr1* gene for the detection and characterization of *Metarhizium* strains. Mycol Res 101: 257–265

Leland JE, McGuire MR, Grace JA, Jaronski ST, Ulloa M, Park YH, Plattner RD (2005) Strain selection of a fungal entomopathogen, *Beauveria bassiana*, for control of plant bugs (*Lygus* spp.) (Heteroptera: Miridae). Biol Control 35:104–114

Lihme M, Jensen AB, Rosendahl S (2009) Local scale population genetic structure of *Entomophthora muscae* epidemics. Fungal Ecol 2:81–86

Liu WT, Marsh TL, Cheng H, Forney LJ (1997) Characterization of microbial diversity by determining terminal restriction fragment length polymorphisms of genes encoding 16S rRNA. Appl Environ Microbiol 63: 4516–4522

Liu ZY, Liang ZQ, Liu AY, Yao YJ, Hyde KD, Yu ZN (2002) Molecular evidence for teleomorph-anamorph connections in *Cordyceps* based on ITS-5.8S rDNA sequences. Mycol Res 106:1100–1108

Lloyd-Jones G, Hunter DWF (2001) Comparison of rapid DNA extraction methods applied to contrasting New Zealand soils. Soil Biol Biochem 33:2053–2059

Lord NS, Kaplan CW, Shank P, Kitts CL, Elrod SL (2002) Assessment of fungal diversity using terminal restriction fragment (TRF) pattern analysis: comparison of 18S and ITS ribosomal regions. FEMS Microbiol Ecol 42:327–337

Lynch MJM, Thorn RG (2006) Diversity of basidiomycetes in Michigan agricultural soils. Appl Environ Microbiol 72:7050–7056

Manzano M, Cocolin L, Iacumin L, Cantoni C, Comi G (2005) A PCR-TGGE (temperature gradient gel electrophoresis) technique to assess differentiation among enological *Saccharomyces cerevisiae* strains. Int J Food Microbiol 101:333–339

Margulies M, Egholm M, Altman WE, Attiya S, Bader JS, Bemben LA, Berka J, Braverman MS, Chen YJ, Chen ZT, Dewell SB, Du L, Fierro JM, Gomes XV, Godwin BC, He W, Helgesen S, Ho CH, Irzyk GP, Jando SC, Alenquer MLI, Jarvie TP, Jirage KB, Kim JB, Knight JR, Lanza JR, Leamon JH, Lefkowitz SM, Lei M, Li J, Lohman KL, Lu H, Makhijani VB, McDade KE, McKenna MP, Myers EW, Nickerson E, Nobile JR, Plant R, Puc BP, Ronan MT, Roth GT, Sarkis GJ, Simons JF, Simpson JW, Srinivasan M, Tartaro KR, Tomasz A, Vogt KA, Volkmer GA, Wang SH, Wang Y, Weiner MP, Yu PG, Begley RF, Rothberg JM (2005) Genome sequencing in microfabricated high-density picolitre reactors. Nature 437:376–380

Martin-Laurent F, Philippot L, Hallet S, Chaussod R, Germon JC, Soulas G, Catroux G (2001) DNA extraction from soils: old bias for new microbial diversity analysis methods. Appl Environ Microbiol 67:2354–2359

Maurer P, Couteaudier Y, Girard PA, Bridge PD, Riba G (1997) Genetic diversity of *Beauveria bassiana* and relatedness to host insect range. Mycol Res 101:159–164

McGuire MR, Ulloa M, Park YH, Hudson N (2005) Biological and molecular characteristics of *Beauveria bassiana* isolates from California *Lygus hesperus* (Hemiptera: Miridae) populations. Biol Control 33:307–314

Meudt HM, Clarke AC (2007) Almost forgotten or latest practice? AFLP applications, analyses and advances. Trends Plant Sci 12:106–117

Meyling NV, Eilenberg J (2006) Isolation and characterisation of *Beauveria bassiana* isolates from phylloplanes of hedgerow vegetation. Mycol Res 110:188–195

Meyling NV, Eilenberg J (2007) Ecology of the entomopathogenic fungi *Beauveria bassiana* and *Metarhizium anisopliae* in temperate agroecosystems: potential for conservation biological control. Biol Control 43:145–155

Meyling NV, Hajek AE (2009) Principles from community and metapopulation ecology: application to fungal entomopathogens BioControl. doi:10.1007/s10526-009-9246-5 (this SI)

Meyling NV, Lubeck M, Buckley EP, Eilenberg J, Rehner SA (2009) Community composition, host range and genetic structure of the fungal entomopathogen *Beauveria* in adjoining agricultural and seminatural habitats. Mol Ecol 18:1282–1293

Midgley DJ, Saleeba JA, Stewart MI, Simpson AE, McGee PA (2007) Molecular diversity of soil basidiomycete communities in northern-central New South Wales, Australia. Mycol Res 111:370–378

Miller DN, Bryant JE, Madsen EL, Ghiorse WC (1999) Evaluation and optimization of DNA extraction and purification procedures for soil and sediment samples. Appl Environ Microbiol 65:4715–4724

Mitina GV, Mikhailova LA, Yli-Mattila T (2007) RAPD-PCR, UP-PCR and rDNA sequence analyses of the entomopathogenic fungus *Verticillium lecanii* and its pathogenicity towards insects and phytopathogenic fungi. Arch Phytopathol Pflanzenschutz 41:113–128

Mor H, Gindin G, BenZeev IS, Raccah B, Geschtovt NU, Ajtkhozhina N (1996) Diversity among isolates of *Verticillium lecanii* as expressed by DNA polymorphism and virulence towards *Bemisia tabaci*. Phytoparasitica 24:111–118

Morin PA, Luikart G, Wayne RK (2004) SNPs in ecology, evolution and conservation. Trends Ecol Evol 19:208–216

Mozes-Koch R, Edelbaum O, Livneh O, Sztejnberg A, Uziel A, Gerson U, Sela I (1995) Identification of *Hirsutella* species, isolates within a species and intraspecific heterokaryons by random amplified polymorphic DNA (RAPD). J Plant Dis Protect 102:284–290

Mullis K, Falcoma F, Scharf S, Snikl R, Horn G, Erlich H (1986) Specific amplification of DNA in vitro: the polymerase chain reaction. Cold Spring Harbor Symp Quant Biol 51:280

Mullis KB, Ferré F, Gibbs RA (1994) The polymerase chain reaction. Birkhäuser, Boston

Munoz C, Talquenca SG, Volpe ML (2009) Tetra primer ARMS-PCR for identification of SNP in beta-tubulin of *Botrytis cinerea*, responsible of resistance to benzimidazole. J Microbiol Methods 78:245–246

Muyzer G, Smalla K (1998) Application of denaturing gradient gel electrophoresis (DGGE) and temperature gradient gel electrophoresis (TGGE) in microbial ecology. Antonie Van Leeuwenhoek 73:127–141

Nagahama T, Sato H, Shimazu M, Sugiyama J (1995) Phylogenetic divergence of the entomophthoralean fungi: evidence from nuclear 18S ribosomal RNA gene sequences. Mycologia 87:203–209

Neuvéglise C, Brygoo Y, Vercambre B, Riba G (1994) Comparative analysis of molecular and biological characteristics of strains of *Beauveria brongniartii* isolated form insects. Mycol Res 98:322–328

Neuvéglise C, Brygoo Y, Riba G (1997) 28S rDNA group-I introns: a powerful tool for identifying strains of *Beauveria brongniartii*. Mol Ecol 6:373–381

Nielsen C, Sommer C, Eilenberg J, Hansen KS, Humber RA (2001) Characterization of aphid pathogenic species in the genus *Pandora* by PCR techniques and digital image analysis. Mycologia 93:864–874

Nielsen C, Milgroom MG, Hajek A (2005) Genetic diversity in the gypsy moth fungal pathogen *Entomophaga maimaiga* from founder populations in North America and source populations in Asia. Mycol Res 109:941–950

Nikoh N, Fukatsu T (2000) Interkingdom host jumping underground: phylogenetic analysis of entomoparasitic fungi of the genus *Cordyceps*. Mol Biol Evol 17:629–638

O'Brien HE, Parrent JL, Jackson JA, Moncalvo JM, Vilgalys R (2005) Fungal community analysis by large-scale sequencing of environmental samples. Appl Environ Microbiol 71: 5544–5550

Oliveira CA, Sá NMH, Gomes EA, Marriel IE, Scotti MR, Guimarães CT, Schaffert RE, Alves VMC (2009) Assessment of the mycorrhizal community in the rhizosphere of maize (*Zea mays* L.) genotypes contrasting for phosphorus efficiency in the acid savannas of Brazil using denaturing gradient gel electrophoresis (DGGE). Appl Soil Ecol 41:249–258

Oulevey C, Widmer F, Kölliker R, Enkerli J (2009) An optimized microsatellite marker set for detection of *Metarhizium anisopliae* genotype diversity on field and regional scales. Mycol Res 113:1016–1024

Ownley B, Gwinn KD, Vega FE (2009) Endophytic fungal entomopathogens with activity against plant pathogens: ecology and evolution. BioControl. doi:10.1007/s10526-009-9241-x (this SI)

Pantou MP, Mavridou A, Typas MA (2003) IGS sequence variation, group-I introns and the complete nuclear ribosomal DNA of the entomopathogenic fungus *Metarhizium*: excellent tools for isolate detection and phylogenetic analysis. Fungal Genet Biol 38:159–174

Perez T, Albornoz J, Dominguez A (1998) An evaluation of RAPD fragment reproducibility and nature. Mol Ecol 7:1347–1357

Pesaro M, Widmer F (2006) Identification and specific detection of a novel *Pseudomonadaceae* cluster associated with soils from winter wheat plots of a long-term agricultural field experiment. Appl Environ Microbiol 72:37–43

Petrosino JF, Highlander S, Luna RA, Gibbs RA, Versalovic J (2009) Metagenomic pyrosequencing and microbial identification. Clin Chem 55:856–866

Piatti P, Cravanzola F, Bridge PD, Ozino OI (1998) Molecular characterization of *Beauveria brongniartii* isolates obtained from *Melolontha melolontha* in Valle d'Aosta (Italy) by RAPD-PCR. Lett Appl Microbiol 26:317–324

Poussier S, Cheron JJ, Couteau A, Luisetti J (2002) Evaluation of procedures for reliable PCR detection of *Ralstonia solanacearum* in common natural substrates. J Microbiol Methods 51:349–359

Qin X, Evans JD, Aronstein KA, Murray KD, Weinstock GM (2006) Genome sequences of the honey bee pathogens *Paenibacillus* larvae and *Ascosphaera apis*. Insect Mol Biol 15:715–718

Qiu XY, Wu LY, Huang HS, McDonel PE, Palumbo AV, Tiedje JM, Zhou JZ (2001) Evaluation of PCR-generated chimeras: mutations, and heteroduplexes with 16S rRNA gene-based cloning. Appl Environ Microbiol 67:880–887

Rakotonirainy M, Cariou M, Brygoo Y, Riba G (1994) Phylogenetic relationships within the genus *Metarhizium* based on 28S rRNA sequences and isozyme comparison. Mycol Res 98:225–230

Ramette A (2007) Multivariate analyses in microbial ecology. FEMS Microbiol Ecol 62:142–160

Ranjard L, Poly F, Lata JC, Mougel C, Thioulouse J, Nazaret S (2001) Characterization of bacterial and fungal soil communities by automated ribosomal intergenic spacer analysis fingerprints: biological and methodological variability. Appl Environ Microbiol 67:4479–4487

R Development Core Team (2008) R: a language and environment for statistical computing. http://www.R-project.org

Rees GN, Baldwin DS, Watson GO, Perryman S, Nielsen DL (2004) Ordination and significance testing of microbial community composition derived from terminal restriction fragment length polymorphisms: application of multivariate statistics. Antonie Van Leeuwenhoek 86:339–347

Rehner SA, Buckley EP (2003) Isolation and characterization of microsatellite loci from the entomopathogenic fungus *Beauveria bassiana* (Ascomycota: Hypocreales). Mol Ecol Notes 3:409–411

Rehner SA, Buckley E (2005) A *Beauveria* phylogeny inferred from nuclear ITS and EF1-alpha sequences: evidence for cryptic diversification and links to *Cordyceps* teleomorphs. Mycologia 97:84–98

Rehner SA, Evans JD (2009) Microsatellite loci for the fungus *Ascosphaera apis*: cause of honey bee chalkbrood disease. Mol Ecol Resour 9:855–858

Rieseberg L (1996) Homology among RAPD fragments in interspecific comparisons. Mol Ecol 5:99–105

Roh C, Villatte F, Kim BG, Schmid RD (2006) Comparative study of methods for extraction and purification of environmental DNA from soil and sludge samples. Appl Biochem Biotechnol 134:97–112

Rohel E, Couteaudier Y, Papierok B, Cavelier N, Dedryver CA (1997) Ribosomal internal transcribed spacer size variation correlated with RAPD-PCR pattern polymorphisms in the entomopathogenic fungus *Erynia neoaphidis* and some closely related species. Mycol Res 101:573–579

Romanowski G, Lorenz MG, Wackernagel W (1993) Use of polymerase chain-reaction and electroporation of *Escherichia coli* to monitor the persistence of extracellular plasmid DNA introduced into natural soils. Appl Environ Microbiol 59:3438–3446

Ronaghi M, Uhlen M, Nyren P (1998) A sequencing method based on real-time pyrophosphate. Science 281:363–365

Sambrook J, Russell DW (2001) Molecular cloning: a laboratory manual. Cold Spring Harbor Laboratory Press, Cold Spring Harbor

Schadt CW, Martin AP, Lipson DA, Schmidt SK (2003) Seasonal dynamics of previously unknown fungal lineages in tundra soils. Science 301:1359–1361

Schneider S, Enkerli J, Widmer F (2009) A generally applicable assay for the quantification of inhibitory effects on PCR. J Microbiol Methods 78:351–353

Schwarzenbach K, Enkerli J, Widmer F (2007a) Objective criteria to assess representativity of soil fungal community profiles. J Microbiol Methods 68:358–366

Schwarzenbach K, Widmer F, Enkerli J (2007b) Cultivation-independent analysis of fungal genotypes in soil by using simple sequence repeat markers. Appl Environ Microbiol 73:6519–6525

Schwarzenbach K, Enkerli J, Widmer F (2009) Effects of biological and chemical insect control agents on fungal community structures in soil microcosms. Appl Soil Ecol 42:54–62

Schwieger F, Tebbe CC (1998) A new approach to utilize PCR-single-strand-conformation polymorphism for 16S rRNA gene-based microbial community analysis. Appl Environ Microbiol 64:4870–4876

Sessitsch A, Hackl E, Wenzl P, Kilian A, Kostic T, Stralis-Pavese N, Sandjong BT, Bodrossy L (2006) Diagnostic microbial microarrays in soil ecology. New Phytol 171(4): 719–735

Smit E, Leeflang P, Glandorf B, van Elsas JD, Wernars K (1999) Analysis of fungal diversity in the wheat rhizosphere by sequencing of cloned PCR-amplified genes encoding 18S rRNA and temperature gradient gel electrophoresis. Appl Environ Microbiol 65:2614–2621

Sosa-Gomez DR, Humber RA, Hodge KT, Binneck E, da Silva-Brandao KL (2009) Variability of the mitochondrial SSU rDNA of *Nomuraea* species and other entomopathogenic fungi from Hypocreales. Mycopathologia 167:145–154

Sudisha J, Kumar SA, Niranjana SR, Shetty NP, Shetty HS (2009) Cloning and development of pathotype-specific SCAR marker associated with *Sclerospora graminicola* isolates from pearl millet. Australas Plant Pathol 38:216–221

Sugimoto M, Koike M, Hiyama N, Nagao H (2003) Genetic, morphological, and virulence characterization of the entomopathogenic fungus *Verticillium lecanii*. J Invertebr Pathol 82:176–187

Suzuki M, Rappe MS, Giovannoni SJ (1998) Kinetic bias in estimates of coastal picoplankton community structure obtained by measurements of small-subunit rRNA gene PCR amplicon length heterogeneity. Appl Environ Microbiol 64:4522–4529

Swofford DL (2002) PAUP*: phylogenetic analysis using parsimony (* and other methods). Version 4.0b10. Sinauer Associates, Sunderland

Takatsuka J (2007) Specific PCR assays for the detection of DNA from *Beauveria bassiana* F-263, a highly virulent strain affecting the Japanese pine sawyer, *Monochamus alternaturs* (Coleoptera: Cerambycidae), by a sequence-characterized amplified region (SCAR) marker. Appl Entomol Zool 42:619–628

Taylor JW, Fisher MC (2003) Fungal multilocus sequence typing—it's not just for bacteria. Curr Opin Microbiol 6:351–356

Tebbe CC, Vahjen W (1993) Interference of humic acids and DNA extracted directly from soil in detection and transformation of recombinant-DNA from bacteria and a yeast. Appl Environ Microbiol 59:2657–2665

Thomsen L, Jensen AB (2002) Application of nested-PCR technique to resting spores from the *Entomophthora muscae* species complex: implications for analyses of host-pathogen population interactions. Mycologia 94:794–802

Tigano MS, Adams B, Maimala S, Boucias D (2006) Genetic diversity of *Hirsutella thompsonii* isolates from Thailand based on AFLP analysis and partial β-tubulin gene sequences. Genet Mol Biol 29:715–721

Tominaga T (2006) Rapid discrimination of *Listeria monocytogenes* strains by microtemperature gradient gel electrophoresis. J Clin Microbiol 44:2199–2206

Tringe SG, Rubin EM (2005) Metagenomics: DNA sequencing of environmental samples. Nat Rev Genet 6:805–814

Tun CC, Ikenaga M, Asakawa S, Kimura M (2002) Community structure of bacteria and fungi responsible for rice straw decomposition in a paddy field estimated by PCR-RFLP analysis. Soil Sci Plant Nutri 48:805–813

Tymon AM, Pell JK (2005) ISSR, ERIC and RAPD techniques to detect genetic diversity in the aphid pathogen *Pandora neoaphidis*. Mycol Res 109:285–293

Tymon AM, Shah PA, Pell JK (2004) PCR-based molecular discrimination of *Pandora neoaphidis* isolates from related entomopathogenic fungi and development of species-specific diagnostic primers. Mycol Res 108:419–433

van Elsas JD, Duarte GF, Keijzer-Wolters A, Smit E (2000) Analysis of the dynamics of fungal communities in soil via fungal-specific PCR of soil DNA followed by denaturing gradient gel electrophoresis. J Microbiol Methods 43:133–151

Vandenkoornhuyse P, Baldauf SL, Leyval C, Straczek J, Young JPW (2002) Evolution—extensive fungal diversity in plant roots. Science 295:2051

Vargas LRB, Rossato M, Ribeiro RTD, de Barros NM (2003) Characterization of *Nomuraea rileyi* strains using polymorphic DNA, virulence and enzyme activity. Braz Arch Biol Technol 46:13–18

Vega FE, Goettel M, Blackwell M, Chandler D, Jackson MA, Keller S, Koike M, Maniania NK, Monzon A, Ownley BH, Pell J, Rangel DEN, Roy HE (2009) Fungal entomopathogens: new insights on their ecology. Fungal Ecol 2:149–159

Vekemans X, Beauwens T, Lemaire M, Roldan-Ruiz I (2002) Data from amplified fragment length polymorphism (AFLP) markers show indication of size homoplasy and of a relationship between degree of homoplasy and fragment size. Mol Ecol 11:139–151

Velásquez VB, Carcamo MP, Merino CR, Iglesias AF, Duran JF (2007) Intraspecific differentiation of Chilean isolates of the entomopathogenic fungi *Metarhizium anisopliae* var. *anisopliae* as revealed by RAPD, SSR and ITS markers. Genet Mol Biol 30:89–99

Versalovic J, Schneider M, DeBruijn FJ, Lupski JR (1994) Genomic fingerprinting of bacteria using repetitive sequence-base polymerase chain reaction. Methods Mol Cell Biol 5:25–40

Viebahn M, Veenman C, Wernars K, van Loon LC, Smit E, Bakker P (2005) Assessment of differences in ascomycete communities in the rhizosphere of field-grown wheat and potato. FEMS Microbiol Ecol 53:245–253

Vos P, Hogers R, Bleeker M, Reijans M, van de Lee T, Hornes M, Frijters A, Pot J, Peleman J, Kuiper M, Zabeau M (1995) AFLP: a new technique for DNA fingerprinting. Nucleic Acids Res 23:4407–4414

Wada S, Horita M, Hirayae K, Shimazu M (2003) Discrimination of Japanese isolates of *Beauveria brongniartii* (Deuteromycotina: Hyphomycetes) by RFLP of the rDNA-ITS regions. Appl Entomol Zool 38:551–557

Wagner-Dobler I, Lunsdorf H, Lubbehusen T, von Canstein HF, Li Y (2000) Structure and species composition of mercury-reducing biofilms. Appl Environ Microbiol 66:4559–4563

Wang CS, St Leger RJ (2007) The MAD1 adhesin of *Metarhizium anisopliae* links adhesion with blastospore production and virulence to insects, and the MAD2 adhesin enables attachment to plants. Eukaryotic Cell 6:808–816

Wang CS, Li ZZ, Typas MA, Butt TM (2003a) Nuclear large subunit rDNA group I intron distribution in a population of *Beauveria bassiana* strains: phylogenetic implications. Mycol Res 107:1189–1200

Wang CS, Shah FA, Patel N, Li ZZ, Butt TM (2003b) Molecular investigation on strain genetic relatedness and population structure of *Beauveria bassiana*. Environ Microbiol 5:908–915

Wang C, Fan M, Li Z, Butt TM (2004) Molecular monitoring and evaluation of the application of the insect-pathogenic fungus *Beauveria bassiana* in southeast China. J Appl Microbiol 96:861–870

Wang CS, Hu G, St Leger RJ (2005a) Differential gene expression by *Metarhizium anisopliae* growing in root exudate and host (*Manduca sexta*) cuticle or hemolymph reveals mechanisms of physiological adaptation. Fungal Genet Biol 42:704–718

Wang SB, Miao XX, Zhao WG, Huang B, Fan MZ, Li ZZ, Huang YP (2005b) Genetic diversity and population structure among strains of the entomopathogenic fungus, *Beauveria bassiana*, as revealed by inter-simple sequence repeats (ISSR). Mycol Res 109:1364–1372

Watson RJ, Blackwell B (2000) Purification and characterization of a common soil component which inhibits the polymerase chain reaction. Can J Microbiol 46:633–642

Welsh J, McClelland M (1990) Fingerprinting genomes using PCR with arbitrary primers. Nucleic Acids Res 18:7213–7218

White TJ, Bruns T, Lee S, Taylor J (1990) Amplification and direct sequencing of fungal ribosomal RNA genes for phylogenetic studies. In: Innis M, Gelfand DH, Sninsky JJ, White TJ (eds) PCR protocols—a guide to methods and applications. Academic Press, San Diego, pp 315–322

Whitehouse CA, Hottel HE (2007) Comparison of five commercial DNA extraction kits for the recovery of *Francisella tularensis* DNA from spiked soil samples. Mol Cell Biochem 21:92–96

Widmer F, Seidler RJ, Watrud LS (1996) Sensitive detection of transgenic plant marker gene persistence in soil microcosms. Mol Ecol 5:603–613

Widmer F, Fliessbach A, Laczko E, Schulze-Aurich J, Zeyer J (2001) Assessing soil biological characteristics: a comparison of bulk soil community DNA-, PLFA-, and Biolog (TM)-analyses. Soil Biol Biochem 33:1029–1036

Widmer F, Hartmann M, Frey B, Kölliker R (2006) A novel strategy to extract specific phylogenetic sequence information from community T-RFLP. J Microbiol Methods 66:512–520

Williams JK, Kubelik AR, Livak KJ, Rafalski JA, Tingey SV (1990) DNA polymorphism amplified by arbitrary primers are useful as genetic markers. Nucleic Acids Res 18: 6531–6535

Wilson IG (1997) Inhibition and facilitation of nucleic acid amplification. Appl Environ Microbiol 63:3741–3751

Xu JP, Guo H, Yang ZL (2007) Single nucleotide polymorphisms in the ectomycorrhizal mushroom *Tricholoma matsutake*. Microbiology 153:2002–2012

Yergeau E, Filion M, Vujanovic V, St-Arnaud M (2005) A PCR-denaturing gradient gel electrophoresis approach to assess *Fusarium* diversity in asparagus. J Microbiol Methods 60:143–154

Yergeau E, Vujanovic V, St-Arnaud M (2006) Changes in communities of *Fusarium* and arbuscular mycorrhizal fungi as related to different asparagus cultural factors. Microb Ecol 52:104–113

Zare R, Kouvelis VN, Typas MA, Bridge PD (1999) Presence of a 20 bp insertion/deletion in the ITS1 region of *Verticillium lecanii*. Lett Appl Microbiol 28:258–262

Zhou J, Bruns MA, Tiedje JM (1996) DNA recovery from soils of diverse composition. Appl Environ Microbiol 62: 316–322

Zhou G, Whong W-Z, Ong T, Chen B (2000) Development of a fungus-specific PCR assay for detecting low-level fungi in an indoor environment. Mol Cell Probes 14:339–348

Zietkiewicz E, Rafalskim A, Labuda D (1994) Genome fingerprinting by simple sequence repeat (SSR)-anchored polymerase chain reaction amplification. Genomics 20:176–183

Author Biographies

Jürg Enkerli is senior scientist and leads research activities on fungal ecology in the group for molecular ecology at the Swiss Federal Research Station ART. His research is focused on ecological aspects of fungal pathogens either used in insect biocontrol or in control of plant pathogens. He develops and applies genetic diagnostics in order to gain information on the genetic resources present in agricultural systems and on how they may be affected by various impacts.

Franco Widmer is heading the Molecular Ecology group at the Swiss Federal Research Station ART. His research interest focuses mainly on molecular soil microbial ecology. The interactions of soil microbial communities and various environmental and anthropogenic factors represent the main objects of his research. He develops specific molecular genetic diagnostics and downstream analytical procedures for assessing changes in soil microbial communities.

BioControl (2010) 55:39–54
DOI 10.1007/s10526-009-9246-5

Principles from community and metapopulation ecology: application to fungal entomopathogens

Nicolai V. Meyling · Ann E. Hajek

Received: 26 June 2009 / Accepted: 15 October 2009 / Published online: 10 November 2009

Abstract Fungal entomopathogens are often studied within the context of their use for biological control, yet these natural enemies are also excellent subjects for studies of ecological interactions. Here, we present selected principles from community ecology and discuss these in relation to fungal entomopathogens. We discuss the relevance of apparent competition, food web construction, intraguild predation and density-mediated and trait-mediated indirect effects. Although current knowledge of community interactions involving fungal entomopathogens are limited, fungal entomopathogens can be important, interactive members of communities and the activities of fungal entomopathogens should be evaluated in the context of ecological principles. We also discuss aspects of metapopulation ecology and the application of these principles to fungal entomopathogens. Knowledge of ecological interactions is crucial if we are to understand and predict the effects of fungal entomopathogens on host populations and understand the interactions among fungal entomopathogens and other organisms in the communities in which they occur.

Keywords Community ecology · Apparent competition · Food webs · Trait-mediated indirect effects · Metapopulation ecology · Host specificity · Fungal entomopathogens

Handling Editor: Dr. Helen Roy.

N. V. Meyling (✉)
Faculty of Life Sciences, Department of Agriculture and Ecology, University of Copenhagen, 1871 Frederiksberg C, Denmark
e-mail: nvm@life.ku.dk

A. E. Hajek
Department of Entomology, Cornell University, Cornstock Hall, Ithaca, NY 14853, USA

Introduction

Fungal entomopathogens are often studied within the context of biological control, i.e., with the objective of exploiting the pathogen to decrease population sizes of specific arthropod pests. In addition, recent studies present impacts of fungal entomopathogens on plant pathogens (e.g., Kim et al. 2007, 2008). In these contexts, ecological principles are implicit, e.g., the population size of the pathogen should increase (either immediately, as for inundation biological control, or after some time, as for inoculation biological control, *sensu* Eilenberg et al. 2001) and the host–pathogen interaction should lead to a reduction in the population density of the host (the pest), usually benefiting the host resource (e.g., the plant being eaten by the host). Fundamental ecological studies investigating the basis for biological control have often included predators or parasitoids but rarely pathogens. In an ecological context, fungal entomopathogens can broadly be included in the definition of parasites as "symbionts that cause harm to another organism, the host, which the parasites utilize as habitat" (Raffel et al. 2008). Likewise, Hatcher et al. (2006) define parasites and pathogens collectively as "organisms that feed on a host individual, usually living on or in it and often causing harm but not immediate death". Indeed, the term 'microparasites' has traditionally been applied to pathogenic microbes within epidemiology (Anderson and May 1981). We emphasize these definitions here because there is an increasing amount of literature on the ecology of parasites, primarily within the context of community ecology. In ecological terms, however, microparasites, or pathogens, are often considered distinct from macroparasites, or 'true' parasites. Pathogens are typically intensity independent, meaning that a single infection event can lead to high within-host reproduction of the pathogen, resulting in characteristic host pathology. In contrast, 'true' parasites (macroparasites) are intensity dependent, as their impact on host individuals is more dependent on increasing infection events (Lafferty et al. 2008).

Concerning fungal entomopathogens, several life history specializations make this group distinctive from other types of pathogens. The fungi are unusual among entomopathogens because they infect through the cuticle of their host (Hajek and Leger 1994), usually not having to be ingested or enter through other natural openings in the host's body. Mostly, fungal entomopathogens must kill their host to produce new infective spores in order to be transmitted to new hosts. Exceptions are few but include some interesting biologies such as active spore discharge from living fly hosts in *Strongwellsea* spp. (Eilenberg 2002). Among the fungal entomopathogens, host specificity ranges from extreme host specialization [one host species, e.g., specific clades of *Entomophthora muscae* s.s. (Cohn) Fresenius (Entomophthoramycotina: Entomophthorales) infecting individual species of flies (Jensen et al. 2001)], through intermediate specialization [particular systematic host groups, e.g., *Pandora neoaphidis* (Remaudière and Hennebert) Humber (Entomophthoramycotina: Entomophthorales) infecting several aphid species (Ekesi et al. 2005)], to pathogens with broad host ranges [e.g., *Beauveria bassiana* (Balsamo) Vuillemin (Ascomycota: Hypocreales) infecting many insect species from different orders (Meyling et al. 2009)]. These adaptations to different breadths of host ranges impact the ecological context and principles to consider, as well as the possible resulting ecological effects due to various types of fungal entomopathogens. This will be emphasized in the following sections.

Here, we present examples of interactions in communities, i.e., among organisms assembled within a specific area, and discuss entomopathogenic fungi as interactive members of communities. The systems and principles we present include interactions and processes within closed communities as well as within and among open communities, the latter covered by the principles of metapopulation ecology.

Direct and indirect effects in community ecology

A community is an assemblage of populations of different species that co-occur in the same habitat or area, and interact. Interactions are traditionally considered to be direct, meaning that two species engage in a direct confrontation such as a trophic interaction (e.g., one consumes the other) or a competitive interaction (e.g., interference competition). Species also interact with each other indirectly when a specific interaction is mediated by a third party. For example, if two species consume a common resource but never encounter each other and thus do not interact directly, they can have an indirect interaction with each other by one species reducing the amount of the resource that is available to the other species (exploitative or scramble competition). In this case, the interaction is mediated by the shared resource. Trophic interactions within a community can be visualized by constructing food webs linking consumers and resources, in which species that consume others are placed at higher trophic levels than their resources. However, the total sum of all interactions, direct and indirect, will increase complexity of the food web. For simplicity, the complex architecture of the whole community can be broken into modules (community modules, sensu Holt 1997) and, to study specific interactions and their effects, attention is usually given to particular community modules of interest (Holt 1997; Holt and Dobson 2006; Hatcher et al. 2006). Community modules consist of a few (three–six) species that interact as "multi-species extensions of basic pair-wise interactions" (Holt and Hochberg 2001; e.g., Figs. 1, 2). However, more complex modules embedded in a food web involving more species with which interactions may also occur can be considered (e.g., by adding more species or trophic levels, such as host plants, to Figs. 1, 2). Below, we present examples from selected community modules, including direct and indirect interactions among species in the modules, and we discuss their relevance for fungal entomopathogens. Although some fungal entomopathogens are known to interact with plants by being endophytes (Vega et al. 2008) and rhizosphere colonizers (Hu and Leger 2002; Bruck 2009), here we are discussing fungal entomopathogens strictly as natural enemies of arthropod hosts and the ecological implications of such interactions.

Imagine a simple food chain in which an intermediate species H feeds on a resource R (e.g., a plant) and H is itself consumed by a natural enemy P (e.g., pathogen). Each species in this chain reduces the abundance of the species on the trophic level immediately below through consumption. This leads to direct positive effects on higher trophic levels by nutrient gain and direct negative effects on lower trophic levels (solid lines in Fig. 1). In addition, the species in the top level also affects the lowest level indirectly by limiting the effect of the intermediate species on its resource. Thus, P has an indirect positive effect on R (broken line in Fig. 1). The effect is indirect in the sense that it is mediated by a third species, H. This effect leads to a trophic cascade, i.e., the effects of trophic links are cascading along the food chain, fitting the phrase "the enemy of my enemy is my friend" (Holt 2000). Achieving indirect effects through trophic cascades is the objective of traditional biological control, when the abundance of a natural enemy of a pest is manipulated to reduce the density and/or impact of the pest. However, studies of direct effects of fungal entomopathogens on their host populations rarely include whether this interaction affects the resource of the host, although protecting the host resource is usually the overall aim. Most studies are limited to the host–pathogen interaction alone and do not include the host's resource. Considering the host–pathogen interaction from the ecological perspective of community modules (see Fig. 1) in future studies would provide more insight into the indirect effects of fungal entomopathogens on host resources, mediated by the host.

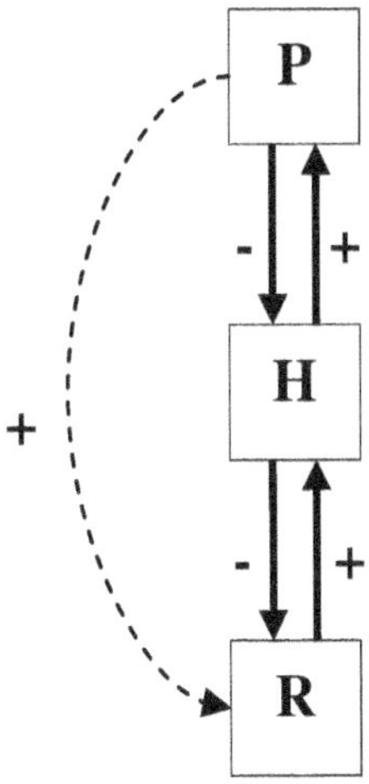

Fig. 1 A simple three-species food chain indicating direct and indirect effects. The basal species R is consumed by H, creating a positive direct effect on H and a negative direct effect on R. Likewise, P consumes H yielding a positive direct effect for itself while H suffers. By consuming H, P also has a positive indirect effect on R (broken line) by reducing the direct negative effect of H on R. H is therefore mediating the effect of P on R. The indirect effect may be both density-mediated (DMIE) or trait-mediated (TMIE), see text for details

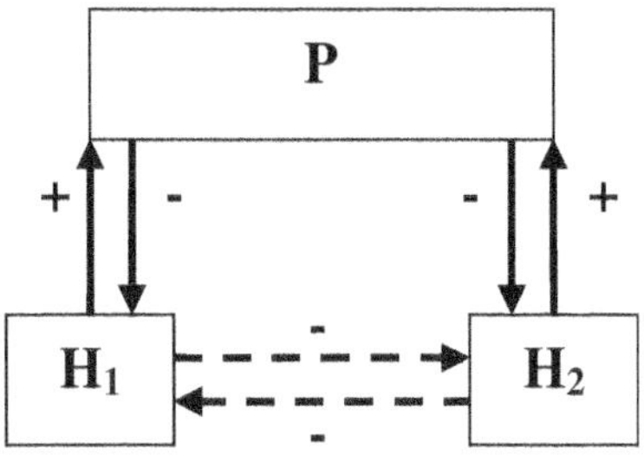

Fig. 2 Two species, H_1 and H_2, live from separate resources but share a natural enemy, P. Through consumption, P has direct negative effects on H_1 and H_2, and thus the two latter species have positive direct effects on P. By sharing a common natural enemy H_1 and H_2 have indirect negative effects on each other through apparent competition (broken lines)

Species that do not consume a common resource may affect each other indirectly by sharing a natural enemy. In Fig. 2, two herbivorous species (H_1 and H_2) that live on separate host plants may share a natural enemy (P) that consumes both H_1 and H_2, and P consequently increases in density. P has direct negative effects on H_1 and H_2 while these two hosts have direct positive effects on P (solid lines in Fig. 2). The two host species have an indirect negative effect on each other (broken lines in Fig. 2) although they do not compete for a resource, because they both provide resources for P and thereby both contribute to the increase in abundance of P. This kind of indirect effect due to sharing an enemy is termed apparent competition (Holt 1977; van Veen et al. 2006a). In Fig. 2, apparent competition is reciprocally negative. The interaction can also be more or less unidirectionally negative, if, for example, H_1 is less susceptible to P than H_2, then H_1 is negatively affecting H_2 to a larger extent than vice versa. As an ultimate consequence, H_2 might be excluded from the community.

Apparent competition mediated by shared predators and parasitoids has been quite extensively studied for herbivorous insect communities [see reviews by van Veen et al. (2006a, b). For examples concerning 'true' parasites see reviews by Hatcher et al. (2006), Raffel et al. (2008) and Lefevre et al. (2009)]. However, apparent competition mediated by fungal entomopathogens has been studied to a very limited extent. Apparent competition is immediately applicable to pest management, as the principle is implicitly embedded within central parts of the strategy known as 'conservation biological control' (Barbosa 1998). This strategy is defined as "modification of the environment or existing practices to protect and enhance specific natural enemies or other organisms to reduce the effect of pests" (Eilenberg et al. 2001). If resources, such as prey or hosts, for a particular natural enemy are increased by environmental manipulation, then populations of this natural enemy can increase and, consequently, the density of the target pest species would decrease. It is implicit in this intended manipulation that the natural enemy is shared among 'reservoir' species (i.e., alternate host(s) occurring in the environment) and the target pest. Apparent competition will be mediated through the shared natural enemy and, as a result, the target pest population will be reduced by this indirect effect to a larger extend than if the reservoir species was not present. The time scale in which to consider apparent competition can be separated into 'short term' and 'long term' based on the duration of the interactions, in comparison to the generation time of the natural enemy (Holt and Lawton 1994; van Veen et al. 2006a).

Aphids (Hemiptera: Aphididae) are important agricultural pests, and apparent competition mediated by aphid natural enemies has been experimentally investigated in a number of studies. Müller and Godfray (1997) studied population dynamics of experimental colonies of nettle aphids, *Microlophium carnosum* (Buckton), next to colonies of grass aphids, *Rhopalosiphum padi* (L.) on separate host plants. Densities in colonies of grass aphids were increased by fertilizing pots and this resulted in earlier population declines of nettle aphids compared to unfertilized controls (Müller and Godfray 1997). The declines in *M. carnosum* densities were caused by increased predation by ladybirds, which were initially attracted by the high numbers of *R. padi*. This experiment therefore demonstrates the principle of short term apparent competition mediated by a shared predator due to manipulation of the environment, here fertilization of grass plots.

Apparent competition between aphids mediated by their parasitoids is also of interest within conservation biological control, but so far demonstration of this mechanism has not been clear (Müller and Godfray 1999). This may be caused by the fact that many aphid parasitoids are relatively specialized, limiting the number of hosts that share the same enemy (Müller and Godfray 1999; van Veen et al. 2008).

As aphid species are often infected by fungal entomopathogens from the order Entomophthorales, apparent competition can also be mediated by fungal entomopathogens if these are shared by more than one aphid species within an aphid community. This was experimentally tested by Pope et al. (2002) in the field. Neighbouring colonies of pea aphids (*Acyrthosiphon pisum* [Harris]) and nettle aphids (*M. carnosum*) were established with one or the other species initially infected with the fungus *P. (=Erynia) neoaphidis* or not, and the inoculated species was called the reservoir species. Population dynamics of colonies of the uninoculated target species were then monitored, for treatments with and without (control) fungal infections in the reservoir species. Infections spread to the neighbouring target colonies from the infected reservoirs, but the infections did not significantly reduce the target populations compared to the uninfected controls. Thus, apparent competition could not be demonstrated (Pope et al. 2002). According to the authors, the indirect effect mediated by a fungal entomopathogen in this case may be strongly affected by weather conditions, thus making the experimental results of the trials inconclusive (Pope et al. 2002). However, the fact that *P. neoaphidis* is shared among different aphid species makes it a suitable candidate for conservation biological control, and aspects of this strategy involving *P. neoaphidis* have been investigated in several further studies (see Ekesi et al. 2005; Pell et al. 2009).

Potential for apparent competition mediated by fungal entomopathogens in insect communities: construction of food webs

In order to establish whether apparent competition mediated by fungal entomopathogens potentially contributes to the structure of an insect community, construction of food webs can provide valuable perspectives (van Veen et al. 2006b, 2008). In its simplest form, a food web is constructed of two trophic layers, usually portrayed horizontally, each comprised of a number of species in the community. The trophic interactions are illustrated by linking the consumer species in the top layer with its resource species in the bottom layer. Links can be purely qualitative (i.e., present or absent) or they can be quantitative (i.e., the number of specific consumer-resource associations out of the total number of associations recorded). Based on the food web, one can assess which prey/hosts on the bottom layer share natural enemies on the top layer.

For apparent competition to potentially occur, some level of generalization in the prey/host range must characterize the natural enemy in question. In aphid communities, predators can be the most likely group of natural enemies for mediating apparent competition, as they attack a broader range of prey while parasitoids tend to be more specialized (van Veen et al. 2006a, 2008). Moreover, predators will congregate at a population of one prey species (e.g., based on a numerical response) and the same individuals will later move onto another neighbouring prey species on a separate host plant. Fungal entomopathogens attacking aphids are considered to be intermediate in their potential to mediate short-term apparent competition (Pope et al. 2002; van Veen et al. 2008).

Studies of food webs can contribute to our knowledge of how a particular group of natural enemies might mediate apparent competition within a community by illustrating which resource species are shared among consumer species, e.g., the level of host/prey specialization. Once a food web has been constructed, a measure of connectance can be calculated based on the number of realized associations divided by the maximum number of potential associations (van Veen et al. 2008). If all potential associations are realized, then connectance equals 1, while the lowest possible connectance ($C_{\min}$) corresponds to the highest level of specialization, i.e., single associations only (the more taxa sampled the lower the $C_{\min}$). In a study by van Veen et al. (2008), values of connectance were calculated for three groups of aphid natural enemies: predators, parasitoids and fungal entomopathogens from the Order Entomophthorales. The predators showed the highest connectance (0.20–0.28), parasitoids the lowest (0.07–0.10) and aphid pathogenic fungi displayed intermediate values (0.16) (van Veen et al. 2008). Although the study did not indicate that aphid pathogenic fungi were the most obvious group of natural enemies to mediate apparent competition among aphid species in the community, it strongly suggested that the nettle aphid, *M. carnosum*, shared pathogens with most other aphids and that the most commonly shared pathogen was *P. neoaphidis*. Potentially, *P. neoaphidis* should therefore be expected to mediate apparent competition between *M. carnosum* and pest aphids in agricultural systems, and the *M. carnosum–P. neoaphidis* association has indeed been investigated for use in conservation biological control (Shah and Pell 2003; Ekesi et al. 2005; Baverstock et al. 2008; Pell et al. 2009).

Adapting the principles of food webs to other groups of fungal entomopathogens may reveal to what degree they have potential for mediating apparent competition in communities of arthropods besides aphids. First, associations between naturally occurring hosts and pathogens within a community must be established, i.e., the ecological host range of the pathogen (Onstad and Carruthers 1990). Consideration must be given to which potential hosts should be sampled and how pathogen infections should be assessed. The major challenge in characterization of host–pathogen associations is based on definitions of species and their identification. Diagnostic features of some taxa are ambiguous and molecular methods for characterization can reveal cryptic species (Rehner 2005; Rehner and Buckley 2005; Bischoff et al. 2006), so an in depth understanding of species identities is lacking for many groups of fungal entomopathogens. The identification of genotypic groups of fungal entomopathogens, coupled with thorough sampling efforts, rather than traditional grouping by morphological characters, may provide new insights into host–pathogen links that will become the foundations for associations within food webs. In a study of host–pathogen associations between fungal isolates from the *Entomophthora muscae* species complex and their fly hosts, Jensen et al. (2001) showed that each of four fly species were infected by separate genotypic groups of *E. muscae*, as illustrated in Fig. 3 (Jensen et al. 2001). Assessment of the *E. muscae* complex by molecular methods therefore revealed a higher degree of host specificity than had been reported by morphological identification alone. Therefore, species within the *E. muscae* complex are unlikely to mediate apparent competition between co-occurring fly species.

The fungal genus *Beauveria* has been reported to have significant cryptic diversification based on molecular characterization (Rehner 2005; Rehner and Buckley 2005). It is often cited that the cosmopolitan morpho-species *B. bassiana* infects more than 700 host species (Inglis et al. 2001), indicating that *B. bassiana*, as identified by morphology, has an extremely broad host range. Identification of individual isolates based on molecular methods has confirmed the view that individual clades of *B. bassiana* have not evolved host specialization (Rehner and Buckley 2005; Meyling et al. 2009). (Below, we refer to individual phylogenetic species of the morpho-species *B. bassiana* as members or species within the *B. bassiana* complex). Thus, infections by members in the *B. bassiana* complex should be expected to be shared among insect species and this group should therefore have the potential to mediate apparent competition. However, for members of the *B. bassiana* complex to mediate apparent competition the pathogen must be shared among insect hosts within a specific community. Until recently, it was not known whether such pathogen-sharing occurs within insect communities, as host ranges have previously been evaluated based on collections of infected insects at larger spatial scales. However, Meyling et al. (2009) demonstrated that several phylogenetic species within *B. bassiana* were shared among insect species on different host plants within a hedgerow community (Fig. 4). Some species in the *B. bassiana* complex were only found rarely and their actual host range therefore cannot be assessed, but the two most common species, *B. bassiana* Eu_1 and Eu_4, infected seven and six host species out of twelve, respectively, on the three host plants utilized by the insects in the community. Thus, at least these two pathogen species have the potential to mediate apparent competition among insect species. Although the data sets of the two studies are not as extensive in the number of links as in the webs presented by van Veen et al. (2008), we have here calculated the connectance for each of the host–pathogen webs. Based on the data from Jensen et al. (2001), connectance in the *E. muscae*-fly host web was 0.36 ($C_{\min} = 0.36$) and in the web of species in the *B. bassiana* complex and their insect hosts, connectance was 0.606 ($C_{\min} = 0.212$). In comparison,

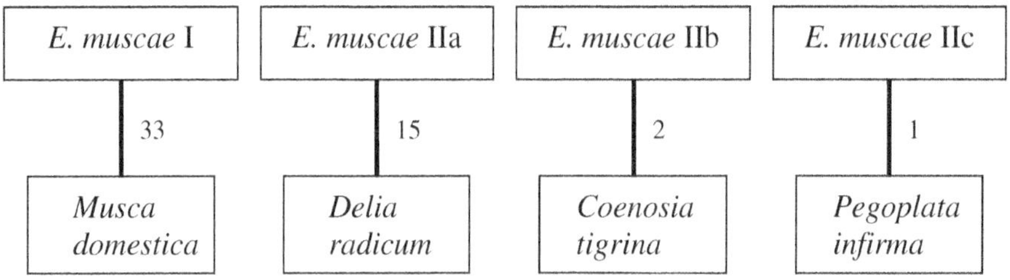

Fig. 3 Host-pathogen web illustrating the trophic association between genotypic groups of *Entomophthora muscae* and their dipteran hosts. The four boxes in the upper panel illustrate the four genotypes identified within *E. muscae*, and the lower boxes represent the four fly hosts from which the pathogens were isolated. The numbers to the right of the connecting lines denote the numbers of individual host–pathogen associations that have been identified. Based on data from Jensen et al. (2001)

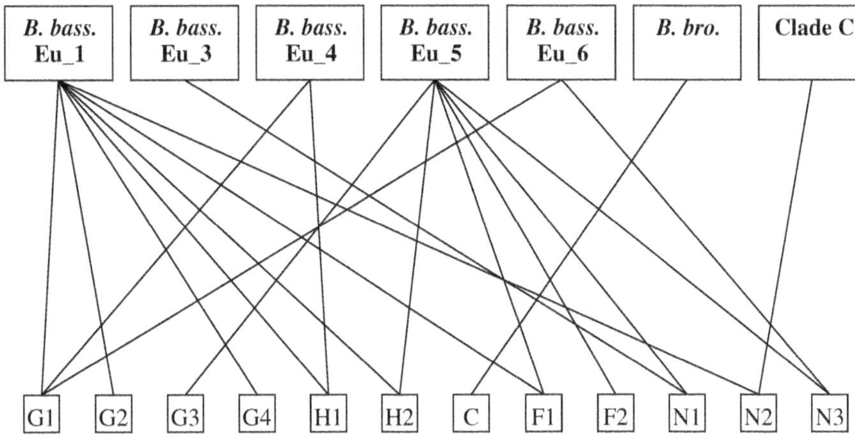

Fig. 4 Host-pathogen web modified from Meyling et al. (2009). The top panel illustrates the species of the *B. bassiana* complex identified by DNA-sequencing which were found to infect insect hosts within a single hedgerow. Five phylogenetic species of *B. bassiana* were identified, Eu_1, Eu_3, Eu_4, Eu_5 and Eu_6, as were *B. brongniartii* (*B. bro.*) and a separate *Beauveria* species, Clade C, which morphologically resemble *B. bassiana*. The twelve host species are presented in the lower panel and host plant is indicated when known. Host plants were: G = grasses, H = hawthorn and N = nettle; the number after host plant indicate separate species from this particular host plant. C denotes the carabid beetle *Nebria brevicollis*, and F1 and F2 denote two anthomyiid fly species. Links were established between each fungus species and the hosts they were found to infect in the insect community

connectance for the Entomophthorales-aphid web in a "typical" season was 0.16 ($C_{\min} = 0.043$) (van Veen et al. 2008). Our calculations are based on semi-quantitative data and give only a rough estimate of linkage in the systems studied. However, connectance of the *B. bassiana* complex—insect system indicates that the degree of linkage is relatively high while in the *E. muscae*-fly system connectance equals $C_{\min}$, emphasizing the high degree of host specialization by this latter entomopathogen complex. The study by Meyling et al. (2009) thus highlights that species within the *B. bassiana* complex are potential candidates for conservation biological control (Meyling and Eilenberg 2007).

Trophic placement and intraguild interaction of fungal entomopathogens

Future studies that focus on the ecological roles of fungal entomopathogens in insect communities will provide important insights into the impact of this group of natural enemies on species distribution and abundance. Placing fungal entomopathogens in the context of food webs will further make our understanding of interactions among insects and their natural enemies more complete. Recently, Lafferty et al. (2008) called for the inclusion of infectious disease agents in food webs as this may provide the 'full' ecological picture of species interactions in communities. However, at which trophic level should fungal entomopathogens be placed? If we look at the well-studied aphid-enemy system, inclusion of several types of natural enemies complicates interactions greatly. For example, in Fig. 5 a community module with four species is presented: one aphid, one predatory ladybird beetle, *Coccinella septempunctata* L. (Coleoptera: Coccinellidae), one parasitoid, *Aphidius ervi* (Halliday) (Hymenoptera: Braconidae) and one fungal entomopathogen (*P. neoaphidis*). The three aphid enemies are members of the same guild, i.e., species that consume a shared resource, but the guild members can also interact with each other by engaging in intraguild predation (IGP) (Polis et al. 1989; Polis and Holt 1992). The ladybird will consume the aphids, the parasitoid larvae within living aphids and parasitoid pupae within mummified aphids (Rosenheim et al. 1995; Brodeur and Rosenheim 2000) as well as aphids infected by *P. neoaphidis* (Roy et al. 1998). The ladybird itself will not become infected by the fungus (Roy et al. 2001). Although *P. neoaphidis* does not infect *A. ervi*, the fungus will outcompete the parasitoid within an aphid harbouring both natural enemies (Powell et al. 1986).

In this latter case, competitive exclusion of one enemy by another is either direct [interference competition, i.e., the pathogen and the parasitoid confront each other within the host; solid line c) in Fig. 5] or indirect [exploitation competition, i.e., the effect is mediated through a shared resource, the aphid; broken line c) in Fig. 5]. The intraguild (IG) predator must be a generalist to some degree (here, the ladybird) and must engage in IGP when consuming hosts harbouring parasitoids or entomopathogens, so-called coincidental intraguild predation (Polis et al. 1989). In constructing a community module, the ladybird could therefore be placed at the highest trophic level (top predator) and the fungus and parasitoid at an intermediate level (intermediate "predators", leading to unidirectional or asymmetric IGP). However, it has also been argued that when other guild members consume uninfected hosts they interact indirectly through exploitation competition for a common resource (Borer et al. 2007) [broken lines a) and b) in Fig. 5]. By consuming intermediate natural enemies, the ladybird may cause an indirect positive effect on the aphid [broken line d) in Fig. 5], especially if the IG prey (fungus or parasitoid) are more effective in consuming the shared prey than the ladybird. This principle was generally found in a meta-analysis of empirical studies of IGP (Vance-Chalcraft et al. 2007) and is also predicted from equilibrium models (Holt and Polis 1997).

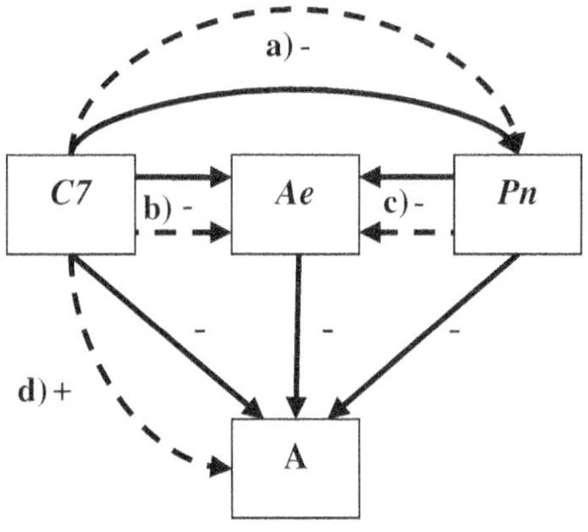

Fig. 5 Direct and indirect effects in a guild of natural enemies of an aphid (*A*), including the ladybird *Coccinella septempunctata* (*C7*), the parasitoid *Aphidius ervi* (*Ae*) and the aphid pathogenic fungus *Pandora neoaphidis* (*Pn*). All three enemies have direct negative effects on the aphid by consuming it (solid lines) but they have also direct negative effects on each other through intraguild predation (see text for details). The intraguild interactions can also be considered to be negative and indirect by resource competition for uninfected hosts, then mediated by the aphid, i.e., broken lines a) and b). Likewise, the fungus can have a negative indirect effect on the parasitoid through exploitation competition mediated through their shared resource, i.e., broken line c). By consuming infected or parasitized aphids the ladybird may have a positive indirect effect on the aphid, i.e., broken line d)

It has been suggested that predation can reduce prevalence of a pathogen in the host population (Packer et al. 2003), but predicting population level effects caused by all of these interactions, whether direct or indirect, and whether the effects will be positive or negative, may be difficult. For example, although *C. septempunctata* can eat fungal-infected hosts, it can also vector *P. neoaphidis* conidia to new hosts (Roy et al. 2001), thus potentially increasing pathogen transmission. Furthermore, other herbivorous insects that are not attacked by the aphid natural enemies, may also impact pathogen dispersal (Baverstock et al. 2008, 2009). Adding just a single additional aphid species to the community module also greatly increases complexity.

Meyling et al. (2009) illustrated phylogenetic species of the *B. bassiana* complex as belonging to one trophic level above the insect host level (as presented in Fig. 4). However, individual species of the *B. bassiana* complex were found to infect both herbivorous and predatory insects in the insect community that was sampled. Thus, the insect hosts themselves can belong to several trophic levels. For example, species of the *B. bassiana* complex can infect both ladybirds and their prey (Roy and Cottrell 2008), as has been shown for other fungal entomopathogens, such as *Isaria fumosorosea* Wize (Ascomycota: Hypocreales) (Pell and Vandenberg 2002). Such fungal entomopathogens with broad host ranges can thus complicate interactions in predator–prey or parasitoid-host systems by acting as co-incidental IG 'predators', just as *C. septempunctata* in Fig. 5. In biological control studies of interactions between fungal entomopathogens and predators, the focus has mostly been on so-called non-target effects, when the pathogen infects predators as well as the target host. As indicated above, these direct interactions among natural enemies of a common prey/host may release the latter from regulation. Thus, the infection of a predator by a fungal entomopathogen can cause a positive indirect effect on the prey/host. Studies of non-target effects of *B. bassiana* and *Metarhizium anisopliae* (Metschnikoff)

Sorokin (Ascomycota: Hypocreales) are reviewed by Zimmermann (2007a, b).

Trait-mediated indirect effects

So far, the indirect effects mentioned have been implicitly based on principles of population ecology, i.e., effects are caused by changes in densities of the species in a community through altered mortality rates. Such effects are defined as density-mediated indirect effects (DMIE), as they result from changes in the density of the species that is mediating the effect. Indirect effects may also be approached from the perspective of evolutionary ecology, with the expectation that prey or host species have evolved adaptations to reduce their risk of being preyed upon, parasitized or infected. Natural selection should favour the individuals that are better at avoiding predation or infection, as they would be more successful in contributing to the next generation. Such adaptations would potentially also affect community structure through indirect effects, but in this case would not be mediated by changes in density but rather by changes in specific traits, such as behaviour. Consequently, these effects are termed trait-mediated indirect effects (TMIE). In recent years, many ecologists have studied TMIE as a mechanism for understanding structures of communities and their influence on ecosystems (Schmitz et al. 2004; van Veen et al. 2005; Okuyama and Bolker 2007; Bukovinszky et al. 2008; Schmitz 2008). In Fig. 1, the indirect effect of P on R, mediated by H, could be realized by a reduced density of H (DMIE) as discussed previously, but could also be realized if individuals of H change the expression of a specific trait, such as a behavioural trait that influences predator avoidance (Schmitz et al. 2004). For example, if the presence of P causes H to spend more time avoiding P and, consequently, less time consuming R, or P causes H to move away from R, these alterations will benefit R, making the indirect effect positive and trait-mediated (TMIE). TMIEs have not yet been studied explicitly for systems involving fungal entomopathogens, but in systems including other natural enemies of insects such as predators, TMIEs have been shown to have profound effects on the ecosystem. Recently, Schmitz (2008) demonstrated that the presence of a sit-and-wait predator (an ambush spider) caused its prey, a grasshopper, to move away from its preferred host plant and seek refuge on a safer plant species. This shift caused a positive TMIE on the original host plant by reducing herbivory and a negative TMIE on the alternate, safer host plant. In addition, the grasshopper host plant shift impacted the functioning of the ecosystem by altering primary productivity and nitrogen mineralization (Schmitz 2008). It remains to be demonstrated whether fungal entomopathogens can cause TMIEs, but it should be expected that potential insect hosts might have evolved adaptations (TMIEs) that would reduce risk of infection by pathogens.

Alterations of behavioural traits in insect hosts to decrease the risk of exposure to fungal entomopathogens have been documented. Observations of searching behaviour of the generalist predatory bug *Anthocoris nemorum* L. (Hemiptera: Anthocoridae) in arenas containing nettle leaves inoculated with *B. bassiana* showed that the presence of the pathogen caused the predator to move away from these patches (Meyling and Pell 2006). Likewise, the predator withdrew instantly upon contact with cadavers sporulating with *B. bassiana* although these predators readily consumed freeze-killed cadavers. Meyling and Pell (2006) demonstrated this change in a behavioural trait in the host in response to the pathogen, but did not estimate the ecological effect that this change might cause. If we transfer the study system of Meyling and Pell (2006) to Fig. 1, P will represent *B. bassiana*, H would be the predatory bug *A. nemorum*, and R would be its prey, the nettle aphid, *M. carnosum*. We would then predict that the presence of *B. bassiana* could cause a positive TMIE toward *M. carnosum* because aphid predation would be reduced because *A. nemorum* would spend more time avoiding *B. bassiana*. The nettle aphid is not within the ecological host range of *B. bassiana* in Northern Europe, but it is important prey for *A. nemorum* (Perrin 1976). Thus, *B. bassiana* is expected to have an indirect and not a direct interaction with *M. carnosum*. If we are to select systems for further experimentation in which TMIE might occur, insect predators and their fungal entomopathogens would be appropriate organisms to choose.

Infection and subsequent mortality due to *Beauveria* spp. or *Metarhizium* spp. is dependent on inoculum density (e.g., Hughes et al. 2004). Moreover, a threshold density of inoculum is often necessary to initiate infection. In such cases, hosts would then be expected to avoid the pathogen to keep exposure as low as possible and close to the infection threshold, in order to increase fitness. It could also be expected that such adaptations would be expressed more in predator species that hunt actively, as they would encounter fungal entomopathogens in their environment more frequently than predators that sit and wait for their prey. Indeed, the actively hunting ladybird, *C. septempunctata*, avoids patches containing *B. bassiana*, both on leaves and on soil (Ormond 2007). Other predatory insects are exposed to fungal entomopathogens in their habitats and therefore could potentially have evolved mechanisms to avoid infection. Predators, that have been documented to be infected by fungal entomopathogens, include staphylinid beetles (Steenberg et al. 1995) and carabid beetles (Vestergaard and Eilenberg 2000). Also hemipteran predators that have been studied from the perspective of non-target effects of fungal entomopathogens (Poprawski et al. 1997; Todorova et al. 2002; Dunkel and Jaronski 2003) could constitute potential species for studying TMIE. In other well-studied systems involving specialist fungal entomopathogens, such as the aphid—*P. neoaphidis* system, we may not expect TMIE mediated by predators or parasitoids. In Fig. 5, the pathogen does not infect the ladybird and, as would be predicted, no avoidance mechanisms toward the fungus have been shown (Pell et al. 1997). It has been demonstrated that the parasitoid, *A. ervi*, also does not show avoidance behaviour in response to *P. neoaphidis* (Baverstock et al. 2005, 2009), although the fungus is mostly successful in colonizing aphids that have already been parasitized (Powell et al. 1986). It is possible that the fungus does not pose a strong selection pressure to the parasitoid and thus mechanisms to avoid the fungus have not evolved. In contrast, aphid parasitoids have evolved adaptations to detect cues from ladybirds, thus avoiding patches with foraging ladybirds that may consume aphids containing parasitoid larvae (Nakashima et al. 2004). It could be predicted that patches with fungus-infected prey would indicate poor quality resources to predators or parasitoids, but to our knowledge this has not been investigated.

Metapopulations of fungal entomopathogens

Up to this point, we have focused on interactions among species in localized communities, i.e., the species involved are present simultaneously within the same area. However, species are rarely limited to closed communities, as some degree of dispersal to and from individual locations will occur. Furthermore, populations of individual species are spatially distributed and locally they exhibit individual dynamics, even becoming extinct locally. In open communities, colonization or recolonization are possible through immigration. As such, spatially separated populations are connected by dispersal, creating larger interchanging groups of populations called metapopulations (Fig. 6).

Populations of fungal entomopathogens are dynamic and it is well established that disease epizootics can cause rapid reductions in populations of arthropod hosts (Fuxa and Tanada 1997; Tanada and Kaya 1993). However, the mechanisms behind host–pathogen relations that result in epizootics are largely not understood. An important missing

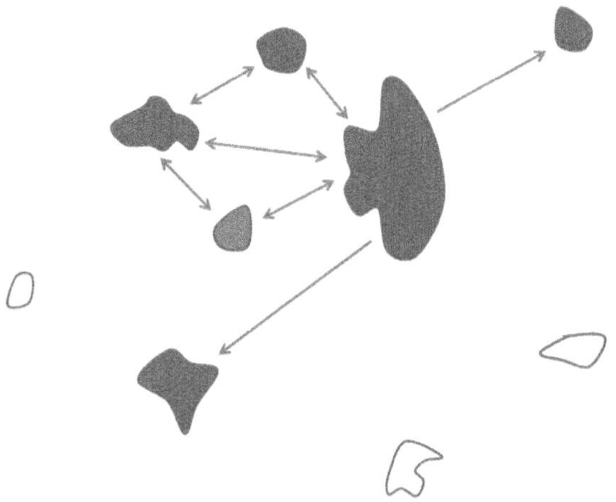

Fig. 6 A hypothetical model of insect host/entomopathogen metapopulations. Filled areas Host populations colonized by the entomopathogen. Empty areas Host populations not colonized by the entomopathogen. Arrows Dispersal of the entomopathogen between isolated host populations. The entomopathogen would disperse among host populations and habitats where transmission and persistence occur frequently (double direction arrows) but host populations living in marginal habitats, where transmission or persistence of the entomopathogen is variable, can act as sinks and the pathogen may not persist (single direction arrows). The entomopathogen never successfully disperses to, infects hosts or persists in some host populations (empty areas)

ingredient in our understanding of insect–pathogen dynamics is the spatial dimension, an area that, as Levin (1992) pointed out, is becoming central to nearly every problem in ecology. However, while there have been important pioneering strides made by theoretical ecologists in exploring the role of space in predator–prey (e.g., Hastings 1977; Hassell et al. 1991) and competitive interactions (e.g., Hastings 1980; Lehman and Tilman 1997), there has been much less attention given to the role of space in insect–pathogen systems.

Arthropod hosts are usually patchily distributed in time and space but, to infect hosts, fungal entomopathogens must be present and active at the same locations as host populations when hosts are present (Fig. 6). Temporal variability in activity of fungal entomopathogens, in part often due to seasonality, has been the focus of many studies that often have investigated relationships between infection or fungal activity and abiotic conditions (e.g., Wraight et al. 2007). Questions regarding spatial variability in pathogen presence and activity are a more recent focus of interest.

Previous interest in the spatial ecology of fungal entomopathogens has often first focused on mechanisms of dispersal (Andreadis 1987). Fungal dispersal can range from passive to active, with species in the large group of anamorphs of Hypocreales, such as *M. anisopliae* and *B. bassiana*, principally employing passive means of dispersal. As examples of passive mechanisms of spread, fungal entomopathogens can be moved by infected arthropods (e.g., Feng et al. 2007) as well as by arthropod members of the community that vector the pathogen (Dromph 2003; Bruck and Lewis 2002a), even including transmission of conidia by mates (e.g., Quesada-Moraga et al. 2008). Studies have also shown that insect predators can act as pathogen vectors (e.g., Roy et al. 2001). The soil is often considered to act as a reservoir for fungal entomopathogens and both aphid prey and their hemipteran predators are known to vector fungal inocula from the soil to the phylloplane (Meyling et al. 2006). Conidia of hypocrealean anamorphs are also moved passively by rain (e.g., Fernández-García and Fitt 1993; Bruck and Lewis 2002b) and wind (e.g., Shimazu et al. 2002).

In contrast to the Hypocreales, the other major taxonomic groups of fungal entomopathogens, the Entomophthorales and Chytridiomycetes, can employ active means for dispersal. Aquatic chytrids have flagellated zoospores that locate hosts. Entomophthoralean species actively eject conidia from cadavers (and, in at least one case, from living insects). While in the laboratory most conidia discharged from flies killed by species in the *E. muscae* species complex land <3.75 cm from cadavers (Six and Mullens 1996), in nature many conidia escape boundary layers and become airborne. Concentrations of conidia in the air can increase dramatically, resulting in temporally variable conidia clouds in the air over crops (Steinkraus et al. 1999; Hemmati et al. 2001) or within forests (Hajek et al. 1999). Aerial dispersal of conidia is hypothesized as being responsible for spread that has been documented from point sources where the entomopathogen *Entomophaga maimaiga* Humber, Shimazu & Soper infecting gypsy moth, *Lymantria dispar* (L.) (Lepidoptera: Lymantriidae) had been released (Hajek et al. 1996).

Whether dispersal is active or passive, how far can conidia of fungal entomopathogens travel and remain alive? While we do not have a direct answer, we can look at changes in the distribution of *E. maimaiga* as this pathogen dispersed across the area populated by the gypsy moth in the northeastern United States, between 1989 and 1992 (Hajek et al. 1995). Results from a spatial mathematical model suggest that there are two scales of movement by *E. maimaiga* conidia: a smaller (localized) scale and long-distance dispersal of conidia on wind currents above the forest canopy (Dwyer et al. 1998).

We have much yet to learn about larger scales of spatial organization and metapopulation processes of fungal entomopathogens. The theory of metapopulation processes states that distance dependent dispersal and isolation drive colonization-extinction processes, so that local interconnected populations become extinct and are later recolonized through dispersal (Hanski and Simberloff 1997). There are certainly strong suggestions that fungal entomopathogens at individual sites within an area disperse among sites, which would create a metapopulation. A model by Weseloh (2004) predicted levels of infection by *E. maimaiga* most similar to observed infection levels in nature when conidial dispersal occurred equally among plots. However, at present, studies have not specifically addressed whether individual populations of entomopathogenic fungi become extinct and are then recolonized. Fungal entomopathogens can persist in nature through the presence of long-lived spores, and probably also fungal stages within cadavers, yet the questions of metapopulation dynamics concern whether the fungus is active in localized areas and not present in dormant or quiescent stages.

There has been tremendous progress in understanding the spatial dynamics of the interactions between humans and their diseases (e.g., Mugglin et al. 2000; Grenfell et al. 2001) as well as progress in understanding the spatial dynamic relationships between fungal plant pathogens and their hosts (e.g., Thrall et al. 2003; Antonovics 2004). Many plant pathogens are well known for their ability to disperse, with obligately biotrophic species (those requiring living plant tissue for survival, e.g., rusts and powdery mildews) using wind dispersal to reach new hosts (Brown and Hovmøller 2002). Studies of the rust fungus *Uromyces valerianae* Fuckel (Basidiomycota: Uredinales) infecting *Valeriana salina* Pleijel have shown that extinction and recolonization can be affected by host population size, prevalence of disease the previous year and proximity of neighbouring populations the current year (Ericson et al. 1999). For a powdery mildew, *Podosphaera plantaginis* (Castagne) U. Braun & S. Takem (Ascomycota: Erysiphales), infecting plantain (*Plantago lanceolata* L.), disease incidence was affected by host density, proximity to a road (possibly facilitating dispersal) and proximity to the coast (presumably affecting microclimate) (Laine and Hanski 2006). A rust fungus, *Triphragmium ulmariae* (DC) Link (Basidiomycota: Uredinales), infecting meadowsweet (*Filipendula ulmaria* (L.) Maxim.) on islands never infected 43% of the 129 populations studied over four years while 37% of host plant populations were consistently infected and there was a weak relationship between disease presence and habitat type (Burdon et al. 1995). Metapopulations of a rust fungus, *Melampsora lini* (Ehrenb.) Lev. (Basidiomycota: Pucciniales), infecting wild flax (*Linum marginale* A. Cunn ex Planch.) have been shown to differ in dynamics based on spatial isolation, with more isolated patches exhibiting lower levels of disease during epidemic peaks. Extinction of pathogen genotypes was positively related to severity of disease during epidemic peaks but negatively related to the level of disease present prior to overwintering (Thrall et al. 2003). Laine and Hanski (2006) state that the high turnover rate of pathogens, distance-dependent dispersal, and regional variation in encounter rates between host and pathogen can also influence coevolutionary dynamics. Thus, spatial population structure impacts fungal pathogens infecting plants in a diversity of ways.

Future studies of spatial dynamics of fungal entomopathogens could aid in understanding host–pathogen dynamics, yet such studies remain to be conducted. Perhaps studies of spatial dynamics can help to clarify relationships between fungal entomopathogens and host density. While results from some studies have suggested that the activity of fungal entomopathogens through time is density dependent, both in outbreak and non-outbreak host populations (e.g., Kamata 2000), other examples have not found density dependence between insect hosts and fungal entomopathogens (e.g., Monzón et al. 2008). As mentioned earlier, explicit molecular characterization of fungal entomopathogens will aid in identifying host–pathogen associations as well as in elucidating whether local extinction of individual pathogen genotypes occurs. The extent to which factors such as host density, infection the previous year and the degree of isolation of both host and fungal entomopathogen are associated with infection must be investigated to begin gaining insights into the extent to which spatial population structure can help us to understand dynamics of insect diseases.

Conclusions

Fungal entomopathogens are involved in a wealth of interactions in the environments in which they occur. They may have both direct and indirect effects on their hosts and other species within communities and they can disperse among communities of hosts. Adapting the general principles from disciplines such as community ecology and metapopulation ecology will provide us with new insights into the role that fungal entomopathogens play in nature. For example, ecological approaches can be used to more fully understand the less direct ways in which fungal entomopathogens interact with both insect hosts and plants (Vega et al. 2009). Moreover, molecular tools will aid in defining explicit delimitations of fungal entomopathogen populations regardless of whether these entities are termed clades or species (Enkerli and Widmer 2009). This knowledge will help us to

better understand the interactions among fungal entomopathogens and their arthropod hosts and thereby improve predictions of the outcome of such interactions. We urge invertebrate pathologists working with fungal entomopathogens to address research questions investigating ecological principles. We also encourage ecologists to include fungal entomopathogens in their research.

References

Anderson RM, May RM (1981) The population dynamics of microparasites and their invertebrate hosts. Phil Trans R Soc Lond Ser B Biol Sc 291:451–524

Andreadis TG (1987) Transmission. In: Fuxa JR, Tanada Y (eds) Epizootiology of insect diseases. Wiley, UK, pp 159–178

Antonovics J (2004) Long-term study of a plant-pathogen metapopulation. In: Hanski I, Gaggiotti OE (eds) Ecology, genetics, and evolution of metapopulations. Elsevier, Amsterdam, pp 471–488

Barbosa P (1998) Conservation biological control. Academic Press, San Diego

Baverstock J, Alderson PG, Pell JK (2005) Influence of the aphid pathogen *Pandora neoaphidis* on the foraging behaviour of the aphid parasitoid *Aphidius ervi*. Ecol Entomol 30:665–672

Baverstock J, Baverstock KE, Clark SJ, Pell JK (2008) Transmission of *Pandora neoaphidis* in the presence of co-occurring arthropods. J Invertebr Pathol 98:356–359

Baverstock J, Roy HE, Pell JK (2009) Entomopathogenic fungi and insect behaviour: from unsuspecting hosts to targeted vectors. BioControl. doi:10.1007/s10526-009-9238-5 (this SI)

Bischoff JF, Rehner SA, Humber RA (2006) *Metarhizium frigidum* sp. nov.: a cryptic species of *M. anisopliae* and a member of the *M. flavoviride* complex. Mycologia 98:737–745

Borer ET, Briggs CJ, Holt RD (2007) Predators, parasitoids, and pathogens: a cross-cutting examination of intraguild predation theory. Ecology 88:2681–2688

Brodeur J, Rosenheim JA (2000) Intraguild interactions in aphid parasitoids. Ent Exp Appl 97:93–108

Brown JK, Hovmøller MS (2002) Aerial dispersal of pathogens on the global and continental scales and its impact on plant disease. Science 297:537–541

Bruck DJ (2009) Fungal entomopathogens in the rhizosphere. BioControl. doi:10.1007/s10526-009-9236-7 (this SI)

Bruck DJ, Lewis LC (2002a) *Carpophilus freeman* (Coleoptera: Nitidulidae) as a vector of *Beauveria bassiana*. J Invertebr Pathol 80:188–190

Bruck DJ, Lewis LC (2002b) Rainfall and crop residue effects on soil dispersion and *Beauveria bassiana* spread to corn. Appl Soil Ecol 20:183–190

Bukovinszky T, van Veen FJF, Jongema Y, Dicke M (2008) Direct and indirect effects of resource quality on food web structure. Science 319:804–807

Burdon JJ, Ericson L, Müller WJ (1995) Temporal and spatial changes in a metapopulation of the rust pathogen *Triphragmium ulmariae* and its host, *Filipendula ulmaria*. J Ecol 83:979–989

Dromph KM (2003) Collembolans as vectors of entomopathogenic fungi. Pedobiologia 47:245–256

Dunkel FV, Jaronski ST (2003) Development of a bioassay system for the predator, *Xylocoris flavipes* (Heteroptera: Anthocoridae), and its use in subchronic toxicity/pathogenicity studies of *Beauveria bassiana* strain GHA. J Econ Entomol 96:1045–1053

Dwyer G, Elkinton JS, Hajek AE (1998) Spatial scale and the spread of a fungal pathogen of gypsy moth. Am Nat 152:485–494

Eilenberg J (2002) Biology of fungi from the order Entomophthorales. DSc thesis. The Royal Veterinary and Agricultural University, Denmark

Eilenberg J, Hajek A, Lomer C (2001) Suggestions for unifying the terminology in biological control. BioControl 46:387–400

Ekesi S, Shah PA, Clark SJ, Pell JK (2005) Conservation biological control with the fungal pathogen *Pandora neoaphidis*: implications of aphid species, host plant and predator foraging. Agric Forest Entomol 7:21–30

Enkerli J, Widmer F (2009) Molecular ecology of fungal entomopathogens: molecular genetic tools and their applications in population and fate studies. BioControl. doi:10.1007/s10526-009-9251-8 (this SI)

Ericson L, Burdon JJ, Müller WJ (1999) Spatial and temporal dynamics of epidemics of the rust fungus *Uromyces valerianae* on populations of its host *Valeriana salina*. J Ecol 87:649–658

Feng MG, Chen C, Shang SW, Ying SH, Shen ZC, Chen XX (2007) Aphid dispersal flight disseminates fungal pathogens and parasitoids as natural control agents of aphids. Ecol Entomol 32:97–104

Fernández-García E, Fitt BDL (1993) Dispersal of the entomopathogen *Hirsutella cryptosclerotium* by simulated rain. J Invertebr Pathol 61:39–43

Fuxa JR, Tanada Y (1997) Epizootiology of insect diseases. Wiley, NY

Grenfell BT, Bjørnstad ON, Kappey J (2001) Travelling waves and spatial hierarchies in measles epidemics. Nature 414:716–723

Hajek AE, Leger RJ (1994) Interactions between fungal pathogens and insect hosts. Annu Rev Entomol 39:293–322

Hajek AE, Humber RA, Elkinton JS (1995) The mysterious origin of *Entomophaga maimaiga* in North America. Am Entomol 41:31–42

Hajek AE, Elkinton JS, Witcosky JJ (1996) Introduction and spread of the fungal pathogen *Entomophaga maimaiga* (Zygomycetes: Entomophthorales) along the leading edge of gypsy moth (Lepidoptera: Lymantriidae) spread. Environ Entomol 25:1235–1247

Hajek AE, Olsen CH, Elkinton JS (1999) Dynamics of airborne conidia of the gypsy moth (Lepidoptera: Lymantriidae) fungal pathogen *Entomophaga maimaiga* (Zygomycetes: Entomophthorales). Biol Contr 16:111–117

Hanski I, Simberloff D (1997) The metapopulation approach, its history, conceptual domain, and application to conservation. In: Hanski IA, Gilpin ME (eds) Metapopulation

biology: ecology, genetics and evolution. Academic Press, San Diego, pp 5–26
Hassell MP, Comins HN, May RM (1991) Spatial structure and chaos in insect populations. Nature 353:255–258
Hastings A (1977) Spatial heterogeneity and the stability of predator-prey systems. Theor Pop Biol 12:37–48
Hastings A (1980) Disturbance co-existence, history, and competition for space. Theor Ecol 18:363–373
Hatcher MJ, Dick JTA, Dunn AM (2006) How parasites affect interactions between competitors and predators. Ecol Lett 9:1253–1271
Hemmati F, Pell JK, McCartney HA, Deadman ML (2001) Airborne concentrations of conidia of *Erynia neoaphidis* above cereal fields. Mycol Res 105:485–489
Holt RD (1977) Predation, apparent competition, and structure of prey communities. Theor Pop Biol 12:197–229
Holt RD (1997) Community modules. In: Gange AC, Brown VK (eds) Multitrophic interactions in terrestrial systems. Blackwell, London, pp 333–350
Holt RD (2000) Trophic cascades in terrestrial ecosystems. Reflections on Polis et al. Trends Ecol Evol 15:444–445
Holt RD, Dobson AP (2006) Extending the principles of community ecology to address the epidemiology of host-pathogen systems. In: Collinge SK, Ray C (eds) Disease ecology—community structure and pathogen dynamics. Oxford University Press, Oxford, pp 6–27
Holt RD, Hochberg ME (2001) Indirect interactions, community modules and biological control: a theoretical perspective. In: Wajnberg E, Scott JK, Quimby PC (eds) Evaluating indirect effects of biological control. CABI, Wallingford, pp 13–38
Holt RD, Lawton JH (1994) The ecological consequences of shared natural enemies. Annu Rev Ecol Syst 25:495–520
Holt RD, Polis GA (1997) A theoretical framework for intraguild predation. Am Nat 149:745–764
Hu G, Leger RJ (2002) Field studies using a recombinant mycoinsecticide (*Metarhizium anisopliae*) reveal that it is rhizosphere competent. Appl Env Microbiol 68:6383–6387
Hughes WOH, Petersen KS, Ugelvig LV, Pedersen D, Thomsen L, Poulsen M, Boomsma JJ (2004) Density-dependence and within-host competition in a semelparous parasite of leaf-cutting ants. BMC Evol Biol 4:45. doi: 10.1186/1471-2148-4-45
Inglis GD, Goettel MS, Butt TM, Strasser H (2001) Use of hyphomycetous fungi for managing insect pests. In: Butt TM, Jackson C, Magan N (eds) Fungi as biocontrol agents. Progress, problems and potential. CABI, Wallingford, pp 23–69
Jensen AB, Thomsen L, Eilenberg J (2001) Intraspecific variation and host specificity of *Entomophthora muscae sensu stricto* isolates revealed by random amplified polymorphic DNA, universal primed PCR, PCR-restriction fragment length polymorphism, and conidial morphology. J Invertebr Pathol 78:251–259
Kamata N (2000) Population dynamics of the beech caterpillar, *Syntypistis punctatella*, and biotic and abiotic factors. Popul Ecol 42:267–278
Kim JJ, Goettel MS, Gillespie DR (2007) Potential of *Lecanicillium* species for dual microbial control of aphids and the cucumber powdery mildew fungus, *Sphaerotheca fuliginia*. Biol Contr 40:327–332
Kim JJ, Goettel MS, Gillespie DR (2008) Evaluation of *Lecanicillium longisporum*, Vertalec for simultaneous suppression of cotton aphid, *Aphis gossypii*, and cucumber powdery mildew, *Sphaerotheca fuliginea*, on potted cucumbers. Biol Contr 45:404–409
Lafferty KD, Allesina S, Arim M, Briggs CJ, De Leo G, Dobson AP, Dunne JA, Johnson PTJ, Kuris AM, Marcogliese DJ, Martinez ND, Memmott J, Marquet PA, McLaughlin JP, Mordecai EA, Pascual M, Poulin R, Thieltges DW (2008) Parasites in food webs: the ultimate missing links. Ecol Lett 11:533–546
Laine A-L, Hanski I (2006) Large-scale spatial dynamics of a specialist plant pathogen in a fragmented landscape. J Ecol 94:17–226
Lefevre T, Lebarbenchon C, Gauthier-Clerc M, Misse D, Poulin R, Thomas F (2009) The ecological significance of manipulative parasites. Trends Ecol Evol 24:41–48
Lehman CL, Tilman D (1997) Competition in spatial habitats. In: Tilman D, Kareiva P (eds) Spatial ecology: the role of space in population dynamics and interspecific interactions. Princeton University Press, Princeton, pp 185–203
Levin SA (1992) The problem of pattern and scale in ecology. Ecology 73:1943–1967
Meyling NV, Eilenberg J (2007) Ecology of the entomopathogenic fungi *Beauveria bassiana* and *Metarhizium anisopliae* in temperate agroecosystems: potential for conservation biological control. Biol Contr 43:145–155
Meyling NV, Pell JK (2006) Detection and avoidance of an entomopathogenic fungus by a generalist insect predator. Ecol Entomol 31:162–171
Meyling NV, Pell JK, Eilenberg J (2006) Dispersal of *Beauveria bassiana* by the activity of nettle insects. J Invertebr Pathol 93:121–126
Meyling NV, Lubeck M, Buckley EP, Eilenberg J, Rehner SA (2009) Community composition, host range and genetic structure of the fungal entomopathogen *Beauveria* in adjoining agricultural and seminatural habitats. Mol Ecol 18:1282–1293
Monzón AJ, Guharay F, Klingen I (2008) Natural occurrence of *Beauveria bassiana* in *Hypothenemus hampei* (Coleoptera: Curculionidae) populations in unsprayed coffee fields. J Invertebr Pathol 97:34–141
Mugglin AS, Cressie N, Gemmell I (2000) Hierarchical statistical modeling of influenza-epidemic dynamics in space and time. Stat Med 21:2703–2721
Müller CB, Godfray HCJ (1997) Apparent competition between two aphid species. J Anim Ecol 66:57–64
Müller CB, Godfray HCJ (1999) Indirect interactions in aphid-parasitoid communities. Res Popul Ecol 41:93–106
Nakashima Y, Birkett MA, Pye BJ, Pickett JA, Powell W (2004) The role of semiochemicals in the avoidance of the seven-spot ladybird *Coccinella septempunctata* by the aphid parasitoid *Aphidius ervi*. J Chem Ecol 30:1103–1116
Okuyama T, Bolker BM (2007) On quantitative measures of indirect interactions. Ecol Lett 10:264–271
Onstad DW, Carruthers RI (1990) Epizootiological models of insect diseases. Annu Rev Entomol 35:399–419
Ormond E (2007) Overwintering interactions between *Coccinella septempunctata* and *Beauveria bassiana*. PhD thesis. Anglia Ruskin University, Cambridge, UK

Packer C, Holt RD, Hudson PJ, Lafferty KD, Dobson AP (2003) Keeping the herd healthy and alert: implications of predator control for infectious disease. Ecol Lett 6:797–802

Pell JK, Vandenberg JD (2002) Interactions among the aphid *Diuraphis noxia*, the entomopathogenic fungus *Paecilomyces fumosoroseus* and the coccinellid *Hippodamia convergens*. Biocontr Sci Technol 12:217–224

Pell JK, Pluke R, Clark SJ, Kenward MG, Alderson PG (1997) Interactions between two aphid natural enemies, the entomopathogenic fungus *Erynia neoaphidis* Remaudière and Hennebert (Zygomycetes: Entomophthorales) and the predatory beetle *Coccinella septempunctata* L. (Coleoptera: Coccinellidae). J Invertebr Pathol 69:261–268

Pell JK, Steinkraus D, Hannam, J (2009) Conservation biological control using fungal entomopathogens. BioControl. doi:10.1007/s10526-009-9245-6 (this SI)

Perrin RM (1976) The population dynamics of the stinging nettle aphid, *Microlophium carnosum* (Bukt.). Ecol Entomol 1:31–40

Polis GA, Holt RD (1992) Intraguild predation—the dynamics of complex trophic interactions. Trends Ecol Evol 7:151–154

Polis GA, Myers CA, Holt RD (1989) The ecology and evolution of intraguild predation—potential competitors that eat each other. Annu Rev Ecol Syst 20:297–330

Pope T, Croxson E, Pell JK, Godfray HCJ, Muller CB (2002) Apparent competition between two species of aphid via the fungal pathogen *Erynia neoaphidis* and its interaction with the aphid parasitoid *Aphidius ervi*. Ecol Entomol 27:196–203

Poprawski TJ, Carruthers RI, Speese J, Vacek DC, Wendel LE (1997) Early-season applications of the fungus *Beauveria bassiana* and introduction of the hemipteran predator *Perillus bioculatus* for control of Colorado potato beetle. Biol Control 10:48–57

Powell W, Wilding N, Brobyn PJ, Clark SJ (1986) Interference between parasitoids [Hym, Aphidiidae] and fungi [Entomophthorales] attacking cereal aphids. Entomophaga 31:293–302

Quesada-Moraga E, Martin-Carballo I, Garrido-Jurado I, Santiago-Alvarez C (2008) Horizontal transmission of *Metarhizium anisopliae* among laboratory populations of *Ceratitis capitata* (Wiedemann) (Diptera: Tephritidae). Biol Control 47:115–124

Raffel TR, Martin LB, Rohr JR (2008) Parasites as predators: unifying natural enemy ecology. Trends Ecol Evol 23:610–618

Rehner SA (2005) Phylogenetics of the insect pathogenic genus *Beauveria*. In: Vega FE, Blackwell M (eds) Insect-fungal associations—ecology and evolution. Oxford University Press, New York, pp 3–27

Rehner SA, Buckley EP (2005) A *Beauveria* phylogeny inferred from nuclear ITS and EF1-alpha sequences: evidence for cryptic diversification and links to *Cordyceps* teleomorphs. Mycologia 97:84–98

Rosenheim JA, Kaya HK, Ehler LE, Marois JJ, Jaffee BA (1995) Intraguild predation among biological control agents—theory and evidence. Biol Control 5:303–335

Roy HE, Cottrell TE (2008) Forgotten natural enemies: Interactions between coccinellids and insect-parasitic fungi. Eur J Entomol 105:391–398

Roy HE, Pell JK, Clark SJ, Alderson PG (1998) Implications of predator foraging on aphid pathogen dynamics. J Invertebr Pathol 71:236–247

Roy HE, Pell JK, Alderson PG (2001) Targeted dispersal of the aphid pathogenic fungus *Erynia neoaphidis* by the aphid predator *Coccinella septempunctata*. Biocontrol Sci Tech 11:99–110

Schmitz OJ (2008) Effects of predator hunting mode on grassland ecosystem function. Science 319:952–954

Schmitz OJ, Krivan V, Ovadia O (2004) Trophic cascades: the primacy of trait-mediated indirect interactions. Ecol Lett 7:153–163

Shah PA, Pell JK (2003) Entomopathogenic fungi as biological control agents. Appl Microbiol Biotech 61:413–423

Shimazu M, Sato H, Maehara N (2002) Density of the entomopathogenic fungus, *Beauveria bassiana* Vuillemin (Deuteromycotina: Hyphomycetes) in forest air and soil. Appl Entomol Zool 37:19–26

Six DL, Mullens BA (1996) Distance of conidial discharge of *Entomophthora muscae* and *Entomophthora schizophorae* (Zygomycotina: Entomophthorales). J Invertebr Pathol 67:253–258

Steenberg T, Langer V, Esbjerg P (1995) Entomopathogenic fungi in predatory beetles (Col: Carabidae and Staphylinidae) from agricultural fields. Entomophaga 40:77–85

Steinkraus DC, Howard MN, Hollingsworth RG, Boys GL (1999) Infection of sentinel cotton aphids (Homoptera: Aphididae) by aerial conidia of *Neozygites fresenii* (Entomophthorales: Neozygitaceae). Biol Control 14:131–135

Tanada Y, Kaya HK (1993) Insect pathology. Academic Press, San Diego

Thrall PH, Godfree R, Burdon JJ (2003) Influence of spatial structure on pathogen colonization and extinction: a test using an experimental metapopulation. Plant Pathol 52:350–361

Todorova SI, Cloutier C, Cote JC, Coderre D (2002) Pathogenicity of six isolates of *Beauveria bassiana* (Balsamo) Vuillemin (Deuteromycotina, Hyphomycetes) to *Perillus bioculatus* (F) (Hem., Pentatomidae). J Appl Entomol 126:182–185

van Veen FJF, van Holland PD, Godfray HCJ (2005) Stable coexistence in insect communities due to density- and trait-mediated indirect effects. Ecology 86:1382–1389

van Veen FJF, Memmot J, Godfray HCJ (2006a) Indirect effects, apparent competition and biological control. In: Brodeur J, Boivin G (eds) Trophic and guild interactions in biological control. Springer, The Netherlands, pp 145–170

van Veen FJF, Morris RJ, Godfray HCJ (2006b) Apparent competition, quantitative food webs, and the structure of phytophagous insect communities. Annu Rev Entomol 51:187–208

van Veen FJF, Müller CB, Pell JK, Godfray HCJ (2008) Food web structure of three guilds of natural enemies: predators, parasitoids and pathogens of aphids. J Anim Ecol 77:191–200

Vance-Chalcraft HD, Rosenheim JA, Vonesh JR, Osenberg CW, Sih A (2007) The influence of intraguild predation on prey suppression and prey release: a meta-analysis. Ecology 88:2689–2696

Vega FE, Posada F, Aime MC, Pava-Ripoll M, Infante F, Rehner SA (2008) Entomopathogenic fungal endophytes. Biol Control 46:72–82

Vega FE, Goettel MS, Blackwell M, Chandler D, Jackson MA, Keller S, Koike M, Maniania NK, Monzon A, Ownley BH, Pell JK, Rangel DEN, Roy HE (2009) Fungal entomopathogens: new insights on their ecology. Fungal Ecol 2:1–11

Vestergaard S, Eilenberg J (2000) Persistence of released *Metarhizium anisopliae* in soil and prevalence in ground and rove beetles. In: Proceedings of the 7th European meeting of the IOBC/WPRS working group: insect pathogens and insect parasitic nematodes, entitled 'Capturing the potential of biological control', vol 23, pp 181–185, Vienna, Austria. 22–26 March 1999

Weseloh RM (2004) Effect of conidial dispersal of the fungal pathogen *Entomophaga maimaiga* (Zygomycetes: Entomophthorales) on survival of its gypsy moth (Lepidoptera: Lymantriidae) host. Biol Control 29:138–144

Wraight SP, Inglis GD, Goettel MS (2007) Fungi. In: Lacey LA, Kaya HK (eds) Field manual of techniques in invertebrate pathology. Springer, Dordrecht, pp 223–248

Zimmermann G (2007a) Review on safety of the entomopathogenic fungi *Beauveria bassiana* and *Beauveria brongniartii*. Biocontrol Sci Tech 17:553–596

Zimmermann G (2007b) Review on safety of the entomopathogenic fungus *Metarhizium anisopliae*. Biocontrol Sci Tech 17:879–920

Author Biographies

Nicolai V. Meyling is an associate professor in the Department of Agriculture and Ecology, University of Copenhagen. He teaches courses in zoology, diversity and biological control. His research focuses on host–pathogen interactions, host range and fundamental ecology of fungal entomopathogens in managed ecosystems.

Ann E. Hajek is a professor studying insect pathology in the Department of Entomology, Cornell University, Ithaca, New York. She teaches undergraduate courses in biological control and invasive species and a graduate course in invertebrate pathology. Her research focuses on the epizootiology of insect diseases, ecology and evolution of entomopathogens and use of entomopathogens for control of insects, particularly invasive insect species.

BioControl (2010) 55:55–73
DOI 10.1007/s10526-009-9249-2

Challenges in modelling complexity of fungal entomopathogens in semi-natural populations of insects

H. Hesketh · H. E. Roy · J. Eilenberg · J. K. Pell · R. S. Hails

Received: 2 October 2009 / Accepted: 19 October 2009 / Published online: 28 November 2009

Abstract The use of fungal entomopathogens as microbial control agents has driven studies into their ecology in crop ecosystems. Yet, there is still a lack of understanding of the ecology of these insect pathogens in semi-natural habitats and communities. We review the literature on prevalence of fungal entomopathogens in insect populations and highlight the difficulties in making such measurements. We then describe the theoretical host-pathogen models available to examine the role that fungal entomopathogens could play in regulating insect populations in semi-natural habitats, much of the inspiration for which has been drawn from managed systems, particularly forests. We further emphasise the need to consider the complexity, and particularly the heterogeneity, of semi-natural habitats within the context of theoretical models and as a framework for empirical studies. We acknowledge that fundamental gaps in understanding fungal entomopathogens from an ecological perspective coupled with a lack of empirical data to test theoretical predictions is impeding progress. There is an increasing need, especially under current rapid environmental change, to improve our understanding of the role of fungi in insect population dynamics beyond the context of forestry and agriculture.

Handling Editor: Eric Wajnberg.

H. Hesketh (✉)
NERC Centre for Ecology & Hydrology, Mansfield Road, Oxford, Oxfordshire OX1 3SR, UK
e-mail: hhesketh@ceh.ac.uk

H. E. Roy
NERC Centre for Ecology & Hydrology, Maclean Building, Benson Lane, Crowmarsh Gifford, Oxfordshire OX10 8BB, UK

J. Eilenberg
Department of Agriculture and Ecology, Faculty of Life Sciences, University of Copenhagen, Thorvaldsensvej 40, 1871 Frederiksberg C, Denmark

J. K. Pell
Department of Plant and Invertebrate Ecology, Rothamsted Research, Harpenden, Hertfordshire AL5 2JQ, UK

R. S. Hails
NERC Centre for Ecology & Hydrology, Mansfield Road, Oxford, Oxfordshire, UK

Keywords Pathogen population dynamics · Theoretical modelling · Epizootiology · Fungal entomopathogens · Entomophthorales · Hypocreales · Non-pest insects

Introduction

Fungal entomopathogens are diverse and globally ubiquitous natural enemies of arthropods. There has been considerable research focus on their potential as microbial control agents (e.g. Goettel et al. 2005, 2008; Pell 2007; Vega et al. 2009; Hajek and Delalibera 2009; Jaronski 2009; Pell et al. 2009). Indeed, they are considered to have been instrumental in the advent of modern microbial control (Steinhaus 1949; Burges and Hussey 1971; Krassilstchik 1888; Vega 2008) and our knowledge of fungal ecology in crop ecosystems has largely been driven by applied studies assessing their potential for microbial control. There is no doubt that such studies have provided fundamental information on the host-fungus relationship. However, there are still profound gaps in our understanding of their ecology particularly in semi-natural or minimally managed systems both in terrestrial and aquatic environments (Johnson et al. 2006; Stentiford et al. 2001; Roy and Cottrell 2008; Roy et al. 2009). The role of pathogens as natural enemies of non-pest insects, including those of conservation value, is seldom considered beyond their context as 'non-targets' of microbial control agents (Roy et al. 2009). It is possible that fungal pathogens are playing a hitherto undetected role in documented declines of some insect species (Balmford et al. 2005; Roy and Cottrell 2008). In this review we consider the literature on prevalence of fungal entomopathogens in populations of insects in crop and semi-natural habitats. We then examine the insights provided by ecological models in exploring the role fungi may play in regulating host populations. As much of the inspiration for this work is drawn from forest ecosystems and, due to the paucity of data available from semi-natural ecosystems, we highlight those studies in managed systems that are also particularly relevant to insect populations in semi-natural habitats.

There are over 700 species of fungal entomopathogens and these are broadly found within two main groups: phylum Ascomycota (subkingdom Dikarya) and the order Entomophthorales (Hibbett et al. 2007; Humber 2008; Blackwell 2009). Within the Ascomycota there are two major orders that contain entomopathogens: Hypocreales (class: Sordariomycetes; subclass: Hypocreomycetidae) and Laboulbeniales (class: Laboulbeniomycetes) (Hibbett et al. 2007). The Hypocreales have both sexual (teleomorph) and asexual (anamorph) forms although most research has focused on the anamorphs. Anamorphic hypocrealean fungi are considered to be generalist pathogens with broad host ranges and even switching between pathogenic and saprophytic lifestyles (Blackwell 2009). The Laboulbeniales (Ascomycota: Laboulbeniaceae) are a group of obligate ectoparasitic fungi that are mainly associated with Coleoptera and do not cause death of their hosts (Weir and Hammond 1997; Roy and Cottrell 2008). The Entomophthorales are all obligate arthropod pathogens historically placed within the phylum Zygomycota but likely to be elevated to a distinct subphylum named Entomophthoramycotina pending resolution of clades from the Zygomycota (Hibbett et al. 2007). Microsporidia are now known to be highly specialised obligate intracellular fungi, closely aligned to the Entomophthorales (Keeling and Fast 2002; Humber 2008) and infecting a diverse array of vertebrate and invertebrate hosts. However, the Laboulbeniomycetes are all associated with insects or other arthropods mostly as biotrophic parasites (Blackwell 2009). There are a number of detailed studies examining the ecological interactions of microsporidia with their hosts particularly in forest Lepidoptera systems (Hoch et al. 2000, 2008; Pilarska et al. 2006; Solter 2006; Solter and Becnel 2007). We will only consider microsporidia superficially in this manuscript; the recent literature on this group is worthy of an entire ecological review. However there are undoubtedly conceptual parallels between this intriguing group of fungi and the others that are described in detail here.

Although the taxonomy of fungal entomopathogens is undergoing significant change, their basic biology and general life history attributes are well understood. All fungal entomopathogens produce infective conidia (spores) that attach to, germinate, and penetrate the cuticle (or digestive tract) of their host. Inside the host they proliferate as single- or multi-celled structures (protoplasts, blastospores, hyphal bodies), usually killing the host and producing either more infective conidia for immediate transmission or resting structures (sexual or asexual resting spores, chlamydospores, mummified hosts) for persistence in the environment (Roy et al. 2006; Pell et al. 2001; Table 1; Fig. 1).

Abiotic and biotic conditions strongly influence key components of fungal activity and fitness including transmission efficiency and persistence within and outside the host (Fuxa and Tanada 1987; Fig. 1). Humidity in excess of 90% in the microenvironment surrounding fungi is required for germination, infection, and sporulation (e.g. Inglis et al. 2001; Wilding 1969) and is considered to be the most critical environmental factor influencing the development of epizootics (Fuxa and Tanada 1987; Hall and Papierok 1982). Ambient temperatures affect speed of germination, growth and kill. There is an inverse relationship

Table 1 Broad comparison of attributes of the Entomophthorales and Hypocreales (modified from Pell et al. 2001). There is considerable variability between species within these families. Some attributes are host dependent

Attribute	Entomophthorales	Hypocreales	Comments	References
Families	Ancylistaceae Completoriaceae Entomophthoraceae Meristacraceae Neozygitaceae	Clavicipitaceae Cordycipitaceae Ophiocordycipitaceae		Hibbett et al. (2007) Humber (2008)
Spore size	Usually >10 μm	<10 μm		Balazy (1993) Samson et al. (1988)
Reproductive output (spores per cadaver)	Few (10^4)	Many (10^7–10^9)		Arthurs and Thomas (2001) Eilenberg (1987) Hua and Feng (2003) Posada and Vega (2005)
Sporulation rate	Fast (hours)	Slow (days)		Eilenberg (1987) Arthurs and Thomas (2001) Sierotzki et al. (2000)
Germination rate	Fast (hours)	Slow (days) but sometimes fast		Oduor et al. (1996) Posada and Vega (2005)
Life cycle	Fast (few days)	Slow (several days or even weeks)		Posada and Vega (2005) Roy et al. (2006)
Higher order production of spores (primary, secondary, etc.)	Always	Rarely	The hypocrealean genus *Aschersonia* produces higher order conidia	Scholte et al. (2004) Shah and Pell (2003)
LC_{50} (spores mm^{-2} or conidia ml^{-1})	Low (10^0–10^4)	High (10^2–10^9)	Host dependent: *P. neoaphidis* to *A. pisum* LC_{50} of 19 conidia mm^{-2}; *P. neoaphidis* to *U. jaceae* LC_{50} of 10^4 conidia mm^{-2} (Ekesi et al. 2005) Isolate dependent: Hypocreales to *Aphis fabae* LC_{50} range 1.62×10^2 - 2.95×10^7 conidia ml^{-1} (Hesketh et al. 2008)	Eilenberg (1987) Ekesi et al. (2005) Hesketh et al. (2008) Roy et al. (2008) Ugine et al. (2005) Xu and Feng (2000)
Active discharge	In most cases	Only in sexual stages	Active discharge is not known for species within the entomophthoralean genus *Massospora* Active discharge is known for some *Cordyceps* (Hypocreales)	Scholte et al. (2004) Wongsa et al. (2005)

Table 1 continued

Attribute	Entomophthorales	Hypocreales	Comments	References
Mucous coated spores?	Often	Rarely	There are exceptions where mucous coated conidia are produced by some Hypocreales for example *Verticillium*, *Hirsutella*, *Aschersonia*	Roy et al. (2006) Shah and Pell (2003)
Resting spores	Common	Rare	Entomophthorales: resting spores are mostly sexual. *Cordyceps* species also produce sexual spores but not resting spores. Hypocreales: *Sorosporella* spp. produce resting spores and *Beauveria* spp. produce microsclerotia	Roy et al. (2006) Shah and Pell (2003) Scholte et al. (2004)
Rhizoids	Present or absent	Absent		Roy et al. (2006)
Host range	Narrow (one host or taxonomically related host species)	Wide (hosts may belong to taxonomically distant groups)	At the species level hypocrealean fungi have broad host ranges but isolates can be more specific. Furthermore, species complexes are known for a number of species in both groups.	Shah and Pell (2003) Scholte et al. (2004)
Epizootics	Common	Common		Scholte et al. (2004)
Common transmission mode	Aerial by wind and rain Host to host	Rain splash Host to host		Roy and Pell (2000) Scholte et al. (2004)
Pre-death sporulation?	Rare	Rare	Entomophthorales: *Entomophthora thripidium* *Strongwellsea* species *Massospora* species Hypocreales: *Lecanicillium* species	Shah and Pell (2003) Roy et al. (2006)
Modification of host behaviour	Common	Rare but occurs in *Cordyceps* species		Roy and Pell (2000) Roy et al. (2006) Pontoppidan et al. (2009)
Primary reservoir	Host	Soil		Roy et al. (2009)
Primary environment	Mostly foliar (resting spores in soil)	Both in soil and foliar		Shah and Pell (2003)
Toxin production	Known for *Conidiobolus* species	Known		Strasser et al. (2000) Shah and Pell (2003)
Saprophytic life strategies	Rare	Common	Species of the entomophthoralean genus *Conidiobolus* can be saprophytic	Shah and Pell (2003)
Primary biological control strategies	Conservation Classical Inocolulation	Augmentation Inundation Classical		Eilenberg et al. (2001) Shah and Pell (2003) Pell (2007)

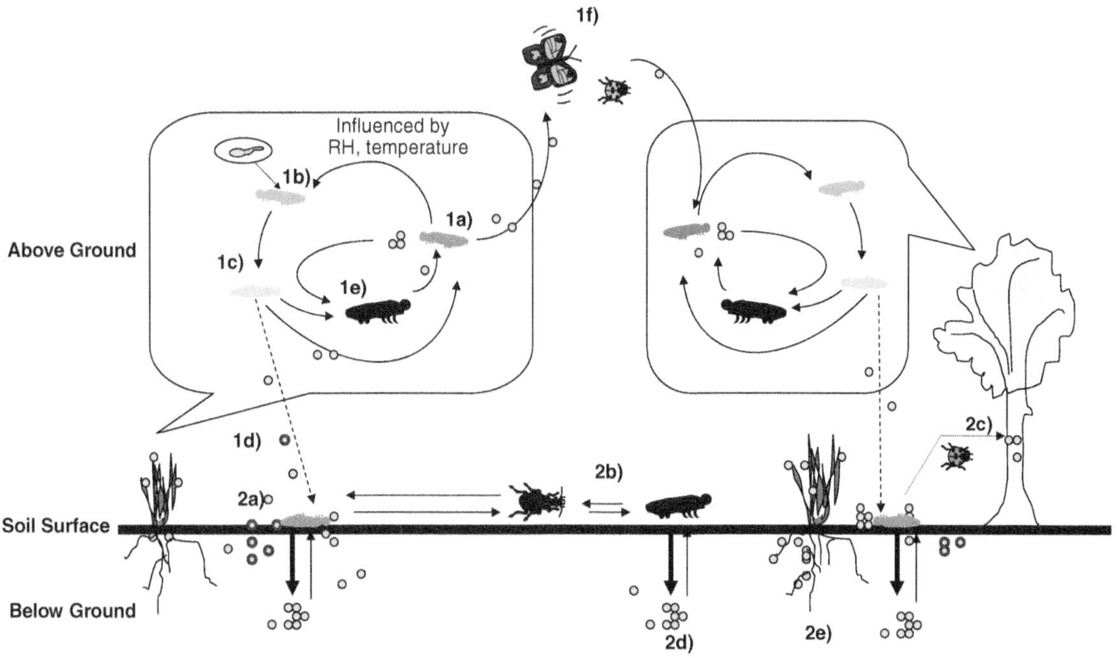

Fig. 1 Entomopathogenic fungi and their hosts exist in a complex landscape influenced by multi-trophic relationships within the community and modulated by abiotic factors. Environmental change, particularly the arrival of new species (either host or fungus), climate change, habitat fragmentation and/or alteration will have differential effects across this community. Arrow size indicates the direction of interaction that is likely to be greatest in semi-natural habitats. 1a) an insect host contacts infective spores which b) germinate and penetrate the host eventually killing it, c) the sporulating cadaver releases spores for further cycles or d) often (in the case of Entomophthoralean fungi) when the number of susceptible hosts decreases resting spores are produced which survive in the soil and produce infective spores under favourable conditions and e) alternative hosts, often taxonomically related to the primary host, may be infected and produce spores that also infect primary hosts f) Conidia can be transported in wind currents, in infected hosts and on the surfaces of non-host invertebrates to other habitats. 2a) Conidia/resting spore distribution and persistence at the soil surface will be influenced by abiotic factors such as rainfall that influence horizontal transmission by promoting conidium formation on cadavers, mechanically dispersing conidia and potentially increasing vectoring by other invertebrates b) Epigeal predators can also remove inoculum by consuming cadavers but may also vector infective stages to new hosts and habitats at the soil surface and c) into foliar environments. Persistence in the soil profile is affected by d) soil type, soil moisture and pore size and by e) interactions in the rhizosphere with soil microbes, root exudates and secondary plant compounds. Within the soil profile conidia may also be dispersed by species such as *Collembola*

between speed of kill and temperature although overall mortality may not be affected (Ekesi et al. 1999; Thomas and Blanford 2003). Solar radiation is detrimental to persistence, particularly on the phylloplane where fungi can be rapidly deactivated (e.g. Fargues et al. 1996; Furlong and Pell 1997).

The detailed ecology, physiology and life cycles of each species within these groups can be exceedingly varied reflecting adaptations to ensure survival and transmission despite the environmental constraints (Pell et al. 2001; Roy et al. 2006; Table 1). However, it is possible to generalise for taxonomically related species/groups. Entomophthoralean fungi demonstrate a continuum of adaptations for dispersal and transmission. They exhibit dispersive, actively discharged conidia produced externally after host death through to sporulation from living hosts prior to host death which is particularly noted in species which require continued host activity to ensure conidia dispersal (Pell et al. 2001; Roy et al. 2006; Table 1; Fig. 1). Species in the Entomophthorales do not generally produce toxins (secondary metabolites) as part of the infection cycle but are characteristically biotrophic with a narrow host range and are common among foliar arthropods (Pell et al. 2001; Shah et al. 2004; Table 1). Eilenberg and Pell (2007) list a number of host-pathogen systems in which the ecology of Entomophthorales has been discussed.

The anamorphic Hypocreales are generally considered to be opportunistic with broad host ranges and most commonly associated with soil-inhabiting arthropods in temperate regions. They are characteristically hemibiotrophic, switching from a parasitic, biotrophic phase in the haemocoel (sometimes producing toxins) to a saprophytic phase colonizing the host after death. Conidia are produced on the cadaver but, unlike Entomophthorales, are not actively discharged. Both Entomophthorales and Hypocreales produce resting structures for persistence in the absence of new hosts or under adverse environmental conditions. Often assumed to be generalists, they are usually considered to be less well adapted to a parasitic life style than entomophthoralean fungi. However, recent research is demonstrating that the challenges of exploiting a wide range of potential hosts requires adaptations that are just as elegant as those required for a specialist life style (Humber 2008). Furthermore, while Hypocreales tend to be considered as generalists and Entomophthorales as specialists, there is considerable variability amongst species within these orders and this is highlighted in Table 1.

The genus *Cordyceps* (Ascomycota: Hypocreales) is, perhaps, the most studied teleomorph within the Ascomycota and the most common fungus encountered in association with arthropods in tropical forests (Evans 1981). Most *Cordyceps* appear to have a very restricted host range (in contrast to their anamorphic counterparts). This has been clearly demonstrated for ants. Sanjuan et al. (2001) documented the importance of host association in the distribution and incidence of *Cordyceps* in forest systems. Number of ants parasitized by *Cordyceps* was greater in disturbed forests compared to near pristine forests and this was closely correlated to the presence of host species. The taxonomy of these fungi is only just being resolved. Indeed it is only recently that the teleomorph and anamorph states have been linked as one species rather than being assigned to separate divisions. Phylogenetic analysis suggests that the *Cordyceps* are not monophyletic but occur in three families: Clavicipitaceae, Cordycipitaceae and Ophiocordycipitaceae (Sung et al. 2007; Blackwell 2009). There are more than 400 species of *Cordyceps* and a number of studies are emerging on the ecology of a few of these (Chee-Sanford 2008; Sanjuan et al. 2001). It is fascinating to consider that the same fungal species can differ so fundamentally in ecology depending on sexual state and not surprising that many basic questions remain unanswered such as: why do teleomorphic ascomycetes not occur so widely in temperate habitats? What is driving the host specificity of the sexual stages? Are the telemorphic ascomycetes utilising the functional niches in the tropics that are occupied by the Entomophthorales in temperate zones?

The anamorphic (asexual) states of the Ascomycota have generally been used as inundative biopesticides. In contrast, research on the Entomophthorales has focused on conservation and inoculation biological control. Accordingly, ecological understanding of the Entomophthorales is more advanced than for the Hypocreales. However, recent research efforts are beginning to address this imbalance (Bidochka et al. 2001; Meyling and Eilenberg 2006a, b; Roy et al. 2009). Studies on the anamorphic states of species within the Ascomycota dominate the literature. The teleomorphic (sexual) states are poorly understood but are undoubtedly critical to our ecological understanding of fungal entomopathogens.

Conceptual framework for understanding the role of fungal entomopathogens in host population regulation

The potential of fungi to regulate insect populations will depend on their abundance in the host population (prevalence) as well as their abundance and persistence in the surrounding environment. Whether or not insect populations are regulated by fungi, our first challenge is to accurately quantify how common fungi are in both hosts and the surrounding environment.

Prevalence in host populations

Accurate measurement of prevalence without biased sampling of either uninfected or diseased insects can be difficult and some challenges are specific to fungal entomopathogens (Fig. 2). A truly accurate assessment of prevalence can only be achieved by sampling all stages of the host in a life table analysis but this is rarely possible. Two methods are usually employed to estimate prevalence (1) sampling living individuals only, followed by laboratory incubation until death when infection can be confirmed by phenotypic characteristics and (2) sampling both living, dead and dying individuals, followed by laboratory incubation and identification (Fig. 2). The choice of sampling

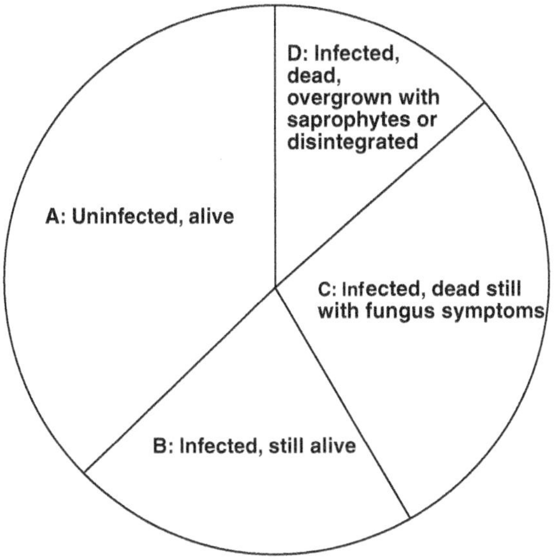

Fig. 2 (Adapted from Eilenberg and Pell 2007). A diagram of the composition of a natural population of an insect species in relation to infection by a fungus pathogen. A: The population of uninfected individuals; B: The population of living, infected individuals. Fungus prevalence will be documented upon sampling these individuals and incubating them in the laboratory. C: Recently killed fungus-infected individuals located in the field; D: Individuals overgrown with saprophytic fungi for which diagnosis is not possible without molecular probing. Example of prevalence assessment: If living individuals are sampled and incubated alongside scoring of any dead individuals in the field then prevalence is assessed as (B+C)/(A+B+C)

process and the life-stages sampled will be dictated by the practicalities of sampling. The most obvious challenge, however, is ensuring the sample is representative of the entire population (Fuxa and Tanada 1987; Fig. 2). Some insects have behavioural characteristics that cause aggregation at specific locations such as late instar larvae of *Lymantria dispar* L. (Lepidoptera: Lymantriidae) moving off trees onto soil (Hajek 2001) or exhibit behavioural changes due to infection such as increased movement in aphids (Roy et al. 2006; Roditakis et al. 2008). Some life stages cannot be easily located, such as small instars, increasing sampling bias towards the larger late instars. Furthermore, host development time could be altered by infection (Hoch et al. 2000) and this could lead to an inaccurate measure of prevalence. However, it is critical that all juvenile (and in some cases also adult) stages are sampled as insects may demonstrate stage specific resistance to fungal infection (Roy et al. 2008) and in some cases, differential susceptibility based on the life stage exposed (Dromph et al. 2002). Ideally, sampled insects should be incubated separately to avoid transmission within the sample and under conditions that do not favour infection as this could lead to overestimation of prevalence.

In recent years, molecular techniques have been developed to detect the presence of fungal pathogens in field collected insect samples. Such methods offer opportunities for more rapid assessment in the future and examples include: enzyme-linked immunosorbent assay (ELISA) to detect *Entomophaga maimaiga* Humber, Shimazu and Soper (Entomophthoramycotina: Entomophthorales) in *L. dispar* (Hajek et al. 1991); DNA probes to confirm *L. dispar* deaths due to *Entomophaga aulicae* (Reichardt in Bail) (Zygomycota: Entomophthorales) Humber or *E. maimaiga* (Hajek et al. 1996); PCR detection of *Pandora neoaphidis* (Remaudière & Hennebert) Humber (Entomophthoramycotina: Entomophthorales) in aphids (Fournier et al. 2008; Tymon et al. 2004). Most examples of prevalence assessments using the two methods described above are for pest insects in managed systems but the methods are appropriate in semi-natural systems (see examples in Table 2).

Abundance in the environment

Fungal propagules can persist outside the host on soil and phylloplanes and in the air where they can act as reservoirs of inoculum. Their abundance can be measured directly (conidia capture) and indirectly (baiting) in these habitats (e.g. Bidochka et al. 2001; Bruck 2004; Hemmati et al. 2001; Klingen et al. 2002; Meyling and Eilenberg 2006a; Wilding and Perry 1980). Soil samples are generally incubated with laboratory reared susceptible bait insects such as wax moth *Galleria melonella* L. (Lepidoptera: Pyralidae) and the frequency of insect infection is used as a measure of fungal abundance. Conidia capture in the aerial environment has been measured using selective media (Shimazu et al. 2002) or microscope slides (Steinkraus et al. 1996) exposed to the air above or within plant canopies. More precise measurements are made using volumetric spore traps e.g. Burkhard traps and rotorod samplers that capture conidia on adhesive materials to determine conidia density at specific locations (e.g. Hajek et al. 1999; Hemmati et al. 2001) Occurrence studies, such as these, are useful measurements of fungal reservoirs within a habitat that may have the potential to infect a

Table 2 Examples of prevalence studies using two methods: collection of live hosts only and collection of both living and dead hosts

Method	Fungus species	Host species	References
Living hosts only	*Entomophthora schizophorae*	*Chamaepsila rosae*	Eilenberg and Philipsen (1988)
	Entomophaga maimaiga, *Isaria farinosus*, *Lecanicillium sp.*	*Lymantria dispar*	Hajek (1997)
	Strongwellsea castrans	*Delia radicum* and other diptera	Eilenberg and Michelsen (1999)
	Beauveria bassiana	*Hypothenemus hampei*	Monzon et al. (2008)
		Musca domestica	Siri et al. (2005)
	Lecanicillium spp., *Beauveria bassiana*, *Metarhizium anisopliae*, *Isaria farinosa*	*Taeniothrips inconsequens*	Brownbridge et al. (1999)
	Pandora neoaphidis	*Sitobion avenae*	Feng et al. (2004)
			Dean and Wilding (1973)
	Entomophthora planchoniana, *Neozygites fresenii*	*Elatobium abietinum*	Nielsen et al. (2001)
Living and dead hosts	*Neozygites fresenii*	*Aphis gossypii*	Steinkraus et al. (1995)
	Pandora neoaphidis, *Entomophthora planchoniana*, *Entomophthora thaxteriana*	*Metopolophium dirhodum*, *Sitobion avenae*	Dean and Wilding (1971)
	Pandora neoaphidis, *Entomophthora planchoniana*, *Neozygites fresenii*, *Beauveria bassiana*, *Lecanicillium sp.*	Cereal aphids including *Diuraphis noxia*	Hatting et al. (1999)
	Neozygites floridana	*Tetranychus urticae*	Klingen et al. (2008)
	Entomophthora planchoniana, *Pandora neoaphidis*, *Neozygites sp.*	*Monella caryella*	Ekbom and Pickering (1990)
	Pandora neoaphidis, *Conidiobolus thromboides*, *Entomophthora chromaphidis*, *Zoophthora occidentalis*, *Neozygites fresenii*, *Lecanicillium sp.*	*Aphids glycines*	Nielsen and Hajek (2005)

particular insect species, although these studies are again generally focused on insects of economic importance (Bruck 2004; Sookar et al. 2008).

Fungal traits

Research on traits of fungal entomopathogens have largely focused on a single trait: the ability of an isolate or species to cause mortality in the host. In part, this reflects the interest of many pathologists to develop fungi for microbial control and there are, therefore, few studies on non-pest hosts. Examples from non-pest hosts mostly consist of descriptive or observational studies on single or a few species (Humber 1976; Keller 1987b) or books on diversity and occurrence of fungal entomopathogens (Bałazy 1993; Samson et al. 1988). These researchers use the qualitative term pathogenicity to describe "the quality or state of being pathogenic" whilst they use the quantitative term virulence for "the disease producing power of an organism, i.e. the degree of pathogenicity within a group" (Shapiro-Ilan et al. 2005). Both pathogenicity and virulence are frequently measured in laboratory bioassays (see Navon and Ascher 2000 for examples). Within the field of fungal insect pathology, virulence is expressed as the lethal dose (LD_{50}) or lethal concentration (LC_{50}) causing mortality of 50% of test insects. In this way, a fungus may be highly virulent if only a few conidia are required to cause a lethal infection. The definitions of pathogenicity and virulence vary within and between disciplines and depending on the type of pathogen concerned. Cross-disciplinary consensus regarding these definitions is required but this will require wide consultation and is beyond the scope of this paper.

In the general epidemiological literature virulence is defined as a measure of the impact of a pathogen on host fitness, and may be expressed as a reduction in either fecundity or survival of infected hosts compared to uninfected hosts (Solter 2006). It is a biological property of the pathogen that may be altered through abiotic and biotic impacts and thus may vary during the progression of an epizootic. Current studies of fungal entomopathogens often only consider isolates and species that are highly virulent and therefore almost invariably cause host mortality. However, we know that there are fungal isolates that have low virulence and do not generally cause high host mortality (Shah et al. 2004). In these cases, and also for virulent isolates, there are additional effects on the host through other mechanisms such as reduced fecundity (Baverstock et al. 2006; Furlong et al. 1997; Roy et al. 2008; Xu and Feng 2002). Only recently with improved molecular techniques are we becoming aware of the previously underestimated role that covert infections may play in insect populations (Burden et al. 2003). Covert viral infections are increasingly considered as important in infection dynamics (Boots et al. 2003) but as yet, there is no evidence to suggest fungi harbour similar covert infections although these may be more likely in the microsporidia.

To what extent do fungal entomopathogens play a role in regulating populations of insects in semi-natural ecosystems?

This question would be best answered by classic life table studies of host populations, yet few such studies exist as previously mentioned. Examples in the ecological literature tend to focus on insect hosts and their parasitoids (Hawkins et al. 1997; Paniaqua et al. 2009), reflecting perhaps the technical difficulties in detecting pathogens in the field as we highlighted earlier. The best examples emanate from the USA, where populations of Lepidoptera are monitored in forests for economic reasons. Although these studies are from forests that are managed monocultures allowing the host species in question to reach high population densities (Dwyer et al. 2004), they still provide the best empirical and theoretical examples of populations to date in which pathogen prevalence has been monitored over time, and illustrate how theoretical models may be used to unravel the relative contributions of different entomopathogens in the control and regulation of their hosts. In the case of invasive non-native insects, a special situation may occur if the invasive species has escaped from its specialized natural enemies and for that reason, significantly increased in population size. This hypothesis, termed 'natural enemy release', (Torchin et al. 2003; Roy et al. 2008) needs confirmation for host specific entomopathogenic fungal species or isolates.

Potential of specialist fungal entomopathogens to regulate host populations

The earliest host pathogen models established the principle that pathogens with persistent stages external to their hosts have the ability to regulate their hosts if sufficiently persistent in the environment (Anderson and May 1981). These models also assume that insect hosts do not acquire immunity to their pathogens and therefore do not include a resistant class of hosts immune to further infection (Grenfell and Dobson 1995). Indeed, it was illustrated that such specialist pathogens (or parasitoids) could be responsible for population cycles in which the period extends over many host generations. These principles were established using models in which, quite deliberately, the host was not influenced by any other form of population regulation, including intra-specific density dependence. The features of the host-pathogen interaction that resulted in population cycles included the density dependent nature of horizontal transmission which is well recorded for fungal entomopathogens (Johnson et al. 2006; Thomas et al. 1995), and the persistent nature of the external infectious stages (Baverstock et al. 2008; Weseloh and Andreadis 1997; Table 1). The density dependence of horizontal transmission ensures that the prevalence of the fungus in susceptible hosts rises as host population density rises, so checking the exponential growth of the host population.

The best studies that demonstrate insect population regulation by fungal entomopathogens are those involving pest insects in agroecosystems (e.g. Klubertanz et al. 1991; Nielsen and Hajek 2005; Smitley et al. 1986) but there are also examples from aquatic systems (Burns 1979; Johnson et al. 2006). External infectious stages ensure that the fungi persist during periods of low host population density, when horizontal transmission is insufficient to maintain the prevalence in the host population (Filotas and Hajek 2004; Hajek et al. 2004). Thus early theoretical work established that pathogens with these life history characters could potentially both regulate, and cause cycles in host populations. This caused considerable interest at the time, because forest insect pests displayed such cycles with no convincing explanation for them and this has been an area of much research and debate ever since (e.g. Abbott et al. 2008; Abbott and Dwyer 2007; Buntgen et al. 2009; Liu et al. 2007; Myers 1988). As fungal entomopathogens possess these life history attributes (Table 1) it is likely that they also have the potential to regulate host populations.

Two important concepts arose from these early models, one of these being the basic reproductive rate of the pathogen (R_0). This is defined as the number of new infections that arise from one primary infection in a wholly susceptible population. This must be greater than one for the pathogen to persist and spread, and so defines the conditions under which the pathogen could invade the host population. Due to the density dependent nature of transmission, host populations with higher densities of susceptible hosts will have higher contact rates with infective conidia, and so will give rise to higher values of R_0. The second concept, the host density threshold (H_T) is related to R_0, and is the density at which $R_0 = 1$. Thus H_T is a critical threshold below which prevalence will decline and above which it will rise. Comparing how assumptions alter the expressions for R_0 and H_T provides a convenient way of comparing different models.

Potential of generalist fungal entomopathogens to regulate host populations

Theoretical models exploring the potential for natural enemies to regulate populations usually consider specialist natural enemies, the densities of which are tightly coupled to the host populations. However, any density dependent relationship may regulate or stabilize a population through heterogeneity in attack rates, and even density independent patterns may do this if there is sufficient heterogeneity in risk from patch to patch (Hassell and May 1988). Consequently, generalist natural enemies may also hold the potential to be prime regulating factors, even though their dynamics may be uncoupled to some degree from the host species.

Both manipulative empirical and theoretical studies have illustrated that attack rates from generalist natural enemies are usually high enough at low host population densities to prevent population outbreaks. For example, the white footed mouse, *Peromyscus leucopus* Rafinesque (Rodentia: Neotominae), is a generalist predator of the gypsy moth *L. dispar*. Empirical data is consistent with this predator regulating the moth at low densities and a nucleopolyhedrovirus regulating the moth at high densities (Elkinton et al. 1996). In another study of the forest tent caterpillar *Malacosoma disstria* Hübner (Lepidoptera: Lasiocampidae), generalist avian predation was found to be the dominant mortality factor, in strong contrast to five specialist parasitoid species (Parry et al. 1997). A review of two herbivore species (the autumnal moth, *Epirrita autumnata* Borkhausen (Lepidoptera: Geometridae) and voles of the genus *Microtus* and *Clethrionomys*) concluded that the population cycles observed in northern Fennoscandia were likely to be caused by specialist natural enemies and the more stable dynamics on the south to be caused by an increase in the density and diversity of generalist natural enemies (Klemola et al. 2002).

Although this evidence is drawn largely from generalist predators, generalist fungal entomopathogens possess the life history characteristics to fill this role very effectively; host-fungus interactions exhibit heterogeneity in attack rate, and they can increase in abundance rapidly in response to the presence of hosts (Kamata 2000). However, the degree to which they cause mortality in populations, outside of the context of crop systems, is virtually unknown.

Combined effects of specialist and generalist natural enemies on host populations

It has long been established by theory and observation that host populations exhibit many different equilibrium states, and consequently it is unlikely that any one natural enemy is responsible for regulating a host species (Henson et al. 2009). Indeed food webs consist of both specialist and generalist natural enemies, with fungal entomopathogens playing both of these roles (Roy and Pell 2000; van Veen et al. 2008), and it is important to consider the combined effect of the suite of natural enemies present; no one pathogen acts in isolation. Furthermore, models in which only one natural enemy is responsible for regulating a host population frequently fail to capture the observed dynamics, with the time between host outbreaks being far more irregular than traditional models would predict. More recently, in a few cases, models are now being developed to include more than one natural enemy, with the stochastic influence of abiotic factors causing the host population to move between a low, stable, equilibrium which may be maintained by generalist natural enemies, and more cyclic dynamic behaviour which is the classic hallmark of a specialist natural enemy (Dwyer et al. 2004). These more complex models can produce behaviour which is more consistent with field observations, namely irregular outbreaks separated by long intervals during which the host is present at low densities, and represent a significant step forward in our understanding of the potential role of complexes of natural enemies in the regulation of herbivores and is applicable to fungal entomopathogens.

Making the models more realistic

Pathogen reservoirs

Clearly host populations are likely to be influenced by constraints on resources, or other factors that will act in a density dependent manner on population growth. Host density dependence has been incorporated into early models in more than one way, and one case concluded that cyclical behaviour occurs over great regions of parameter space (Dwyer 1994), and in another case that cycles were less likely, with the cycle period depending upon parameter values (Bowers et al. 1993). The range of parameter values considered and the form of the density dependence is likely to be influential in determining which outcome is most probable. Perhaps a more significant extension of theory was to include the concept of a 'pathogen reservoir', in which pathogens cannot infect hosts but where their degradation rate is low (Hochberg 1989). Pathogens may also move out of the reservoir to re-enter the infection cycle, or hosts may 'visit' the reservoir and acquire infection (Fig. 1). The ability of fungal conidia or resting spores in reservoirs to re-enter the infection cycle has been demonstrated (Bitton et al. 1979; Hajek 1999). The presence of such a reservoir has a storage effect, which dampens cycles and increases the likelihood of a stable equilibrium. Environmental reservoirs of fungal entomopathogens have been found in a number of circumstances of which a few examples are: *E. maimaiga* in forest soil (Hajek 1999); *Beauveria bassiana* (Balsamo) Vuillemin (Ascomycota: Hypocreales) on logs (Reay et al. 2007) and within agricultural soils (Meyling and Eilenberg 2006b); *P. neoaphidis* within agricultural soil (Baverstock et al. 2008; Nielsen et al. 2003); *Entomophthora planchoniana* Cornu (Entomophthoramycotina: Entomophthorales) as hyphal bodies on tree trunks or resting spores in soil (Keller 1987a, b); *Neozygites fresenii* (Nowakowski) Batko (Entomophthoramycotina: Entomophthorales) as resting spores on trees (Bitton et al. 1979).

Transmission and disease resistance

In contrast to the experimental literature on viral entomopathogens (Elderd et al. 2008), there are no studies directly examining heterogeneity in transmission rates for fungal entomopathogens. Heterogeneity in transmission is expected, however, due to individual differences in host susceptibility observed in the laboratory (Ferrari et al. 2001; Keller et al. 1999; Roy et al. 2008) and the heterogeneous distribution of infective conidia in the field (Meyling and Eilenberg 2006b; Tscharntke et al. 2008). Such heterogeneity in natural-enemy attack rates is strongly stabilizing (Hassell et al. 1991) and produces stable cycles for a range of parameter values in host-pathogen models (Dwyer et al. 2000).

An element of heterogeneity in host susceptibility has a genetic basis. It has been illustrated that host populations can develop a degree of resistance or at least reduced susceptibility, to some fungal entomopathogens within and between generations (Ferrari et al. 2001; Milner 1982, 1985; Stacey et al. 2003). This phenomenon has also been illustrated in response to other entomopathogens (Boots and Begon 1993; Cooper et al. 2002). Indeed, Stow et al. (2007) suggest that selection by microbial pathogens, and more specifically production of antimicrobial defences, was critical to the evolution of sociality. However, few studies show that host investment in resistance to fungal entomopathogens may change depending upon host density and these are limited to species exhibiting density-dependent phase polyphenism (Wilson et al. 2001, 2002). The hypothesis is that at high host densities, insects are more likely to encounter inoculum due to the density dependent nature of horizontal transmission, and that some species can exhibit a plastic response to this and allocate more of their limited resources to disease defence than at low densities. This has also been illustrated for some species in response to nucleopolyhedroviruses (Reeson et al. 1998, 2000), but for others the reverse pattern is suggested, with susceptibility increasing at high densities, and this has been postulated to be due to stress (Reilly and Hajek 2008). The form of the relationship between disease resistance and density dependence will influence the impact on population dynamics, with the inverse relationship between population density and disease resistance having a stabilizing influence (Reilly and Hajek 2008).

Given that there is a heritable element to resistance in some cases, it is possible that susceptibility to entomopathogens may change during the course of an epizootic, particularly if there is a cost to resistance. Again there is supporting empirical evidence for this in the case of viruses (Cory and Myers 2009) but evidence for fungal entomopathogens is limited (Miller et al. 2009). If natural selection drives rates of transmission through altered host susceptibility, theoretical models suggest that cycles are more likely to be observed even at high rates of heterogeneity in transmission (Elderd et al. 2008). This illustrates the importance of including natural selection in host-pathogen models when attempting to discover the role of entomopathogens in host population dynamics.

Summarising, theory illustrates that host specific fungal entomopathogens could potentially regulate their host populations, but the question remains open as to whether such pathogens really are the prime regulating factor in many cases. There are many details of the host-pathogen interaction that would benefit from further empirical data. It is notable that vertical transmission of fungi has only been demonstrated in very few cases (e.g. Tarrant and Soper 1986). High rates of vertical transmission would make the conditions for regulation less stringent (Anderson and May 1982).

Dispersal: keeping up with your host

Greif and Currah (2007) demonstrated the importance of arthropods in dispersing fungi but highlighted the need for more data comparing patterns among substrates, fungal species and their arthropod carriers. Most ecological studies are conducted at a small spatial scale. More recently, and particularly in the context of arthropod species shifting their ranges as a consequence of climate change, there has been greater focus on the mechanisms and rate at which pathogens spread through host populations. One fundamental constraint on the part of a specialist pathogen is that, when considering the regional scale, it is unlikely to arrive in a new habitat ahead of the host. This has led to the hypothesis that the increased abundance observed at the leading edge of species shifting their ranges is due to the host escaping, albeit temporarily, the regulating influence of some natural enemies (Gaston 2009; Menendez et al. 2008). In some cases, pathogens may hitch a ride with their hosts in the form of covert infections vertically transmitted to offspring (Burden et al. 2003). Covert infections are uncommon in fungal entomopathogens (Tarrant and Soper 1986), however, modern molecular tools may reveal hitherto hidden fungal infections at non lethal levels.

The simplest theoretical models describing pathogen dispersal within a host population are based on the process of diffusion and provide a moderately good description of dispersal at small spatial scales (Dwyer et al. 1998). These relatively simple models assume that conidia obey the laws of diffusion, although the precise shape of the dispersal kernel is unlikely to be Gaussian, and more likely to be 'fat-tailed'. The moderately good fit between models and data suggest that the majority of fungal infection at small spatial scales represented by experimental plots is due to a process akin to diffusion. However, a study of the regional spread of *E. maimaiga* through gypsy moth populations in North America found that rates of spread at the regional scale could not be predicted from diffusion models fitted to data obtained at local scales (Dwyer et al. 1998). Similarly, more detailed simulation models incorporating local abiotic factors such as temperature, rainfall and humidity could only accurately represent patterns of spread over a 3 km area if airborne conidia are allowed to freely disperse over the whole area (Weseloh 2003, 2004). This suggests that dispersal mechanisms, such as wind currents above the forest canopy, which operate at long distances, are crucial in explaining observed patterns of dispersal of conidia independent of their hosts. There are parallels here with studies on the dispersal of seeds, in which models have been developed combining local and long distance dispersal processes (Wichmann et al. 2009), and there is a strong argument that similar theoretical developments, combining local and regional processes in an analytical framework, would be appropriate for fungal entomopathogens (Dwyer et al. 2004).

Spatially heterogeneous environments

Habitat loss through environmental change leads to an increasingly fragmented landscape, with only patches of habitat that are suitable for hosts to persist. How will this influence host-pathogen dynamics, particularly in light of the broad host range of some fungi, and the dispersal ability of conidia discussed above? Hess (1996) developed a host-pathogen model from the classical Levins (1969) metapopulation model to explore the conditions under which hosts and pathogens may persist in a fragmented landscape. This original model was based on direct transmission between infected and susceptible hosts although subsequently, we have explored similar models based on pathogens such as baculoviruses and many fungal entomopathogens which infect by means of free-living infective stages, and the conclusions are not qualitatively different (White and Hails personal communication). Hess (1996) concluded that host dispersal between patches enhanced the spread of disease and thus could lead to host extinction. Fungus-infected hosts have the ability to disperse and to spread disease into new colonies as documented for aphid species (Feng and Chen 2002; Feng et al. 2004). Some specialist fungi such as *Strongwellsea* spp. sporulate from one or two holes on living hosts and conidia are dispersed in this way. Whether hosts themselves are the principle means by which fungal pathogens disperse between patches in a fragmented landscape has yet to be determined.

McCallum and Dobson (2002) further developed this framework to consider a 'generalist' pathogen, the abundance of which is maintained in a second host species which acts as a reservoir. In contrast to Hess (1996), they concluded that greater landscape connectance enhanced the stability of the host-pathogen interaction. Habitat corridors allow host species to disperse and 'escape' pathogens, effectively creating a form of refuge. However, complete connectance is equivalent to a homogenous habitat; and a degree of habitat partitioning actually promotes co-existence of host species by, for example, relaxing apparent competition mediated by a shared natural enemy (Holt 1984). A general principle that emerges from these and other studies is that the spatial complexity of population structure is a source of heterogeneity that can promote the co-existence of hosts and pathogens. However, the precise dynamics will depend upon the spatial distribution of hosts, the productivity of patches (in terms of host growth rates), the life history characteristics of the pathogens and the mobility patterns of hosts and pathogens (Namba et al. 1999; Rodriguez and Torres-Sorando 2001). Consequently, the response of fungal entomopathogens to habitat fragmentation would be best explored in specific host populations using models of intermediate complexity that have been adapted to incorporate species specific information.

Conclusions

Fungal entomopathogens are ubiquitous in semi-natural habitats and play a role in insect population dynamics. There is, however, a scarcity of empirical data available to evaluate their relative importance in controlling and regulating insect populations in semi-natural ecosystems. Even within well studied crop systems such as forest insects, we have a limited understanding of the role of fungal entomopathogens and insect population dynamics. Anticipated changes in disease prevalence due to key anthropogenic drivers (Millennium Ecosystem Assessment 2005) such as climate change and habitat fragmentation as well as the arrival of invasive species are likely to affect the prevalence of all entomopathogens in semi-natural ecosystems (Roy et al. 2009). The effects of such changes in disease prevalence will be relevant to the management of both pest insects and insects of conservation interest (Roy et al. 2009). The practicalities of studying fungal entomopathogens in any system can be challenging; there are limitations imposed by the research tools available and many of the complex multitrophic interactions are yet to be revealed (Cory and Ericsson 2009). However, it is imperative that we drive research effort forward by coupling rigorous research in the field with theoretical modelling in order to unravel the complexity of interactions between fungal entomopathogens and their hosts in semi-natural habitats.

Acknowledgments HH, HER and RSH were funded by the Environmental Change Integrating Fund through the NERC Centre for Ecology & Hydrology. JKP was funded by Department for Environment, Food and Rural Affairs of the United Kingdom (Defra) and the Biotechnology and Biological Sciences Research Council (BBSRC) of the United Kingdom. Rothamsted Research is an Institute of the BBSRC. JE was funded by the University of Copenhagen, Denmark.

References

Abbot KC, Dwyer G (2007) Food limitation and insect outbreaks: complex dynamics in plant-herbivore models. J Anim Ecol 76:1004–1014

Abbot KC, Moriss WF, Gross K (2008) Simultaneous effects of food limitation and inducible resistance on herbivore population dynamics. Theor Popul Biol 73:63–78

Anderson RM, May RM (1981) The population dynamics of microparasites and their invertebrate hosts. Philos Trans R Soc B 291:451–524

Anderson RM, May RM (1982) Coevolution of hosts and parasites. Parasitology 85:411–426

Arthurs S, Thomas MB (2001) Effects of temperature and relative humidity on sporulation of *Metarhizium anisopliae* var. *acridum* in mycosed cadavers of *Schistocerca gregaria*. J Invertebr Pathol 78:59–65

Balazy S (1993) Flora of Poland, Fungi (Mycota), Vol. XXIV Entomophthorales. Polish Academy of Sciences

Balmford A, Bennun L, ten Brink B, Cooper D, Côté IM, Crane P, Dobson A, Dudley N, Dutton I, Green RE, Gregory R, Harrison J, Kennedy ET, Kremen C, Leader-Williams N, Lovejoy T, Mace G, May R, Mayaux P, Phillips J, Redford K, Ricketts TH, Rodriguez JP, Sanjayan M, Schei P, van Jaarsveld A, Walther BA (2005) Science and the convention on biological diversity's 2010 target. Science 307:212–213

Baverstock J, Roy HE, Clark SJ, Alderson PG, Pell JK (2006) Effect of fungal infection on reproductive potential of aphids and their progeny. J Invertebr Pathol 91:136–139

Baverstock J, Clark SJ, Pell JK (2008) Effect of seasonal abiotic conditions and field margin habitat on the activity of *Pandora neoaphidis* inoculum on soil. J Invertebr Pathol 97:282–290

Bidochka M, Kamp AM, Lavender TM, Dekoning J, de Croos JNA (2001) Habitat associated in two genetic groups of the insect-pathogenic fungus *Metarhizium anisopliae*: uncovering cryptic species? Appl Environ Microb 67:1335–1342

Bitton S, Kenneth RG, Ben-Ze'ev I (1979) Zygospore overwintering and sporulative germination in *Triplosporium fresenii* (Entomopthoraceae) attacking *Aphis spriaecola* on citrus in Israel. J Invertebr Pathol 34:295–302

Blackwell M (2009) Fungal evolution and taxonomy. BioControl. doi:10.1007/s10526-009-9243-8 (this SI)

Boots M, Begon M (1993) Trade-offs with resistance to a granulosis virus in the India meal moth, examined by a laboratory evolution experiment. Func Ecol 7:528–534

Boots M, Greenman J, Ross D, Norman R, Hails R, Sait S (2003) The population dynamical implications of covert infections in host-microparasite interactions. J Anim Ecol 72:1064–1072

Bowers RG, Begon M, Hodgkinson DE (1993) Host-pathogen population cycles in forest insects? Lessons from simple models reconsidered. Oikos 67:529–538

Brownbridge M, Adamowitz A, Skinner M, Parker BL (1999) Prevalence of fungal entomopathogens in the life cycle of pear thrips, *Taeniothrips inconsequens* (Thysanoptera: Thripidae), in Vermont sugar maple forests. Biol Control 16:54–59

Bruck D (2004) Natural occurrence of entomopathogens in Pacific Northwest nursery soils and their virulence to the black vine weevil, *Otiorhynchus sulcatus* (F.) (Coleoptera: Curculionidae). Environ Entomol 33:1335–1343

Buntgen U, Frank D, Liebold A, Johson D, Carrer M, Urbinati C, Grabner M, Nicolussi K, Levanic T, Esper J (2009) Three centuries of insect outbreaks across the European Alps. New Phytol 182:929–941

Burden JP, Nixon CP, Hodgkinson AE, Possee RD, Sait SM, King LA, Hails RS (2003) Covert infections as a mechanism for long-term persistence of baculoviruses. Ecol Lett 6:524–531

Burges HD, Hussey NW (eds) (1971) Microbial control of insects and mites. Academic Press Inc., London

Burns CW (1979) Fungal parasitism in a copepod population: the effects of *Aphanomyces* on the population dynamics of *Boeckella dilatata* Sars. J Plankton Res 7:201–205

Chee-Sanford (2008) Weed seeds as nutritional resources for soil *Ascomycota* and characterization of specific associations between plant and fungal species. Biol Fert Soils 44:763–771

Cooper VS, Reiskind MH, Miller JA, Shelton KA, Walther BA, Elkinton JS, Ewald PW (2002) Timing of transmission and the evolution of virulence of an insect virus. Proc R Soc Lond B Biol 269:1161–1165

Cory JS, Ericsson JD (2009) Fungal entomopathogens in a tritrophic context. BioControl. doi:10.1007/s10526-009-9247-4 (this SI)

Cory J, Myers JH (2009) Within and between population variation in disease resistance in cyclic populations of western tent caterpillars: a test of the disease defence hypothesis. J Anim Ecol 78:646–655

Dean GJW, Wilding N (1971) *Entomophthora* infecting cereal aphids *Metopolophium dirhodum* and *Sitobion avenae*. J Invertebr Pathol 18:169–176

Dean GJW, Wilding N (1973) Infection of cereal aphids by fungus *Entomophthora*. Ann Appl Biol 74:133–138

Dromph KM, Pell JK, Eilenberg J (2002) Influence of flight and colour morph on susceptibility of *Sitobion avenae* to infection by *Erynia neoaphidis*. Biocontrol Sci Technol 12:753–756

Dwyer G (1994) Density-dependence and spatial structure in the dynamics of insect pathogens. Am Nat 143:533–562

Dwyer G, Elkinton JS, Hajek AE (1998) Spatial scale and the spread of a fungal pathogen of Gypsy moth. Am Nat 152:485–494

Dwyer G, Dushoff J, Elkinton JS, Levin SA (2000) Pathogen-driven outbreaks in forest defoliators revisited: building models from experimental data. Am Nat 156:105–120

Dwyer G, Dushoff J, Yee SH (2004) The combined effects of pathogens and predators on insect outbreaks. Nature 430:341–345

Eilenberg J (1987) The culture of *Entomophthora muscae* (C.) Fres. (Zygomycetes: Entomophthorales) in carrot flies (*Psila rosae* F.) (Diptera: Psilidae) and the effect of temperature on the pathology of the fungus. Entomophaga 32:425–435

Eilenberg J, Michelsen V (1999) Natural host range and prevalence of the genus *Strongwellsea* (Zygomycota: Entomophthorales) in Denmark. J Invertebr Pathol 73:189–198

Eilenberg J, Pell JK (2007) Ecology. In: Keller S (ed) Arthropod pathogenic Entomophthorales: biology, ecology, identification. Office des Publications Officielles des Communautés Européennes pp 7–26

Eilenberg J, Philipsen H (1988) The occurrence of Entomophthorales on the carrot fly (*Psila rosae* F.) in the field during two successive seasons. Entomophaga 33:135–144

Eilenberg J, Hajek A, Lomer C (2001) Suggestions for unifying the terminology in biological control. BioControl 46: 387–400

Ekbom BS, Pickering J (1990) Pathogenic fungal dynamics in fall populations of the blackmargined aphid (*Monella caryella*). Ent Exp et Appl 57:29–37

Ekesi S, Maniania NK, Ampong-Nyarko K (1999) Effect of temperature on germination, radial growth and virulence of *Metarhizium anisopliae* and *Beauveria bassiana* on *Megalurothrips sjostedti*. Biocontrol Sci Technol 9:177–185

Ekesi S, Shah PA, Clark SJ, Pell JK (2005) Conservation biological control with the fungal pathogen *Pandora neoaphidis*: implications of aphid species, host plant and predator foraging. Agric For Entomol 7:21–30

Elderd BD, Dushoff J, Dwyer G (2008) Host-pathogen interactions, insect outbreaks, and natural selection for disease resistance. Am Nat 172:829–842

Elkinton JS, Healy WM, Buonaccorsi JP, Boettner GH, Hazzard AM, Smith HR, Liebhold AM (1996) Interactions among gypsy moths, white-footed mice, and acorns. Ecology 77:2332–2342

Evans HC (1981) Entomogenous fungi in tropical forest ecosystems: an appraisal. Ecol Entomol 7:47–60

Fargues J, Goettel MS, Smits N, Ouedraogo A, Vidal C, Lacey LA, Lomer CJ, Rougier M (1996) Variability in susceptibility to simulated sunlight of conidia among isolates of entomopathogenic Hyphomycetes. Mycolpathol 135: 171–181

Feng MG, Chen C (2002) Incidences of infected *Myzus persicae* alatae in flight imply place-to-place dissemination of entomophthoralean fungi in aphid populations through migration. J Invertebr Pathol 81:53–56

Feng MG, Chen C, Chen B (2004) Wide dispersal of aphid-pathogenic Entomophthorales among aphids relies upon migratory alates. Enviro Microbiol 6:510–516

Ferrari J, Müller CB, Kraaijeveld AR, Godfray HC (2001) Clonal variation and covariation in aphid resistance to parasitoids and a pathogen. Evolution 55:1805–1814

Filotas MJ, Hajek AE (2004) Influence of temperature and moisture on infection of forest tent caterpillars (Lepidoptera: Lasiocampidae) exposed to resting spores of the entomopathogenic fungus *Furia gastropachae* (Zygomycetes: Entomophthorales). Environ Entomol 33: 1127–1136

Fournier A, Enkerli J, Keller S, Widmer F (2008) A PCR-based tool for the cultivation-independent monitoring of *Pandora neoaphidis*. J Invertebr Pathol 99:49–56

Furlong MJ, Pell JK (1997) The influence of environmental factors on the persistence of *Zoophthora radicans* conidia. J Invertebr Pathol 69:223–233

Furlong MJ, Pell JK, Reddy GVP (1997) Premortality effects of *Zoophthora radicans* (Zygomycetes: Entomophthorales) infection on larval feeding rate and adult fecundity in *Plutella xylostella* (Lepidoptera: Yponomeutidae). J Invertebr Pathol 70:214–220

Fuxa JR, Tanada Y (eds) (1987) Epizootiology of insect diseases. Wiley, New York

Gaston KJ (2009) Geographic range limits of species. Proc R Soc Lond B Biol 276:1391–1393

Goettel MS, Eilenberg J, Glare TR (2005) Entomopathogenic fungi and their role in regulation of insect populations. In: Gilbert LI, Latrou K, Gill S (eds) Comprehensive molecular insect science, vol 6. Elsevier, Oxford, pp 361–406

Goettel MS, Koike M, Kim JJ, Aiuchi D, Shinya R, Brodeur J (2008) Potential of *Lecanicillium* spp. for management of insects, nematodes and plant diseases. J Invertebr Pathol 98:256–261

Greif MD, Currah RS (2007) Patterns in the occurrence of saprophytic fungi carried by arthropods caught in traps baited with rotted wood and dung. Mycologia 99:7–19

Grenfell BT, Dobson AP (eds) (1995) Ecology of infectious diseases in natural populations. Cambridge University Press, UK

Hajek AE (1997) Fungal and viral epizootics in gypsy moth (Lepidoptera: Lymantriidae) populations in central New York. Biol Control 10:58–68

Hajek AE (1999) Pathology and epizootiology of *Entomophaga maimaiga* infections in forest Lepidoptera. Microbiol Mol Biol R 63:814–835

Hajek AE (2001) Larval behaviour in *Lymantria dispar* increases risk of fungal infection. Oecologia 126:285–291

Hajek AE, Delalibera I (2009) Fungal pathogens as classical biological control agents against arthropods. BioControl. doi:10.1007/s10526-009-9253-6 (this SI)

Hajek AE, Butt TM, Strelow LI, Gray SM (1991) Detection of *Entomophaga maimaiga* (Zygomycetes, Entomophthorales) using enzyme-linked-immunosorbent-assay. J Invertebr Pathol 58:1–9

Hajek AE, Butler L, Walsh SRA, Silver JC, Hain FP, Hastings FL, Odell TM, Smitley DR (1996) Host range of the gypsy moth (Lepidoptera: Lymantriidae) pathogen *Entomophaga maimaiga* (Zygomycetes: Entomophthorales) in the field versus the laboratory. Environ Entomol 25:709–721

Hajek AE, Olsen CH, Elkinton JS (1999) Dynamics of airborne conidia of the gypsy moth (Lepidoptera: Lymantriidae) fungal pathogen *Entomophaga maimaiga* (Zygomycetes: Entomophthorales). Biol Control 16:111–117

Hajek AE, Strazanac JS, Wheeler MM, Vermeylen FM, Butler L (2004) Persistence of the fungal pathogen *Entomophaga maimaiga* and its impact on native Lymantriidae. Biol Control 30:466–473

Hall RA, Papierok B (1982) Fungi as biological control agents of arthropods of agricultural and medical importance. Parasitol 84:205–240

Hassell MP, May RM (1988) Spatial heterogeneity and the dynamics of parasitoid-host systems. Ann Zoo Fenn 25:55–61

Hassell MP, Comins HN, May RM (1991) Spatial structure and chaos in insect population dynamics. Nature 353:255–258

Hatting JL, Humber RA, Poprawski TJ, Miller RM (1999) A survey of fungal pathogens of aphids from South Africa, with special reference to cereal aphids. Biol Control 16: 1–12

Hawkins BA, Cornell HV, Hochberg ME (1997) Predators, parasitoids and pathogens as mortality agents in phytophagous insect populations. Ecology 78:2145–2152

Hemmati F, Pell JK, McCartney HA, Deadman ML (2001) Airborne concentrations of conidia of *Erynia neoaphidis* above cereal fields. Mycol Res 105:485–489

Henson KSE, Craze PG, Memmott J (2009) The restoration of parasites, parasitoids and pathogens to heathland communities. Ecology 90:1840–1851

Hesketh H, Alderson PG, Pye BJ, Pell JK (2008) The development and multiple uses of a standardised bioassay method to select hypocrealean fungi for biological control of aphids. Biol Control 46:242–255

Hess G (1996) Disease in metapopulation models: implications for conservation. Ecology 77:1617–1632

Hibbett DS, Binder M, Bischoff JF, Blackwell M, Cannon PF, Eriksson OE, Huhndorf S, James T, Kirk PM, Lücking R, Thorsten Lumbsch H, Lutzoni F, Brandon Matheny P, McLaughlin DJ, Powell MJ, Redhead S, Schoch CL, Spatafora JW, Stalpers JA, Vilgalys R, Aime MC, Aptroot A, Bauer R, Begerow D, Benny GL, Castlebury LA, Crous PW, Dai YC, Gams W, Geiser DM, Griffith GW, Gueidan C, Hawksworth DL, Hestmark G, Hosaka K, Humber RA, Hyde KD, Ironside JE, Kõljalg U, Kurtzman CP, Larsson K-H, Lichtwardt R, Longcore J, Miądlikowska J, Miller A, Moncalvo JM, Mozley-Standridge S, Oberwinkler F, Parmasto E, Reeb V, Rogers JD, Roux C, Ryvarden L, Sampaio JP, Schüßler A, Sugiyama J, Thorn RG, Tibell L, Untereiner WA, Walker C, Wang Z, Weir A, Weiss M, White MM, Winka K, Yao YJ, Zhang N (2007) A higher-level phylogenetic classification of the Fungi. Mycol Res 111:509–547

Hoch G, Schopf A, Maddox JV (2000) Interactions between an entomopathogenic microsporidium and the endoparasitoid *Glyptapanteles liparidis* within their host, the gypsy moth larva. J Invertebr Pathol 75:59–68

Hoch G, D'Amico V, Solter LF, Zubrik M, McManus ML (2008) Quantifying horizontal transmission of *Nosema lymantriae*, a microsporidian pathogen of the gypsy moth, *Lymantria dispar* (Lep., Lymantriidae) in field cage studies. J Invertebr Pathol 99:146–150

Hochberg ME (1989) The potential role of pathogens in biological control. Nature 337:262–265

Holt RD (1984) Spatial heterogeneity, indirect interactions, and the coexistence of prey species. Am Nat 124:377–406

Hua L, Feng M-G (2003) New use of broomcorn millets for production of granular cultures of aphid-pathogenic fungus *Pandora neoaphidis* for high sporulation potential and infectivity to *Myzus persicae*. FEMS Microbiol Lett 227:311–317

Humber RA (1976) Systematics of the genus *Strongwellsea* (Zygomycetes: Entomophthorales). Mycologia 68:1042–1060

Humber RA (2008) Evolution of entomopathogenicity in fungi. J Invertebr Pathol 98:262–266

Inglis GD, Goettel MS, Butt TM, Strasser H (2001) Use of hyphomycetous fungi for managing insect pests. In: Butt TM, Jackson C, Magan N (eds) Fungi as biocontrol agents; progress, problems and potential. CABI Publishing, Wallingford, UK. pp 23–69

Jaronski S (2009) Ecological factors in the inundative use of fungal entomopathogens. doi:10.1007/s10526-009-9248-3 (this SI)

Johnson PTJ, Longcore JE, Stanton DE, Carnegie RB (2006) Chytrid infections of *Daphnia pulicaria*: development, ecology, pathology and phylogeny of *Polycaryum leave*. Freshwater Biol 51:634–648

Kamata N (2000) Population dynamics of the beech caterpillar, *Syntypistis punctatella*, and biotic and abiotic factors. Pop Ecol 42:267–278

Keeling PJ, Fast NM (2002) Microsporidia: biology and evolution of highly reduced intracellular parasites. Ann Rev Microbiol 56:93–116

Keller S (1987a) Observation on the overwintering of *Entomophthora planchoniana*. J Invertebr Pathol 50:333–335

Keller S (1987b) Arthropod-pathogenic Entomophthorales of Switzerland. I. Conidiobolus, Entomophaga and Entomophthora. Sydowia 40:122–167

Keller S, Schweizer C, Shah P (1999) Differential susceptibility of two *Melolontha* populations to infections by the fungus *Beauveria brongniartii*. Biocontrol Sci Technol 9:441–446

Klemola T, Tanhuanpää M, Korpimäki E, Ruohomäki K (2002) Specialist and generalist natural enemies as an explanation for geographical gradients in population cycles of northern herbivores. Oikos 99:83–94

Klingen I, Eilenberg J, Meadow R (2002) Effects of farming system, field margins and bait insect on the occurrence of insect pathogenic fungi in soils. Agric Ecosyst Environ 91:191–198

Klingen I, Wærsted G, Westrum K (2008) Overwintering and prevalence of *Neozygites floridana* (Zygomycetes: Neozygitaceae) in hibernating females of *Tetranychus urticae* (Acari: Tetranychidae) under cold climatic conditions in strawberries. Exp Appl Acarol 46:231–245

Klubertanz TH, Pedigo LP, Carlson RE (1991) Impact of fungal epizootics on the biology and management of the

2-spotted spider mite (Acari, Tetranychidae) in soybean. Environ Entomol 20:731–735
Krassilstchik J (1888) La production industrielle des parasites végétaux pour la destruction des insects nuisibles. Bulletin Biologique de la France et de la Belgique 19:461–472
Levins R (1969) Some demographic and genetic consequences of environmental heterogeneity for biological control. Bull Entomol Soc Am 15:237–240
Liu WC, Bonsall MB, Godfray HCJ (2007) The form of host density-dependence and the likelihood of host-pathogen cycles in forest-insect systems. Theor Popul Biol 72:86–95
McCallum H, Dobson A (2002) Disease, habitat fragmentation and conservation. Proc R Soc Lond B Biol 269: 2041–2049
Menendez R, Gonzalez-Megias A, Lewis OT, Shaw MR, Thomas CD (2008) Escape from natural enemies during climate-driven range expansion: a case study. Ecol Entomol 33:413–421
Meyling NV, Eilenberg J (2006a) Isolation and characterisation of *Beauveria bassiana* isolates from phylloplanes of hedgerow vegetation. Mycol Res 110:188–195
Meyling NV, Eilenberg J (2006b) Occurrence and distribution of soil borne entomopathogenic fungi within a single organic agroecosystem. Agric Ecosyst Environ 113:336–341
Millenium Ecosystem Assessment (2005) Ecosystems and human well-being: biodiversity synthesis. World Resources Institutes, Washington DC
Miller GA, Pell JK, Simpson SJ (2009) Crowded locusts produce hatchlings vulnerable to fungal attack. Biol Lett (in press)
Milner RJ (1982) On the occurrence of pea aphids, *Acyrthosiphon pisum*, resistant to isolates of the fungal pathogen *Erynia neoaphidis*. Entomol Exp Appl 32:23–27
Milner RJ (1985) Distribution in time and space of resistance to the pathogenic fungus *Erynia neoaphidis* in the pea aphid *Acyrthosiphon pisum*. Entomol Exp Appl 37: 235–240
Monzon AJ, Guharay F, Klingen I (2008) Natural occurrence of *Beauveria bassiana* in *Hypothenemus hampei* (Coleoptera: Curculionidae) populations in unsprayed coffee fields. J Invertebr Pathol 97:134–141
Myers JH (1988) Can a general hypothesis explain population cycles of forest Lepidoptera. Adv Ecol Res 18:179–242
Namba T, Umemoto A, Minami E (1999) The effects of habitat fragmentation on persistence of source-sink metapopulations in systems with predators and prey or apparent competitors. Theor Popul Biol 56:123–137
Navon A, Ascher KRS (eds) (2000) Bioassays of entomopathogenic microbes and nematodes. CABI Publishing, Wallingford, UK
Nielsen C, Hajek AE (2005) Control of invasive soybean aphid, *Aphis glycines* (Hemiptera: Aphididae), populations by existing natural enemies in New York State, with emphasis on entomopathogenic fungi. Environ Entomol 34:1036–1047
Nielsen C, Eilenberg J, Harding S, Oddsdottir E, Haldorsson G (2001) Geographical distribution and host range of Entomophthorales infecting the green spruce aphid *Elatobium abietinum* Walker in Iceland. J Invertebr Pathol 78:72–80
Nielsen C, Hajek AE, Humber RA, Bresciani J, Eilenberg J (2003) Soil as an environment for winter survival of aphid-pathogenic Entomophthorales. Biol Control 28: 92–100
Oduor GI, Yaninek JS, van der Geest LPS, Moraes GJ (1996) Germination and viability of capilliconidia of *Neozygites floridana* (Zygomycetes: Entomophthorales) under constant temperature, humidity, and light conditions. J Invertebr Pathol 67:267–278
Paniaqua MR, Medianero E, Lewis OT (2009) Structure and vertical stratification of plant galler-parasitoid food webs in two tropical forests. Ecol Entomol 34:310–320
Parry D, Spence JR, Volney WJA (1997) Responses of natural enemies to experimentally increased populations of the forest tent caterpillar, *Malacosoma disstria*. Ecol Entomol 22:97–108
Pell JK (2007) Ecological approaches to pest management using entomopathogenic fungi; concepts, theory, practice and opportunities. In: Ekesi S, Manianai N (eds) Use of entomopathogenic fungi in pest management. Research Signpost, pp 145–177
Pell JK, Eilenberg J, Hajek AE, Steinkraus D (2001) Biology, ecology and pest management potential of Entomophthorales. In: Butt TM, Jackson C, Magan N (eds) Fungi as biocontrol agents: progress, problems and potential. CABI Publishing, pp 71–153, Wallingford, UK
Pell JK, Hannam J, Steinkraus D (2009) Conservation biological control using fungal entomopathogens. BioControl. doi:10.1007/s10526-009-9245-6 (this SI)
Pilarska DK, Solter LF, Kereselidze M, Linde A, Hoch G (2006) Microsporidian infections in *Lymantria dispar* larvae: interactions and effects of multiple species infections on pathogen horizontal transmission. J Invertebr Pathol 93:105–113
Pontoppidan MB, Himanan W, Hywell-Jones NL, Boomsma JJ, Hughes DP (2009) Graveyyards on the move: the spatio-temporal distribution of *Ophiocordyceps*-infected ants. PlosOne 4(e4835):1–10
Posada F, Vega F (2005) A new method to evaluate the biocontrol potential of single spore isolates of fungal entomopathogens. 10 pp. J Insect Sci 5:37 available online: http://insectscience.org/5.37/
Reay SD, Hachet C, Nelson TL, Brownbridge M, Glare TR (2007) Persistence of conidia and potential efficacy of *Beauveria bassiana* against pinhole borers in New Zealand southern beech forests. Forest Ecol Manag 246:232–239
Reeson AF, Wilson K, Gunn A, Hails RS, Goulson D (1998) Baculovirus resistance in the noctuid *Spodoptera exempta* is phenotypically plastic and responds to population density. Proc R Soc Lond B Biol 265:1787–1791
Reeson AF, Wilson K, Cory JS, Hankard P, Weeks JM, Goulson D, Hails RS (2000) Effects of phenotypic plasticity on pathogen transmission in the field in a Lepidoptera-NPV system. Oecologia 124:373–380
Reilly JR, Hajek AE (2008) Density-dependent resistance of the gypsy moth *Lymantria dispar* to its nucleopolyhedrovirus, and the consequences for population dynamics. Oecologia 154:691–701
Roditakis E, Couzin ID, Franks NR, Charnley AK (2008) Effects of *Lecanicillium longisporum* infection on the

behaviour of the green peach aphid *Myzus persicae*. J Insect Physiol 54:128–136

Rodriguez DJ, Torres-Sorando L (2001) Models of infectious diseases in spatially heterogeneous environments. B Math Biol 63:547–571

Roy HE, Cottrell T (2008) Forgotten natural enemies: interactions between coccinellids and insect-parasitic fungi. Eur J Entomol 105:391–398

Roy HE, Pell JK (2000) Interactions between entomopathogenic fungi and other natural enemies: implications for biological control. Biocontrol Sci Technol 10:737–752

Roy HE, Steinkraus D, Eilenberg E, Pell JK, Hajek A (2006) Bizarre interactions and endgames: entomopathogenic fungi and their arthropod hosts. Ann Rev Entomol 51:331–357

Roy HE, Brown PMJ, Rothery P, Ware RL, Majerus MEN (2008) Interactions between the fungal pathogen *Beauveria bassiana* and three species of ladybird: *Harmonia axyridis*, *Coccinella septempunctata* and *Adalia bipunctata*. BioControl 53:265–276

Roy HE, Hails RS, Hesketh H, Roy DB, Pell JK (2009) Beyond biological control: non-pest insects and their pathogens in a changing world. Insect Conserv Diver 2:65–72

Samson RA, Evans HC, Latge JP (1988) Atlas of entomopathogenic fungi. Springer, Berlin

Sanjuan T, Henao LG, Amat G (2001) Spatial distribution of *Cordyceps* spp. (Ascomycotina: Clavicipitaceae) and its impacts on the ants in forests of the Amazonia Colombian foothill. Rev Biol Trop 49:945–955

Scholte E-J, Knols BGJ, Samson RA, Takken W (2004) Entomopathogenic fungi for mosquito control: a review. 24 pp. J Insect Sci 4:19, Available online: http://www.insectscience.org/4.19/

Shah PA, Pell JK (2003) Entomopathogenic fungi as biological control agents. Appl Microbiol Biotechnol 61:413–423

Shah PA, Clark SJ, Pell JK (2004) Assessment of aphid host range of *Pandora neoaphidis* (Zygomycetes, Entomophthorales). Biol Control 29:90–99

Shapiro-Ilan DI, Fuxa JR, Lacey LA, Onstad DW, Kaya HK (2005) Definitions of pathogenicity and virulence in invertebrate pathology. J Invertebr Pathol 88:1–7

Shimazu M, Sato H, Maehara N (2002) Density of the entomopathogenic fungus, *Beauveria bassiana* Vuillemin (Deuteromycotina: Hyphomycetes) in forest air and soil. Appl Entomol Zool 37:19–26

Sierotzki H, Camastral F, Shah PA, Aebi M, Tuor U (2000) Biological characteristics of selected *Erynia neoaphidis* isolates. Mycol Res 104:213–219

Siri A, Scorsetti AC, Dikgolz VE, Lopez CC (2005) Natural infections caused by the fungus *Beauveria bassiana* as a pathogen of *Musca domestica* in the neotropic. BioControl 50:937–940

Smitley DR, Kennedy GG, Brooks WM (1986) Role of the entomogenous fungus, *Neozygites floridana*, in population declines of the two spotted spider mite, *Tetranychus urticae*, on field corn. Entomol Exp Appl 41:255–264

Solter LF (2006) Transmission as a predictor of ecological host specificity with a focus on vertical transmission of microsporidia. J Invertebr Pathol 92:132–140

Solter LF, Becnel JJ (2007) Entomopathogenic microsporidia. In: Lacey LA, Kaya K (eds) Field manual of techniques in invertebrate pathology. Application and evaluation of pathogens for control of insects and other invertebrate pests. 2nd edn, Springer, pp 199–221

Sookar P, Bhagwant S, Ouna EA (2008) Isolation of entomopathogenic fungi from the soil and their pathogenicity to two fruit fly species (Diptera: Tephritidae). J Appl Entomol 132:778–788

Stacey DA, Thomas MB, Blanford S, Pell JK, Pugh C, Fellowes MDE (2003) Genotype and temperature influence pea aphid resistance to a fungal entomopathogen. Physiol Entomol 28:75–81

Steinhaus E (1949) Principles of insect pathology. McGraw-Hill, NY, p 757

Steinkraus DC, Hollingsworth RG, Slaymaker PH (1995) Prevalence of *Neozygites fresenii* (Entomophtholales: Neozygitaceae) on cotton aphids (Homoptera: Aphididae) in Arkansas cotton. Environ Entomol 24:465–474

Steinkraus DC, Hollingsworth RG, Boys GO (1996) Aerial spores of *Neozygites fresenii* (Entomophthorales: Neozygitaceae): Density, periodicity, and potential role in cotton aphid (Homoptera: Aphididae) epizootics. Environ Entomol 25:48–57

Stentiford GD, Neil DM, Atkinson RJA (2001) The relationship of *Hematodinium* infection prevalence in a Scottish *Nephrops norvegicus* population to season, moulting and sex. ICES J Mar Sci 59:814–823

Stow A, Briscoe D, Gillings M, Holley M, Smith S, Leys R, Silberbauer T, Turnbull C, Beattie A (2007) Antimicrobial defences increase with sociality in bees. Biol Lett 3:422–424

Strasser H, Vey A, Butt TM (2000) Are there any risks in using entomopathogenic fungi for pest control with particular reference to the bioactive metabolites of *Metarhizium*, *Tolypocladium* and *Beauveria* species? Biocontrol Sci Technol 10:717–735

Sung GH, Hywel-Jones NL, Sng JM, Luangsa-Ard JJ, Shrestha B, Spatafora JW (2007) Phylogentic classification of *Cordyceps* and the clavicipitaceous fungi. Stud Mycol 57:5–59

Tarrant CA, Soper R (1986) Evidence for the vertical transmission of *Coelomycidium simulii* (Myceteae (Fungi): Chytridiomycetes). In: Samson RA, Vlak JM, Peters D (eds) Fundamental and applied aspects of invertebrate pathology. Fourth Int. Colloq. Invertebr. Pathol., Wageningen, Netherlands, p 212

Thomas MB, Blanford S (2003) Thermal biology in insect-parasite interactions. Trends Ecol Evol 18:344–350

Thomas MB, Wood SN, Lomer CJ (1995) Biological control of locusts and grasshoppers using a fungal pathogen—the importance of secondary cycling. Proc R Soc Lond B Biol 259:265–270

Torchin ME, Lafferty KE, Dobdon AP, McKenzie VJ, Kuris AM (2003) Introduced species and their missing parasites. Nature 421:628–630

Tscharntke T, Bommarco R, Clough Y, Crist TO, Kleijn D, Rand TA, Tylianakis JM, van Nouhuys S, Vidal S (2008) Conservation biological control and enemy diversity on a landscape scale (Reprinted from Biol. Control, vol 43, pg 294–309, 2007). Biol Control 45:238–253

Tymon AM, Shah PA, Pell JK (2004) PCR-based molecular discrimination of *Pandora neoaphidis* isolates from related entomopathogenic fungi and development of species-specific diagnostic primers. Mycol Res 108:1–15

Ugine TA, Wraight SP, Brownbridge M, Sanderson JP (2005) Development of a novel bioassay for estimation of median lethal concentrations (LC_{50}) and doses (LD_{50}) of the entomopathogenic fungus *Beauveria bassiana*, against western flower thrips, *Frankliniella occidentalis*. J Invertebr Pathol 89:210–218

van Veen FJF, Muller CB, Pell JK, Godfray HCJ (2008) Food web structure of three guilds of natural enemies: predators, parasitoids and pathogens of aphids. J Anim Ecol 77:191–200

Vega FE (2008) Insect pathology and fungal endophytes. J Invertebr Pathol 98:277–279

Vega FE, Goettel MS, Blackwell M, Jackson MA, Keller S, Koike M, Maniania NK, Monzón A, Ownley B, Pell JK, Rangel D, Roy HE (2009) Fungal entomopathogens: new insights on their ecology. Fungal Ecol 2:149–159

Weir A, Hammond PM (1997) Laboulbeniales on beetles: host utilization patterns and species richness of the parasites. Biodivers Conserv 6:701–719

Weseloh RM (2003) Short and long range dispersal in the Gypsy moth (Lepidoptera: Lymantriidae) fungal pathogen, *Entomophaga maimaiga* (Zygomycetes: Entomophthorales). Environ Entomol 32:111–122

Weseloh RM (2004) Effect of conidial dispersal of the fungal pathogen *Entomophaga maimaiga* (Zygomycetes: Entomophthorales) on survival of its gypsy moth (Lepidoptera: Lymantriidae) host. Biol Control 29:138–144

Weseloh RM, Andreadis TG (1997) Persistence of resting spores of *Entomophaga maimaiga*, a fungal pathogen of the gypsy moth, *Lymantria dispar*. J Invertebr Pathol 69:195–196

Wichmann MC, Alexander MJ, Soons MB, Galsworthy S, Dunne L, Gould R, Fairfax C, Niggemann M, Hails RS, Bullock JM (2009) Human-mediated dispersal of seeds over long distances. Proc R Soc B Biol 276:523–532

Wilding N (1969) Effect of humidity on the sporulation of *Entomophthora aphidis* and *E. thaxteriana*. Trans Brit Mycol Soc 53:126–130

Wilding N, Perry JN (1980) Studies on *Entomophthora* in populations of *Aphis fabae* on field beans. Ann Appl Biol 94:367–378

Wilson K, Cotter SC, Reeson AF, Pell JK (2001) Melanism and disease resistance in insects. Ecol Lett 4:637–649

Wilson K, Thomas MB, Blanford S, Doggett M, Simpson SJ, Moore SL (2002) Coping with crowds: density-dependent disease resistance in desert locusts. Proc Natl Acad Sci USA 8:5471–5475

Wongsa P, Tasanatai K, Watts P, Hywel-Jones N (2005) Isolation and in vitro cultivation of the insect pathogenic fungus *Cordyceps unilateralis*. Mycol Res 109:936–940

Xu J-H, Feng M-G (2000) The time–dose–mortality modeling and virulence indices for two Entomophthoralean species, *Pandora delphacis* and *P. neoaphidis*, against the green peach aphid, *Myzus persicae*. BiolControl 17:29–34

Xu J-H, Feng M-G (2002) *Pandora delphacis* (Entomophthorales: Entomophthoraceae) infection affects the fecundity and population dynamics of *Myzus persicae* (Homoptera: Aphididae) at varying regimes of temperature and relative humidity in the laboratory. Biol Control 25:85–91

Author Biographies

Helen Hesketh is an ecologist at the NERC Centre for Ecology & Hydrology (UK). Her research focuses on the ecology of entomopathogenic fungi and baculoviruses with an emphasis on the role of these entomopathogens in regulating insect populations and their use in biological control.

Helen Roy leads zoological research in the Biological Records Centre at the NERC Centre for Ecology & Hydrology (UK). The focus of her research is insect community interactions with particular emphasis on the effects of environmental change. She is an associate editor of BioControl.

Jørgen Eilenberg is a Professor at the Department of Agriculture and Ecology at University of Copenhagen, Denmark. His main interests include ecology of insect pathogenic fungi and other groups of insect pathogens, and their use in biological control.

Judith K. Pell heads the Insect Pathology Group in the Department for Plant and Invertebrate Ecology at Rothamsted Research. She leads research on the ecology of entomopathogenic fungi, to elucidate their role in population regulation and community structure and to inform biological control strategies. Specifically: intraguild interactions; the relationships between guild diversity, habitat diversity and ecosystem function; pathogen-induced host behavioural change.

Rosie Hails is an ecologist at the NERC Centre for Ecology & Hydrology (UK) and a Professor at Oxford Brookes University. Her personal research interests include the persistence and transmission of insect pathogens, exploiting pathogens for biocontrol, the role of pathogens in regulating insect and plant populations, population ecology of feral crop plants and the risk assessment of genetically modified plants and viruses. She was awarded an MBE for services to environmental research in 2000.

BioControl (2010) 55:75–88
DOI 10.1007/s10526-009-9247-4

FORUM PAPER

Fungal entomopathogens in a tritrophic context

Jenny S. Cory · Jerry D. Ericsson

Received: 16 July 2009 / Accepted: 19 October 2009 / Published online: 10 November 2009

Abstract Variation in plant quality has an important impact on insect growth and development and there is considerable evidence that plants can also influence an insect's natural enemies. Here we discuss the potential for plant-mediated effects on fungal entomopathogens. Fungi differ from other insect pathogens in that they infect an insect directly through its cuticle. This means that they are particularly vulnerable to changes in microclimate and properties of the insect cuticle. Potential direct and indirect mechanisms for plant-mediated effects on fungal entomopathogens are discussed. It is clear from these studies that fungal entomopathogens could be affected by plant volatiles and plant surface chemistry. Plant secondary chemicals can also inhibit fungal growth, potentially protecting the insect herbivore. However, the site of action and the mechanism behind these effects in plant-based studies is not always clear. The implications for biocontrol using fungal entomopathogens are discussed.

Keywords Ecology · Trophic cascade · Bodyguard · Direct effects · Allelochemicals

Handling Editor: Helen Roy.

J. S. Cory (✉) · J. D. Ericsson
Department of Biological Sciences, Simon Fraser University, 8888 University Drive, Burnaby, BC V5A 1S6, Canada
e-mail: jsc21@sfu.ca

Introduction

Intra- and interspecific plant variation can strongly influence insect herbivore development and population growth rate and can also have direct and indirect impacts on the natural enemies of herbivores (Inbar and Gerling 2008; Johnson 2008). Plants have sophisticated ways of responding to insect herbivory by releasing volatiles that will attract insect natural enemies, often in response to damage from specific herbivore species. These volatile signals can also influence undamaged neighbouring plants (Arimura et al. 2009). Much of the research on plant responses to damage has focussed on interactions with parasitoids (e.g. DeMoraes et al. 1998; Fatouros et al. 2005; Gols and Harvey 2009). However, increasing evidence suggests that plant-mediated effects can also impact entomopathogens (Cory and Hoover 2006).

Fungal entomopathogens have been rather understudied from an ecological perspective, with the main focus of activities being directed towards their use as biocontrol agents (e.g. Goettel et al. 1995; Scholte et al. 2004). Entomopathogenic fungi are usually applied as inundative sprays with the expectation of short-term pest suppression (Inglis et al. 2001). The potential for rapid multiplication, persistence and passive dispersal of fungal entomopathogens, however, means that they also have potential for longer-term control (Hajek 1997). In addition, they commonly occur in nature and are important mortality factors in some groups of insects (Hajek 1999).

Knowledge of entomopathogen ecology is therefore important for elucidating their role in natural and managed ecosystems and is likely to be important in their successful development as biocontrol agents. Investigations of tritrophic effects may supply some answers as to why some entomopathogens have variable success as biocontrol agents and a greater understanding of these interactions will allow the development of more effective pest management strategies.

Fungal entomopathogens have been isolated from almost all regions on earth ranging from the Arctic Circle in Finland, to islands in the Antarctic, and from almost all soil types (Bidochka et al. 1998). This cosmopolitan distribution is evidence of both their evolutionary success, and their complex and flexible life histories that involve a range of interactions between fungi, plants, insects and other sources of nutrients in their environments. Fungal entomopathogens are distinguished from other entomopathogens in that they are directly transmitted by contact with susceptible hosts, rather than needing to be ingested to initiate infection. This means that they are likely to adopt different strategies for survival that will include potentially novel tritrophic interactions.

Fungal biology in a tritrophic context

One of the most notable characteristics of fungal entomopathogens is that they exhibit a huge range in specificity (Wraight et al. 2007). With other insect pathogens this usually means that they can potentially infect a wide range of host species, but for fungal entomopathogens this means that they can vary from obligate pathogens that are not known to grow outside the host, to others that can include a saprophytic or endophytic life stage and are able to acquire nutrients from sources other than an insect (Vega et al. 2009). This continuum of host range affects other general characteristics of their ecology (Table 1). Obligate pathogens with a narrow host range, such as the Entomophthorales *Pandora neoaphidis* (Remaudière and Hennebert) or *Entomophaga maimaiga* Humber, tend to be more virulent than facultative fungal entomopathogens and often require less inoculum to kill the insect host (Goettel 1995). In evolutionary terms, these highly specific fungal entomopathogens may increase the probability of infection by maximizing virulence towards a small number of species. In contrast, facultative fungal entomopathogens may vary widely in their virulence based on several environmental factors including their prior nutrient history (St. Leger et al. 1997), but will maintain a larger host range. Variation in the ability to use multiple nutrient sources is likely to have a significant impact on their sensitivity to tritrophic interactions. We would predict that fungi that are obligate pathogens with a narrow host range are more likely to be influenced by tritrophic interactions as they need to persist in the insect host's environment and will have a greater opportunity for coevolutionary interactions.

The abundance of obligate fungal entomopathogens in the environment is correlated with host species abundance, and is facilitated by epizootic events (Hajek 1997). The saprophytic and endophytic nature of facultative fungal entomopathogens does not require an insect host for their local establishment (enzootic), but their presence in the environment will increase the probability of contact with a susceptible host. Because the distribution of obligate fungal entomopathogens is more likely to be patchy (following that of their hosts) some species have developed strategies to increase the likelihood of successful infection. For example, infection can be

Table 1 General characteristics of fungal entomopathogens with narrow and wide host ranges (Goettel 1995)

Attribute	Narrow host range	Wide host range
Distribution	Restricted	Ubiquitous
Pathogenicity	Obligate	Facultative
Virulence	High	Low
Epizootiology	Often epizootic	Mostly enzootic
Potential use	Classical/inoculative	Inundative
Non-host habitat	Soil, as resting spores	Soil, as resting saprophytes

promoted by early germination and pre-penetration growth on the insect, or the ability of some Entomophthorales species to propel their spores (Hajek 1997), enabling greater coverage in the search for new hosts. This approach also requires greater sensitivity to environmental cues than facultative fungal entomopathogens, and this could include sensitivity to plant-mediated signals. With facultative fungal entomopathogens, it has been shown that recognition of suitable hosts depends on specific structural and chemical cues found on the insect cuticle (Wang and St. Leger 2006), and thus indirect plant-mediated effects via the insect might be more important, although it is likely that this recognition occurs to some degree in all fungal entomopathogens.

One trait that is shared amongst all fungal entomopathogens is the potential for and the maintenance of different adaptations for growth and survival within and outside the host. In the environment fungal entomopathogens are usually found as spores. This means that fungal entomopathogens can spend a significant period of time on the plant surface and are vulnerable to plant surface characteristics, exposure to damaging ultra-violet radiation and changes in microclimate. When the spore attaches to a host it germinates to form a mycelium, an appresorium, and a germ tube that uses a combination of enzymatic and mechanical forces to breach the insect's cuticle (Hajek 1997). High humidity or free water is usually required for successful germination and infection. Once inside the host, the fungus begins budding into yeast-like hyphal bodies that proliferate, and it is here that plant influences on host condition or sequestered allelochemicals could alter the infection process. Once the nutrients have been exhausted from the host, the hyphal bodies begin to grow as mycelial threads that move towards the surface of the host. The mycelia eventually generate sporogenic structures. Spores are produced in various quantities and can be partially specialized for optimal survival in soil environments (resting spore) and partially specialized for horizontal transmission (primary, secondary and higher order spores). In both obligate and facultative fungi, the infective propagules can vary in function and structure based on the timing of their production. First, primary spores (usually asexual conidia) are produced, which are often specialized for dispersal and infection of a new host and tend to be more virulent (although for some species, principally entomophthoralean fungi, the secondary spore is more virulent than the primary spore). In some species the end of the sporulation cycle results in the production of resting spores which tend to be more stable in the environment, and cause different pathologies in the host (Hajek 1997; Hesketh et al. 2009). Given these complex life histories, some general trends can be anticipated when considering their ecology and potential for tritrophic interactions.

Plant-mediated effects and mechanisms

In general terms, plants could affect fungal entomopathogens either directly or indirectly via changes in the insect hosts, or even other components of the system (Fig. 1). By direct effects we mean anything emanating from or produced by the plant that influences fungal infection of the insect. Indirect effects could occur before or after exposure of the insect to fungus, and would include factors that alter insect condition. Insect condition is a term that usually describes the 'health' of an insect and, depending on the context, can include factors such as weight, the level of energy reserves and the capacity to resist infection (e.g. via innate immunity). As fungal entomopathogens do not need to be ingested to initiate infection, plant-mediated effects are likely to be very different for this group, as compared with pathogens, such as baculoviruses and bacteria, which attack the insect through the mid-gut. Thus fungal entomopathogens are likely to be very sensitive to leaf topology, plant surface chemistry and microclimate. Potential direct plant-mediated effects might include: (1) Plant exudates affecting the conidia directly, (2) herbivore-induced plant volatiles affecting sporulation or germination, (3) leaf topology and surface chemistry, influencing the rate of spore acquisition by the host insect, (4) plant architecture altering spore persistence, and (5) leaf modifications of microclimate and thus spore germination.

With indirect effects it can sometimes be more difficult to demonstrate cause and effect, and the distinction between direct and indirect effects can be blurred, for instance in circumstances where insects sequester plant secondary chemicals which could have a direct effect on fungal entomopathogens or an indirect effect through insect condition. Possible

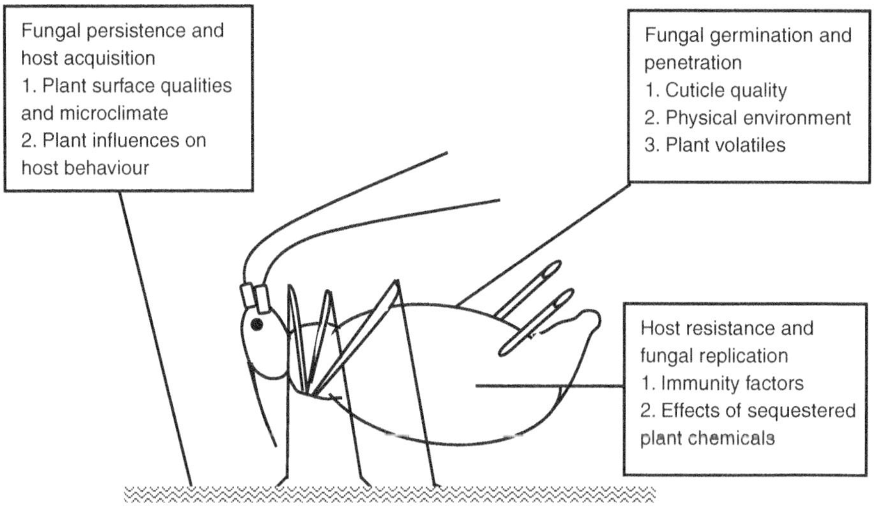

Fig. 1 Potential direct and indirect plant-mediated effects on the efficacy of fungal entomopathogens

indirect effects, discussed below, could include: (1) Plant quality, either allelochemicals or nutrients, altering insect condition (e.g. immunity) and thus disease resistance, (2) nutritional quality altering insect morphology (e.g. cuticle depth) which would influence the infection process, (3) changes in insect growth rate, which might alter the exposure of the insect to fungal entomopathogens, and (4) plant structure altering insect behaviour, and thus fungal encounter rate.

Direct tritrophic effects

Several studies on fungal entomopathogens have pointed to interesting and novel ways in which plants could modify their efficacy (Table 2). A fascinating example considers whether plant volatiles can affect fungal entomopathogens. One of the first studies on this topic showed that green leaf volatiles from the tobacco plant, *Nicotiana tabacum* L., inhibited the germination of *P. neoaphidis* conidia and its in vitro growth rate (Brown et al. 1995). A similar result was found when *P. neoaphidis* was exposed to plants that had been attacked by aphids. However, infection rates did not appear to be affected when the conidia landed directly on the aphid host, leading the authors to propose that the inhibitory response delayed germination until a suitable host could be encountered (Brown et al. 1995). A later study on *P. neoaphidis*, using the broad-bean *Vicia faba* L. as host plant and *Acyrthosiphon pisum* (Harris) as target aphid, measured the effect of volatiles from *A. pisum*-damaged *V. faba* on various stages of the infection process. They found that exposure to the volatiles had no effect on the number of conidia, conidia size or in vitro fungal growth rate (Baverstock et al. 2005). In contrast to Brown et al. (1995), however, percent germination of the conidia was greater on aphids feeding on previously damaged plants, although the resulting infection rate of aphids introduced to damaged and undamaged plants after fungal exposure was not different (Baverstock et al. 2005). The authors suggest that the inhibitory effects recorded by Brown et al. (1995) were the result of toxic effects of chemicals, such as nicotine, released from the tobacco plants, whereas their results indicate a potential positive effect of herbivore-induced plant volatiles which might act as a cue to indicate suitable hosts are near (Baverstock et al. 2005). The lack of increased fungal infectivity did not support this, however, and they go onto conclude that the positive effects on germination that were measured could result from factors other than plant volatiles, such as altered microclimate in aphid-damaged versus undamaged plants.

In a different system, involving the interaction of the cassava green mite, *Mononychellus tanajoa*, (Bondar) (a pest of cassava, *Manihot esculenta* Crantz) and the fungal pathogen *Neozygites tanajoae* Delal., Humber & A. E. Hajek (Entomophthorales: Neozygitaceae), Hountondji et al. (2005) found that green leaf volatiles alone suppressed conidiation,

Table 2 Tritrophic interactions involving fungal entomopathogens (excluding in vitro studies)

Mechanism	Fungi	Insect	Plant(s)	Effect	References
Plant volatiles	*Neozygites tanajoae*	Cassava green mite, *Mononychellus tanajoa*	Cassava, *Manihot esculenta*	Green leaf volatiles inhibit conidiation, herbivore induced plant volatiles enhance conidiation	Hountondji et al. (2005)
Leaf surface wax	*Metarhizium anisopliae*	Mustard beetles *Phaedon cochleariae*	Various crucifers	Dewaxing increased germination. Surface leachates increased germination and virulence.	Inyang et al. (1999a)
	Pandora neoaphidis	Pea aphid, *Acyrthosiphon pisum*	Pea, *Pisum sativum*	Reduced wax increased mortality via increased adhesion and germination of conidia.	Duetting et al. (2003)
Leaf surface topology	*Beauveria bassiana*	Western flower thrip, *Frankliniella occidentalis*	Bean (*Phaseolus vulgaris*) and impatiens (*Impatiens wallerana*)	Acquisition of conidia differed, possibly due to differences in leaf veining.	Ugine et al. (2007)
Leaf microclimate?	*Pandora neoaphidis*	Pea aphid, *Acyrthosiphon pisum*	Broad bean, *Vicia faba*	Germination of conidia greater on previously damaged plants.	Baverstock et al. (2005)
Fumigant effect of sequestered phytochemicals	*Beauveria bassiana* *Metarhizium anisopliae*	Willow leaf beetle, *Phratora vitellinae*	Willows (Salicaceae)	Enhanced susceptibility to entomopathogenic fungi after removal of glandular secretions	Gross et al. (2008)
Plant secondary chemicals?	*Beauveria bassiana*	Whitefly, *Bemisia argentifolii*	Cotton and melon	Lower mortality on cotton due to reduced germination on cuticle	Poprawski and Jones (2000)
		Sweet potato whitefly, *Bemisia tabaci*	Various	Differences in mortality, speed of kill and conidial production	Santiago-Álvarez et al. (2006)

whereas herbivore induced volatiles promoted conidiation. This suggests that the fungus is able to time its production of conidia to the presence of actively feeding hosts and not just solely plant cues. Follow up studies demonstrated that a common component of herbivore-induce plant volatiles, methyl salicylate promoted the production of primary conidia but not their germination into secondary capilliconidia (Hountondji et al. 2006). This response was isolate-dependent and differed from the earlier study, indicating that methyl salicylate is unlikely to be the only factor involved. Thus, while the available studies demonstrate some potentially interesting interactions involving plant volatiles, the data are equivocal. Positive effects on infection rate have yet to be demonstrated.

An area that is likely to have a major impact on the fungal infection process is the leaf surface or phylloplane. Leaf surfaces vary in smoothness, surface structures and the presence of cuticular substances, such as waxes, that could alter the retention of conidia, conidial survival, germination rate or alter insect behaviour, which in turn will change rates of fungal acquisition (Inyang et al. 1998). For example, germination of *Metarhizium anisopliae* (Metchnikoff) Sorokin (Ascomycota: Hypocreales) was higher on dewaxed leaves than intact leaves from a variety of crucifers, and leaf exudates and soluble extracts increased germination and virulence to the mustard beetle *Phaedon cochleariae* (F.) (Inyang et al. 1999a). In another system, pea plants with a reduced waxy bloom promoted better adhesion and germination of the fungus *P. neoaphidis* and thus greater mortality of *A. pisum* (Duetting et al. 2003).

As with other entomopathogens, the focus of most tritrophic studies has been the effect of plant secondary chemicals on fungal growth and infection. In general it is thought that plant secondary chemicals, particularly those that are induced, are produced to deter herbivores, and thus tend to slow herbivore growth and development or promote anti-feeding behaviour. But how do these same chemicals affect fungal entomopathogens? Numerous in vitro studies have investigated the effect of plant secondary chemicals on the growth of fungal entomopathogens. Costa and Gaugler (1989) found that the alkaloid tomatine inhibited colony growth, development of conidiophores and germination of *Beauveria bassiana* (Bals.-Criv.) Vuill. Artificial diet containing tomatine also inhibited growth and sporulation of *Nomuraea rileyi* (Farlow) (Ascomycota: Hypocreales). Tomatine also impaired the growth of the host *Helicoverpa zea* (Boddie), although addition of tomatine to artificial diet did reduce *N. rileyi* induced mortality in *H. zea* (Gallardo et al. 1990). This indicates possible opposing effects of the impact of allelochemicals on herbivores and their fungal pathogens. The speed and degree of in vitro germination of blastospores from *Isaria fumosorosea* (=*Paecilomyces fumosoroseus*) Wize (Ifr) (Ascomycota: Hypocreales) were inhibited by various phenolics (primarily catechol, chlorogenic acid and gallic acid) but not by sugar beet saponin (a triterpenoid) or sinigrin (a glucosinolate) at concentrations ranging from 100 to 1,000 ppm (Vega et al. 1997). These concentrations are well below the levels found in most plant tissues (Vega et al. 1997). Similarly, germination of both aerial conidia and blastospores of *I. fumosorosea* was inhibited by alkaloids (tomatine, camptothecin, solanine), xanthotoxins and tannins, although fungal growth rates were only inhibited by tomatine, camptothecin and xanthotoxin (Lacey and Mercadier 1998). Inyang et al. (1999b) demonstrated that a range of isothiocyanates common in Brassicaceae severely inhibited both germination and growth of *M. anisopliae*, and also reduced fungal-induced mortality on *P. cochleariae*. Overall these studies show that allelochemicals from a diverse range of plants tend to have a negative impact on the growth and germination of fungal entomopathogens. However, what is not clear is whether plants produce these compounds in sufficient quantities to elicit an effect and whether the fungal entomopathogens actually encounter them. Tomatine, for example, has been found to make up as much as 5% of the fresh weight of tomato tissue (Sandrock and VanEtten 1998). However, whether the fungus encounters the secondary chemicals will depend on whether the insect host excretes, detoxifies or sequesters them, or even whether degradation products have activity that also affects fungal entomopathogen growth rate and subsequent sporulation on the cadaver.

Some insect groups, principally Lepidoptera and Coleoptera, are able to utilize plant secondary compounds for their own defence (Kuhn et al. 2004). However, the impact of sequestering these chemicals on entomopathogen-induced mortality has received little attention. Several species of leaf beetle (Chrysomelidae) sequester salicin and saligenin from willow leaves to produce a volatile glandular secretion rich in salicylaldehyde (Kuhn et al. 2004; Gross et al. 2008). These exocrine glandular secretions strongly inhibit both the germination and growth of *M. anisopliae* in vitro when applied directly and as a fumigant (Gross et al. 2002). More importantly, removal of the glandular reservoirs significantly increased larval susceptibility of the willow leaf beetle *Phratora vitellinae* (L.) to two strains of *M. anisopliae* and *B. bassiana* in vivo, supporting the hypothesis that the salicyaldehyde acts as a fumigant for anti-microbial defence (Gross et al. 2008).

Studies have also investigated tritrophic effects on fungal entomopathogens in vivo using plants [discussed below]. One of the drawbacks of these studies, however, is that the effect of the host plant on the condition of the insect through feeding, is often not distinguished from the influence of the host plant on the fungus, for example, the effect of the leaf surface on spore retention and survival. Therefore, while plant-mediated effects may be observed, their cause remains unclear. Results will also be affected by the methodologies used. For example, more natural approaches where the insects acquire conidia by walking on leaf surfaces are likely to yield very different outcomes compared to experiments where insects are exposed to fungal entomopathogens by leaf dipping or spray application. Lacey and Mercadier (1998) suggest that one approach might be to study plant cultivars with varying levels of allelochemicals to tease apart these effects and pinpoint the site(s) of antifungal activity.

In one example, the whitefly *Bemisia argentifolii* Bellows and Perring was clearly found to be less susceptible to both *I. fumosorosea* and *B. bassiana* when it was reared on cotton rather than melon (a preferred host plant). However, as the insects had been both reared and treated on the two plants, the site and cause of this difference was not clear (Poprawski and Jones 2000). The differences in mortality could have been caused by a host plant effect on insect condition prior to fungal challenge, or direct impact of allelochemicals in the insects on fungal infection. Alternatively, plant surface variation could affect the persistence or adhesion of the fungal entomopathogens or change insect behaviour and the rate of spore acquisition by the insect. Further study showed that germination of both fungi was considerably reduced on the cuticle of insects reared on cotton, and the authors suggested that this was the result of a fungistatic response to a secondary chemical produced by cotton (Poprawski and Jones 2000). In vitro assays using the terpenoid, gossypol, a key chemical component of cotton, indicated that both fungi were tolerant to gossypol, which only produced significant inhibition of fungal germination at high concentrations. This suggests that gossypol was not the main cause of reduced germination. In a similar study, the greenhouse whitefly *Trialeurodes vaporariorum* (West.) was found to be much more susceptible to *B. bassiana* and *I. fumosorosea* on cucumber than on tomato (Poprawski et al. 2000). In vitro assays with tomatine showed that it completely inhibited germination of conidia of *I. fumosorosea* at concentrations above 500 ppm but this effect was less marked in *B. bassiana*, indicating that other factors were involved.

Plant-mediated effects can influence other aspects of the fungal-insect interaction in addition to insect mortality. In a study with *B. bassiana* and the sweet potato whitefly, *Bemisia tabaci* (Genn.), the fungus was applied by dipping infested leaves of ten different plant species in a fungal suspension (Santiago-Álvarez et al. 2006). The whiteflies were reared on all but one of the host plants for one generation before the experiment. There was a significant difference in resulting levels of pathogenicity, with mortality being lowest on plants such as cotton and green pepper, and much higher on preferred food plants such as marrow and cucumber (Santiago-Álvarez et al. 2006). Survival time also varied with host plant, tending to increase as mortality levels decreased. Interestingly, conidial production from the cadavers also varied. It was by far the highest on melon, a preferred host plant species and the one on which the main whitefly colony was reared. In general, conidial production had an inverse relationship with survival time. However, it is unclear what caused the effects observed in this study; the insects were only reared for one generation on the different host plants (except melon) and the fungal conidia were not acquired naturally. This would point to host plant chemistry effects, but it is not clear whether it is a direct or indirect interaction.

A more complex experiment was carried out by Ugine et al. (2007) using the western flower thrips,

Frankliniella occidentalis (Pergande) and beans (*Phaseolus vulgaris* L.) and/or impatiens (*Impatiens wallerana* Hook) treated with *B. bassiana*. Levels of mortality were compared depending on which plant the insect was reared on (before challenge effects) and also the plant at the time of infection. In this case, the insects were infected by exposing them to leaf discs already treated with the fungus, so the thrips had to acquire an infective dose from the disc. This is likely to be a more realistic test of the effect host plant on fungal acquisition. Mortality was significantly higher when the dose was administered on bean rather than impatiens, regardless of what plant the thrips had been reared on prior to treatment. Direct mixing of host plant macerates with the fungus had no effect, indicating that it was not a direct chemical interaction. The authors suggest that the difference was due to different rates of dose acquisition; bean plants have external leaf veins and it is possible that fungal spores accumulated more between them. There was also an indication of an opposing effect in this experiment, with insects reared on impatiens being more susceptible to fungal infection. The authors attribute this to the possible effects of host switching but it may also indicate a difference in the condition of the insects before infection.

Indirect tritrophic effects

One of the major ways in which plants could exert a general indirect effect on fungal pathogenicity is via changes in insect condition. For example, caterpillars that were given a diet that was rich in carbohydrate and poor in protein had significantly lower survival when exposed to baculoviruses than insects fed a high protein diet (Lee et al. 2006). Dietary protein levels also altered the level of constitutive immune function. Insects have potent, innate immune systems that are effective against almost all microbial invaders including fungi. The immune system itself is comprised of a population of circulating haemocytes that vary in type, function, and abundance with the insect order and species (Price and Ratcliffe 1974), as well as a humoral component consisting of a range of bacteriolytic enzymes, and antimicrobial (AMP) peptides (Gillespie et al. 1997). These proteins vary in structure and function, but generally attack membrane structures on fungal and bacterial cells, or target other key cellular processes in the microbe. Many fungal entomopathogens are able to survive the phagocytic and encapsulation reactions, and even grow from within partially formed capsules in the haemolymph (Gillespie et al. 2000a). Fungal entomopathogens can apparently develop in their host with little resistance from the host immune system. The hyphal bodies are able to grow within the host's haemolymph unchecked because they produce immune modulating compounds called destruxins, beauverolides, cyclosporins, cytochalasins, and various proteases (Vilcinskas and Gotz 1999). The hyphal bodies also alter their cell wall such that the pathogen associated molecular patterns are not recognized and evasion occurs (Wang and St. Leger 2006). For example, fungal entomopathogens produce several proteases that prevent the expression of host immune genes, degrade the effector molecules themselves, cause paralysis and other toxic symptoms, or induce the apoptosis of the immune cells through anti-haemocyte peptides (Vilcinskas and Gotz 1999; Gillespie et al. 2000a). Although fungal entomopathogens possess an impressive array of immune suppressive tools to enable their infection and replication within a host, the insect is not passive. At early stages in the infection, several changes in immune parameters occur and can prevent the death of the insect (Gillespie et al. 2000b). Thus the probability of surviving infection by a fungal entomopathogen depends on a rapid and potent immune response, that in turn is influenced by the general condition and vigour of the insect.

Host plants could also change insect morphology. For fungal entomopathogens, the insect cuticle is likely to be a particularly important barrier. Cuticle thickness varies widely amongst insects, but in general has been found to be thinner in leaf-feeding beetles than in carnivorous ones (Rees 1986). Because fungal entomopathogens invade directly through the cuticle, if plant quality could alter the surface chemistry of the insect, such as the composition of secreted waxes, or modify the thickness or resilience of its cuticle through altered sclerotization and melanization reactions, or through resorption and alteration of the cuticle's protein composition, this could have consequences for successful infection. It has been shown in *Manduca sexta* L. that cuticle proteins vary in their origin and can even be transported from the haemolymph, and this indicates that the cuticle maintains an intimate, and continuous metabolic interplay with multiple organ systems (Csikos et al. 1999). Thus if plants are of low quality, or alter their quality though the production of protease inhibitors, the cuticle will be thinner and in turn more vulnerable to fungal entomopathogen attack.

Facultative fungal entomopathogens: above and below ground effects?

In facultative fungal entomopathogens like *Metarhizium* spp., saprophytic growth in the rhizosphere may be mediated by root exudates that sustain the fungi in the absence of a susceptible host. This source of nutrients enables the proliferation of *M. anisopliae* within the rhizosphere several centimetres deep in the soil profile (Hu and St. Leger 2002). There is also evidence that treatment of corn seed with *M. anisopliae* improves the fresh weight yield and stand density, but also may reduce herbivory by wireworms (Kabaluk and Ericsson 2007). Regardless of the specific effect, there is evidence that certain fungal entomopathogens can associate with the rhizosphere and survive within this highly specialized niche. It is not clear how fungal populations are regulated in the rhizosphere, because their unchecked growth may reach a level that is harmful to the plant. Thus plants could potentially produce root exudates that regulate the microflora of this critical interface. But the question that emerges is, what is the role of the fungal entomopathogens in this case, and how does the plant benefit from this intimate association? Although mixotrophy, the plants phagocytic consumption of mycorrhizal fungi, has only been shown with plant symbionts (Selosse and Roy 2009), it is possible that a similar mechanism regulates the abundance of fungal entomopathogens in the rhizosphere, or that it facilitates their establishment as an endophyte. For example, various attempts at establishing *B. bassiana* as an endophyte have shown that fungi are transported through the plant and can be isolated from leaf, and stem tissues after inoculation, and can also become established through different routes (Posada and Vega 2005; Vega et al. 2009).

The presence of fungal entomopathogens as endophytes also opens the possibility for more convoluted interactions with plants. Because endophytes obtain nutrition saprophytically from the plants it is reasonable to expect that molecules that are biologically active against insects could be obtained from the plant. In vitro studies performed with both obligate and facultative fungal entomopathogens have shown that the metabolites produced by certain fungi can be vastly different depending on the nutrients available and the growth environment (Isaka et al. 2005). If this is the case for all fungal entomopathogens, then host plant adaptation by facultative fungi could result in a single species producing different compounds possibly active against different insects, and may also partially explain their larger host range of facultative fungal entomopathogens.

Could plants manipulate fungal entomopathogens for their own benefit?

Increasing evidence demonstrates that plants can influence the behaviour of certain groups of natural enemies, particularly parasitoids, to increase herbivore suppression and presumably increase plant fitness. Plants may also influence entomopathogens in a similar way (Elliot et al. 2000) and fungal entomopathogens perhaps offer the best opportunities to become plant bodyguards (Cory and Hoover 2006). Any plant traits that enhanced fungal infection and showed genetically-based variation could theoretically be selected for as an adaptive response. For the fungal entomopathogens to act as a plant bodyguard there must be benefits for the fungi and the plant that are based on co-evolved processes.

Mechanisms could fall into two main categories: the fungus could maintain higher populations on the plant (numerical response) or plants could enhance fungal efficacy in some way (functional response). For the first category, the plant could develop an architecture that favours the fungus and prolongs it persistence. One suggestion is that plants could deliberately maintain fungi as endophytes (Gerson et al. 2008). In this study the authors isolated fungi (*Meira* and *Acaromyces* spp.) from grapefruit, which appeared to have a toxic effect on mites. They failed to isolate the fungus by leaf washing but determined that the fungi were endophytic inside flavedo of fruit's peel and sealed flower buds. The fungi had no effect on the plants, suggesting that it acted as a plant bodyguard (Gerson et al. 2008). However, the mode of transmission of the fungus was not ascertained, nor was it clear how long the fungus would maintain its virulence for insects if maintained in this manner. It should also be added that initiation of endophytic infections of *B. bassiana,* has been suggested as a means for controlling cryptic pests (Akello et al. 2008). A more compelling example of beneficial plant structural changes is the evidence for reduced surface wax on leaves resulting in higher infection of the pea aphid by *P. neoaphidis*, apparently via better retention of conidia (Duetting et al. 2003). Reduced wax bloom is controlled by a single gene in this system, and thus it could readily be envisaged as a trait amenable to selection.

An alternative route for decreasing herbivore attack is to enhance the virulence of fungi. A chemical factor that facilitates fungal germination and increases insect mortality should promote plant fitness. One potential route already discussed is the role of plant volatiles, particularly those produced as a result of herbivore damage (Hountondji et al. 2005, 2006). If a fungus can respond to herbivore-induced plant volatiles in a way that enhances the infection rate, such as by increased germination in the presence of the host, then fungal-induced mortality should also increase. The evidence for several allelochemicals produced by the plant in response to herbivory is that they inhibit fungal infection, releasing the insect from its natural enemy and making these chemicals unlikely to be involved in the recruiting or retention of plant bodyguards.

One of the crucial aspects of acting as a plant bodyguard is that the mechanism must be reliable. A major drawback with fungal entomopathogens is their lack of mobility; they cannot respond to plant cues in the way that predators or parasitoids could. It might be possible to attract infected insects. This, however, has inherent risks in terms of increasing the herbivore problem. Maintaining fungal entomopathogen populations on a plant and being able to transmit them vertically to plant progeny would seem the best way of getting around this problem, assuming that virulence is maintained. Another issue is specificity; some plant volatiles are able to attract specific species of parasitoid. Fungal entomopathogens, however, vary in their host range and given their low mobility, a better approach to maintaining viable populations of bodyguards would be to focus on the wider host range species (which would also tend to be facultative pathogens). Whether fungal entomopathogens can act as plant bodyguards or if the effects seen are simply side effects of plant variability, remains to be seen.

Could plant-mediated effects influence the ecological and evolutionary dynamics of insects and fungal entomopathogens?

Even without the mechanisms being identified, it is clear that inter- and intra-specific variation in plants has the potential to influence key components of insect–fungal entomopathogen interaction, such as mortality levels, speed of action and conidial production. These could in turn influence insect population dynamics. However, there are few, if any, long-term data sets on insect–fungal dynamics that begin to address this issue. Could plant-mediated effects affect the evolution of the insect–fungus interaction? This topic has also received very little attention, although it has obvious repercussions for the efficacy and sustainability of biological control using fungal entomopathogens, in addition to the role it might play in insect population dynamics and structuring communities. For example, does a polyphagous insect species that can complete its development on a range of host plant species respond differently to fungal challenge, and how might this influence the development of host resistance or fungal virulence? If there is heritable variation in both insect resistance to fungal entomopathogens and fungal virulence, then the raw material for selection to take place is available. As far as we are aware, the evolution of resistance to fungal entomopathogens and its potential costs have not been addressed. However, there are studies that indicate that variation in resistance could be important.

Most work has focussed on the pea aphid, *A. pisum*, and *P. neoaphidis*. Comparison of clonal populations of *A. pisum* showed significant variation in their resistance to *P. neoaphidis* (Ferrari et al. 2001). The pea aphid is also a well-studied model system with respect to the study of host race development in insects. Clear evidence shows that this species forms genetically distinct races on different legume species and that trade-offs in performance occur on at least two of these host plants, alfalfa (*Medicago sativa* L.) and red clover (*Trifolium pratense* L.) (Via 1999; Via et al. 2000).

However, is the interaction of the pea aphid races with its fungal entomopathogens the same on the different host plants? The answer seems to be no. Resistance to *P. neoaphidis* was compared on *T. pratense* and *Lotus uliginosus* (*pedunculatus* Cav.) in its native range. Marked differences in susceptibility occurred, with pea aphids isolated from *T. pratense* being almost totally resistant to the fungus, whereas those from *L. uliginosus* were not (Ferrari and Godfray 2003). These data clearly indicate that the specialization of this species on particular host plants can have major effects on the evolution of disease resistance. The mechanisms behind this difference are not clear, although it is unrelated to the host plant on which the assays were conducted and there are indications that it might be related to the possession of an endosymbiont (Scarborough et al. 2005).

The other side of the issue is whether there is genetic variation for virulence in fungal entomopathogens, and whether fungal population structure could be influenced by plant variation and selection. As has already been discussed, the plant surface plays an important role in the ecology of fungal entomopathogens. Unlike other entomopathogens, fungi are not intimately mixed and exposed to plant chemicals in the insect mid-gut as part of their normal infection pathway. Fungi must, however, persist on the phylloplane until acquired by an insect (if not sprayed or showered directly onto it), and as the examples above have shown, plant surface characteristics can have a large influence on this process. Therefore the possibility exists that plants could influence the survival and thus selection of certain fungal strains. There is some evidence in another entomopathogen-insect system that baculoviruses become adapted to the host plant on which their insect host feeds (Cory and Myers 2004). This is particularly likely to occur if the pathogen spends extended time periods on the plant surface between host generations. We are unaware of any studies that have considered this possibility in entomopathogenic fungi. However, if natural fungal populations comprise mixed genotypes that vary in relevant genetically-based traits (e.g. their adhesion to the phylloplane surface), then selection for specific genotypes could occur.

The taxonomy of higher fungi has recently undergone a major revision (Vega et al. 2009) and the taxonomy of several fungal entomopathogens, including *M. anisopliae* and *B. bassiana,* is currently being revised using molecular techniques. These studies have discovered that the genetic variation is much higher than previously thought (Driver et al. 2000), and that some species co-occur as a complex containing cryptic species (Rehner and Buckley 2005). Given the species diversity estimated by sequence data from a single soil sample, however, it is not known what role this diversity has on pathogenicity to one or more hosts, or if the structure of the species complex is more important than the individual genetic contributions.

Characterization of fungal species is a growing area and numerous isolates of the more commonly used fungal entomopathogens with wider host ranges, such as *B. bassiana* and *M. anisopliae,* show variation in activity in a range of hosts and in the same host species from different sites (e.g. Castrillo et al. 2008; Devi et al. 2008; Inglis et al. 2008). However, the identification of individual sub-species and strains and the level of genetic variation at different spatial scales, for example, within an individual cadaver versus a collection of soil samples, is not so clear. Information of this nature is needed if fungal population structure and evolution is going to be investigated over temporal and spatial scales.

Ramifications for biocontrol using fungal entomopathogens

Although the mechanisms are not entirely clear, the available data indicate that variation in host plants can affect fungal efficacy, and in some cases significantly enough to reduce pest control. As such, plant-mediated interactions should certainly be taken into account when planning or assessing a biocontrol program. One little understood aspect is whether feeding on different plant species before the fungus is applied, makes any difference to the resulting levels of mortality. There is strong evidence that the plant surface can affect fungal persistence or rate of acquisition, but the impact on insect condition is far less clear. Part of the relevance of tritrophic interactions to effective biocontrol will depend on whether the target insect is directly contacted by a fungal spray or acquires the spores naturally. Even if an insect can be sprayed directly, some individuals are likely to be missed, and immigration or reproduction will result in uninfected insects. For these to be infected and pest suppression to be continued, secondary transmission of the fungus is needed.

Can performance be improved by better knowledge of plant surface characteristics and insect behaviour? Quite possibly. One of the puzzles associated with the deployment of microbial insecticides like fungal entomopathogens is that their performance in the field is rarely as good as that found in the laboratory. Although there are numerous biotic and abiotic reasons why this might be the case, it seems clear that tritrophic effects originating from the host plant may greatly affect the results of a trial.

Conclusions

Considerable gaps exist in our knowledge of fungal ecology and tritrophic interactions in particular. While a significant body of data on the in vitro effects of plant secondary chemicals clearly shows that they have a negative effect on the growth and germination of fungal entomopathogens, their relevance to the field situation is unclear. Experiments need to be designed that will tease apart the mechanisms behind the observed effects and to see whether pre- and post-challenge plant-mediated effects can be demonstrated. Insect behaviour, in relation to dose acquisition, and plant surface characteristics are likely to be particularly important in unravelling the influence of plants on the outcome of insect–fungal interactions.

Acknowledgments The authors would like to thank Judy Myers for her comments and help with Fig. 1 and three reviewers (Judy Pell, Helen Roy & Fernando Vega) for their comments on the manuscript. The authors also acknowledge NSERC for funding.

References

Akello J, Dubois T, Coyne D, Kyamanywa S (2008) Endophytic *Beauveria bassiana* in banana (*Musa* spp.) reduces banana weevil (*Cosmopolites sordidus*) fitness and damage. Crop Prot 27:1437–1441

Arimura G, Matsui K, Takabayashi J (2009) Chemical and molecular ecology of herbivore-induced plant volatiles: proximate factors and their ultimate functions. Plant Cell Physiol 50:911–923

Baverstock J, Elliott SL, Alderson PG, Pell JK (2005) Response of the entomopathogenic fungus *Pandora neoaphidis* to aphid-induced plant volatiles. J Invertebr Pathol 89:157–164

Bidochka MJ, Kasperski JE, Wild GAM (1998) Occurrence of the fungal entomopathogens *Metarhizium anisopliae* and *Beauveria bassiana* in soils from temperate and near-northern habitats. Can J Botany 76:1198–1204

Brown GC, Prochaska GL, Hildebrand DF, Nordin GL, Jackson DM (1995) Green leaf volatiles inhibit conidial germination of the entomopathogen *Pandora neoaphidis* (Entomophthorales: Entomophthoraceae). Environ Entomol 24:1637–1643

Castrillo LA, Ugine TA, Filotas MJ, Sanderson JP, Vandenberg JD, Wraight SP (2008) Molecular characterization and comparative virulence of *Beaveria bassiana* isolates (Ascomycota: Hypocreales) associated with the greenhouse shore fly, *Scatella tenuicosta* (Diptera: Ephydridae). Biol Control 45:154–162

Cory JS, Hoover K (2006) Plant-mediated effects in insect–pathogen interactions. Trends Ecol Evol 21:278–286

Cory JS, Myers JH (2004) Adaptation in an insect host-plant pathogen interaction. Ecol Lett 7:632–639

Costa SD, Gaugler RR (1989) Sensitivity of *Beauveria bassiana* to solanine and tomatine: plant defensive chemicals inhibit an insect pathogen. J Chem Ecol 15:697–706

Csikos G, Molnar K, Borhegy NH, Talian CS, Sass M (1999) Insect cuticle, an in vivo model of protein trafficking. J Cell Sci 112:2113–2124

DeMoraes CM, Lewis WJ, Pare PW, Alborn HT, Tumlinson JH (1998) Herbivore-infected plant selectively attract parasitoids. Nature 393:570–573

Devi KU, Padmavathi K, Rao CUM, Khan AAP, Mohan MC (2008) A study of host specificity in the entomopathogenic fungus *Beauveria bassiana* (Hypocreales, Clavicipitaceae). BioControl Sci Techn 18:975–989

Driver F, Milner RJ, Trueman JWH (2000) A taxonomic revision of *Metarhizium* based on a phylogenetic analysis of rDNA sequence data. Mycol Res 104:134–150

Duetting PS, Ding HJ, Neufeld J, Eigenbrode SD (2003) Plant waxy bloom on peas affects infection of pea aphids by *Pandora neoaphidis*. J Invertebr Pathol 84:149–158

Elliot SL, Sabelis MW, Janssen A, van der Geest LPS, Beerling EAM, Fransen J (2000) Can plants use entomopathogens as bodyguards? Ecol Lett 3:228–235

Fatouros NE, van Loon JJA, Hordijk KA, Smid HM, Dicke M (2005) Herbivore-induced plant volatiles mediate in-flight host discrimination by parasitoids. J Chem Ecol 31:2033–2047

Ferrari J, Godfray HCJ (2003) Resistance to a fungal pathogen and host plant specialization in the pea aphid. Ecol Lett 6:111–118

Ferrari J, Müller CB, Kraaijeveld AR, Godfray HCJ (2001) Clonal variation and covariation in aphid resistance to parasitoids and a pathogen. Evolution 55:1805–1814

Gallardo F, Boethel DJ, Fuxa JR, Richter A (1990) Susceptibility of *Heliothis zea* (Boddie) larvae to *Nomuraea rileyi* (Farlow) Samson effects of α-tomatine at the third trophic level. J Chem Ecol 16:1751–1759

Gerson U, Gafni A, Paz Z, Sztejnberg A (2008) A tale of three acaropathogenic fungi in Israel: *Hirsutella, Meira* and *Acaromyces*. Exp Appl Acarol 46:183–194

Gillespie J, Kanost MR, Trenczek T (1997) Biological mediators of insect immunity. Annu Rev Entomol 42:611–643

Gillespie JP, Bailey AM, Cobb B, Vilcinskas A (2000a) Fungi as elicitors of insect immune responses. Arch Insect Biochem 44:49–68

Gillespie JP, Burnett C, Charnley AK (2000b) The immune response of the desert locust *Schistocerca gregaria* during mycosis of the entomopathogenic fungus, *Metarhizium anisopliae* var *acridum*. J Insect Physiol 46:429–437

Goettel MS (1995) The utility of bioassays in the risk assessment of entomopathogenic fungi. In: Biotechnology Risk assessment: USEPA/USDA/Environment Canada/Agriculture and Agri-Food Canada "Risk Assessment Methodologies." Proceedings of the Biotechnology Risk Assessment Symposium, June 6-8, Pensacola, Florida, pp 2-8

Goettel MS, Johnson DL, Inglis GD (1995) The role of fungi in the control of grasshoppers. Can J Bot 73:S71–S75

Gols R, Harvey JA (2009) Plant-mediated effects in the Brassicaceae on the performance and behaviour of parasitoids. Phytochem Rev 8:187–206

Gross J, Podsiadlowski L, Hilker M (2002) Antimicrobial activity of exocrine glandular secretion of *Chrysomela* larvae. J Chem Ecol 28:317–331

Gross J, Schumacher K, Schmidtberg H, Vilcinskas A (2008) Protected by fumigants: Beetle perfumes in antimicrobial defense. J Chem Ecol 34:179–188

Hajek AE (1997) Ecology of terrestrial fungal entomopathogens. In: Jones JG (ed) Advances in microbial ecology, vol 15. Plenum Press, New York, pp 193–249

Hajek AE (1999) Pathology and epizootiology of *Entomophaga maimaiga* infections in forest Lepidoptera. Microbiol Mol Biol R 63:814–835

Hesketh H, Roy HE, Eilenberg, J, Pell JK, Hails RS (2009) Challenges in modelling complexity of entomopathogenic fungi in semi-natural populations of insects. BioControl. doi 10.1007/s10526-009-9249-2 (this SI)

Hountondji FCC, Sabelis MW, Hanna R, Janssen A (2005) Herbivore-induced plant volatiles trigger sporulation in fungal entomopathogens: the case of *Neozygites tanajoae* infecting the cassava green mite. J Chem Ecol 31:1003–1021

Hountondji FCC, Hanna R, Sabelis MW (2006) Does methyl salicylate, a component of herbivore-induced plant odour, promote sporulation of the mite-pathogenic fungus *Neozygites tanajoae*? Exp Appl Acarol 39:63–74

Hu G, St. Leger RJ (2002) Field studies using a recombinant mycoinsecticide (*Metarhizium anisopliae*) reveal that it is rhizosphere competent. Appl Environ Microb 68:6383–6387

Inbar M, Gerling G (2008) Plant-mediated interactions between whiteflies, herbivores, and natural enemies. Annu Rev Entomol 53:431–448

Inglis GD, Goettel MS, Butt TM, Strasser H (2001) Use of hyphmycetous fungi for managing insect pests. In: Butt TM, Jackson J, Magan N (eds) Fungi as Biocontrol Agents. CAB International, UK, pp 23–69

Inglis GD, Duke GM, Goettel MS, Kabaluk JT (2008) Genetic diversity of *Metarhizium anisopliae* var. *anisopliae* in southwestern British Columbia. J Invertebr Pathol 98: 101–113

Inyang EN, Butt TM, Ibrahim L, Clark SJ, Pye BJ, Beckett A, Archer S (1998) The effect of plant growth and topography on the acquisition of conidia of the insect pathogen *Metarhizium anisopliae* by larvae of *Phaedon cochleariae*. Mycol Res 102:1365–1374

Inyang EN, Butt TM, Beckett A, Archer S (1999a) The effect of crucifer epicuticular waxes and leaf extracts on the germination and virulence of *Metarhizium anisopliae* conidia. Mycol Res 103:419–426

Inyang EN, Butt TM, Doughty KJ, Todd AD, Archer S (1999b) The effect of isothiocyanates on the growth of the entomopathogenic fungus *Metarhizium anisopliae* and its infection of the mustard beetle. Mycol Res 103:974–980

Isaka M, Kittakoop P, Kirtikara K, Hywel-Jones NL, Thebtaranonth Y (2005) Bioactive substances from insect pathogenic fungi. Acc Chem Res 38:813–823

Johnson MTJ (2008) Bottom-up effects of plant genotype on aphids, ants and predators. Ecology 89:145–154

Kabaluk JT, Ericsson JD (2007) *Metarhizium anisopliae* seed treatment increases yield of field corn when applied for wireworm control. Agron J 99:1377–1381

Kuhn J, Pettersson EM, Feld BK, Burse A, Termonia A, Pasteels JM, Boland W (2004) Selective transport systems mediate sequestration of plant glucosides in leaf beetles: a molecular basis for adaptation and evolution. Proc Natl Acad Sci 101:13808–13813

Lacey LA, Mercadier G (1998) The effect of selected allelochemicals on germination of conidia and blastospores and mycelial growth of the entomopathogenic fungus *Paecilomyces fumosoroseus* (Deuteromycotina: Hyphomycetes). Mycopathologia 142:17–25

Lee KP, Cory JS, Wilson K, Raubenheimer D, Simpson SJ (2006) Flexible diet choice offsets protein costs of pathogen resistance in a caterpillar. Proc R Soc London Ser B 173:823–829

Poprawski TJ, Jones WJ (2000) Host plant effects on activity of the mitosporic fungi *Beauveria bassiana* and *Paecilomyces fumosoroseus* against two populations of *Bemesia* whiteflies (Homoptera: Aleyrodidiae). Mycopathologia 151:11–20

Poprawski TJ, Greenberg SM, Ciomperlik MG (2000) Effect of host plant on *Beauveria bassiana*- and *Paecilomyces fumosoroseus*- induced mortality of *Trialeurodes vaporariorum* (Homoptera: Aleyrodidae]. Environ Ent 29:1048–1053

Posada F, Vega FF (2005) Establishment of the fungal entomopathogen *Beauveria bassiana* (Ascomycota: Hypocreales) as an endophyte in cocoa seedlings (*Theobroma cacao*). Mycologia 97:1195–1200

Price CD, Ratcliffe NA (1974) A reappraisal of insect hemocyte classification by the examination of blood from fifteen insect orders. Cell Tissue Res 147:537–549

Rees CJC (1986) Skeletal economy in certain herbivorous beetles as an adaptation to a poor dietary supply of nitrogen. Ecol Ent 11:221–228

Rehner SA, Buckley EP (2005) A *Beauveria* phylogeny inferred from nuclear ITS and EF1-alpha sequences: evidence for cryptic diversification and links to *Cordyceps* teleomorphs. Mycologia 97:84–98

Sandrock RW, VanEtten HD (1998) Fungal sensitivity to and enzymatic degradation of the phytoanticipin α-tomatine. Phytopathology 88:137–143

Santiago-Álvarez C, Maranhao EA, Maranhao E, Quesada-Moraga E (2006) Host plant influences pathogenicity of

Beauveria bassiana to *Bemisia tabaci* and its sporulation on cadavers. BioControl 51:519–532
Scarborough CL, Ferrari J, Godfray HCJ (2005) Aphid protected from pathogen by endosymbiont. Science 310:1781
Scholte EJ, Knols BGJ, Samson RA, Takken W (2004) Entomopathogenic fungi for mosquito control: a review. J Insect Sci 4:19
Selosse M-A, Roy M (2009) Green plants that feed on fungi: facts and questions about mixotrophy. Trends Plant Sci 14(2):64–70
St. Leger RJ, Joshi L, Roberts DW (1997) Adaptation of proteases and carbohydrates of saprophytic, phytopathogenic and fungal entomopathogens to the requirements of their ecological niches. Microbiology 143:1983–1992
Ugine TA, Wraight SP, Sanderson JP (2007) A tritrophic effect of host plant on susceptibility of western flower thrips to the entomopathogenic fungus *Beauveria bassiana*. J Invertebr Pathol 96:162–172
Vega FE, Dowd PF, McGuire MR, Jackson MA, Nelsen TC (1997) In vitro effects of secondary plant compounds on germination of blastospores of the entomopathogenic fungus *Paecilomyces fumosoroseus* (Deuteromycotina: Hyphomycetes). J Invertebr Pathol 70:209–213
Vega FE, Goettel MS, Blackwell M, Chandler D, Jackson MA, Keller S, Koike M, Maniana NK, Monzon A, Ownley B, Pell JK, Rangel D, Roy HE (2009) Fungal entomopathogens: new insights on their ecology. Fungal Ecology 2: 149–159
Via S (1999) Reproductive isolation between sympatric races of pea aphids. I. Gene flow and restricted habitat choice. Evolution 53:1446–1457
Via S, Bouck AC, Skillman S (2000) Reproductive isolation between divergent races of pea aphids on two hosts II. Selection against migrants and hybrids in the parental environments. Evolution 54:1626–1637
Vilcinskas A, Gotz P (1999) Parasitic fungi and their interactions with the insect immune system. Adv Parasit 43: 267–313
Wang C, St. Leger RJ (2006) A collagenous protective coat enables *Metarhizium anisopliae* to evade insect immune responses. Proc Natl Acad Sci USA 103:6647–6652
Wraight SP, Inglis GD, Goettel MS (2007) Fungi. In: Lacey LA, Kaya HK (eds) Field Manual of Techniques in Invertebrate Pathology, 2nd edn. Dordrecht, The Netherlands, pp 223–248

Author Biographies

Jenny Cory investigates the ecology and evolution of insect pathogens and their development as biocontrol agents. She is particularly interested in multitrophic interactions involving entomopathogens and the interplay between host resistance and pathogen virulence in host-parasite systems.

Jerry Ericsson is a PhD candidate in the Department of Biological Sciences, at Simon Fraser University. His research involves quantifying host-pathogen interactions between fungal and bacterial pathogens and their various insect hosts. His particular focus investigates the role of the insect immune reactions in conferring tolerance, and susceptibility to both general and specialized entomopathogens.

BioControl (2010) 55:89–102
DOI 10.1007/s10526-009-9238-5

Entomopathogenic fungi and insect behaviour: from unsuspecting hosts to targeted vectors

Jason Baverstock · Helen E. Roy · Judith K. Pell

Received: 20 July 2009 / Accepted: 5 October 2009 / Published online: 29 October 2009

Abstract The behavioural response of an insect to a fungal pathogen will have a direct effect on the efficacy of the fungus as a biological control agent. In this paper we describe two processes that have a significant effect on the interactions between insects and entomopathogenic fungi: (a) the ability of target insects to detect and avoid fungal pathogens and (b) the transmission of fungal pathogens between host insects. The behavioural interactions between insects and entomopathogenic fungi are described for a variety of fungal pathogens ranging from commercially available bio-pesticides to non-formulated naturally occurring pathogens. The artificial manipulation of insect behaviour using dissemination devices to contaminate insects with entomopathogenic fungi is then described. The implications of insect behaviour on the use of fungal pathogens as biological control agents are discussed.

Handling Editor: Eric Wajnberg.

J. Baverstock (✉) · J. K. Pell
Department of Plant and Invertebrate Ecology, Rothamsted Research, Harpenden, Hertfordshire AL5 2JQ, UK
e-mail: jason.baverstock@bbsrc.ac.uk

H. E. Roy
Biological Records Centre, NERC Centre for Ecology & Hydrology, Crowmarsh Gifford, Oxfordshire OX10 8BB, UK

Keywords Entomopathogenic fungi · Attraction · Avoidance · Transmission · Vectoring · Autodissemination

Introduction

A co-evolutionary arms race occurs between insects and their pathogens. Whereas selection on the pathogen is for greater exploitation of the host, selection on the host is for greater exclusion of the pathogen (Bush et al. 2001; Roy et al. 2006). The evolution of this behaviour and a description of some of the diverse interactions that occur between arthropods and fungi have recently been described in a review by Roy et al. (2006). Whilst these interactions are of great interest to evolutionary biologists, understanding the fundamental behavioural processes that occur between insects and pathogens is also essential for insect pathologists who wish to exploit fungal entomopathogens as biological control agents. Several species of entomopathogenic fungi are currently available as formulated bio-pesticides, including; Vertalec® (*Lecanicillium longisporum* ((Petch) Zare & Gams Zimmerman)) (Ascomycota: Hypocreales), BotaniGard® (*Beauveria bassiana* (Balsamo) Vuillemin) (Ascomycota: Hypocreales) and Green Muscle® (*Metarhizium anisopliae* var. *acridum* (Metsch.)) (Ascomycota: Hypocreales) (Milner 1997; Shah and Pell 2003). In addition, non-formulated species of entomopathogenic fungi such as

Pandora neoaphidis (Remaudière & Hennebert) Humber (Zygomycota: Entomophthorales) are also being assessed for inclusion as part of integrated pest management schemes. The overall success of using entomopathogenic fungi as biological control agents is affected by numerous abiotic and biotic factors, including the behavioural response of the target insects towards the entomopathogen. This paper describes pre- and post-contact responses of insects to entomopathogenic fungi that are under development as biological control agents. Transmission and vectoring of entomopathogenic fungi to uninfected hosts is then described along with the use of dissemination devices designed to attract and contaminate insects with fungi.

Response of insects to entomopathogenic fungi

Insects and entomopathogenic fungi are under opposing selection pressures. Insects gain a selective advantage from detecting and avoiding fungal pathogens while successful infection of an insect by an entomopathogen requires contact to be made between the host and the pathogen. The behaviour of insects can influence whether contact is made, with changes in activity increasing or decreasing the likelihood of infection (Cory and Hoover 2006). An insect may gain a selective advantage if it is able to detect the risk of attack from entomopathogenic fungi and respond via behavioural avoidance or through post-contact responses such as grooming (Chouvenc et al. 2008). This response may reduce the efficiency of the fungus as a biological control agent. In contrast, fungal pathogens could gain an advantage by attracting or remaining invisible to host insects.

Avoidance of entomopathogenic fungi

The ability of insects to detect and respond to entomopathogenic fungi within the order Hypocreales has been widely assessed, with reports of avoidance of fungi by species within the Coleoptera, Isoptera, Hemiptera and Orthoptera (Chouvenc et al. 2008; Meyling and Pell 2006; Myles 2002; Rath 2000; Staples and Milner 2000; Thompson and Brandenburg 2005; Villani et al. 1994). Termites are a global pest and their behavioural response to entomopathogenic fungi has received considerable attention from insect pathologists. Although various species of termites are susceptible to both *B. bassiana* and *M. anisopliae*, the use of these fungal pathogens as biological control agents is technically challenging due to the lifestyle and behaviour of termites (Staples and Milner 2000). Direct application of entomopathogenic fungi to control subterranean pests such as termites is difficult due to the logistics in applying conidia into colonies (Chouvenc et al. 2008). In addition to this, it has been demonstrated that termites show a variety of behavioural responses towards conspecifics infected with fungal pathogens to reduce transmission of the pathogen between uninfected and infected individuals (Chouvenc et al. 2008; Rath 2000). For infection to occur, direct contact between the termite and the pathogen is required. However, termites such as *Coptotermes lacteus* (Froggatt) displayed an avoidance response to *M. anisopliae* by only making short tunnels into substrates containing the pathogen, which they then seal off preventing further contact with the fungus (Staples and Milner 2000). Furthermore, this response appeared to be dependent on the virulence of the isolate, with a less pronounced response being observed when an isolate of low virulence was applied (Staples and Milner 2000). It may be possible to reduce the repellence of entomopathogenic fungi to termites through the use of baits. When incorporated into a cellulose bait (cellulose powder mixed with the conidial suspension), *M. anisopliae* was not repellent to the termites *Reticulitermes flavipes* (Kollar) and *Coptotermes formosanus* (Shiraki) at inoculum levels of 1.5×10^8 and 3×10^8 conidia respectively (Wang and Powell 2004). The development of a palatable formulation with an appropriate pathogen concentration may therefore be the key to increasing the efficiency of the control agent (Wang and Powell 2004). However, an increase in application efficiency may not be enough to control termites. Indeed, post-contact responses of termites to conspecifics contaminated with fungal pathogens may reduce the control potential. Myles (2002) found that uninfected *R. flavipes* displayed a combination of alarm behaviour, aggregation and defensive reactions towards individuals that were contaminated with *M. anisopliae*. This behaviour persisted for approximately 24 minutes after which it was replaced by grooming, biting, defecation and burial of the infected termite (Myles 2002). Grooming can be an efficient mechanism for removing conidia from the cuticle and preventing infection in termites; *Reticulitermes speratus* (Kolbe) were able to ingest 90% of *M. anisopliae* conidia deposited on their body surface within three hours (Shimizu and Yamaji 2003). Control of termites with entomopathogenic fungi therefore remains challenging. Chouvenc et al. (2008) found that *M. anisopliae* was unable to control *R. flavipes* even when 6.25% of the population was infected with the fungus. It was concluded that successfully controlling a field population where less than 1% of the termites were inoculated with entomopathogenic fungi would be difficult.

By exploiting the repellency of a pathogen, as opposed to its infectivity, it may still be possible to utilise entomopathogenic fungi to control termites and other economically important pests. Sun et al. (2008) found that organic mulches supplemented with *M. anisopliae* significantly repelled foraging *C. formosanus* and reduced mulch consumption by up to 71%. A second example is that of Japanese beetle larvae, *Popillia japonica* (Newman), which avoided soil containing *M. anisopliae* for up to 20 days after applications (Villani et al. 1994). The tawny mole cricket *Scapteriscus vicinus* (Scudder) and the Southern mole cricket *Scapteriscus borellii* (Giglio-Tos) both avoid making contact with *B. bassiana* (Thompson and Brandenburg 2005). Surface tunnelling, vertical tunnels descending into the soil and tunnelling along the perimeter were reduced in containers treated with *B. bassiana* strain DB-2 compared to untreated containers (Thompson and Brandenburg 2005). However, this was not observed when the soil was treated with *B. bassiana* strain 10–22, suggesting that strain selection should be taken into account when assessing the repellency of entomopathogenic fungi towards insects (Thompson and Brandenburg 2005). *Scapteriscus vicinus* has also been shown to avoid making contact with *M. anisopliae* (Villani et al. 2002).

Non-avoidance of entomopathogenic fungi

Although there is evidence of insects being attracted to fungi, this is predominantly to non-entomopathogenic fungi. Female wood living beetles *Malthodes fuscus* (Waltl), *Anaspis marginicollis* (Lindberg) and *Anaspis rufilabris* (Gyllenhall) and the moth *Epinotia tedella* (Clerck) were all attracted to volatiles released from the bracket fungus *Fomitopsis pinicola* (Polyporales: Fomitopsidaceae) (Fäldt et al. 1999) whilst the Deathwatch beetle, *Xestobium rufovillosum* (De Geer), was attracted to volatiles released from wood decaying fungi (Belmain et al. 2002). Hymenopteran parasitoids have also been reported responding to volatiles released from fungi. The Pteromalid *Roptrocerus xylophagorum* (Ratzeburg) and the Braconid *Spathius pallidus* (Ashmead) were attracted to the odour of bark of loblolly pine colonized by blue stain fungus, a fungal associate of the parasitoid's Coleopteran hosts (Sullivan and Berisford 2004). However, the majority of studies to date indicate that insects are either not able to detect entomopathogenic fungi, or can detect the fungus but do not perceive it as being a threat and exhibit no avoidance behaviour. The Colorado potato beetle, *Leptinotarsa decemlineata* (Say) is a serious pest of potatoes and has developed resistance to many insecticides. Although *L. decemlineata* is susceptible to *B. bassiana* and can be contaminated with the fungus when pre-pupae and adults move across soil and make contact with conidia deposited on either the substrate or on infected beetle cadavers (Long et al. 2000), *B. bassiana* cadavers had no effect on orientation by the beetle nor was there avoidance of areas containing *B. bassiana*-sporulating cadavers (Klinger et al. 2006). Mortality of the Western flower thrips, *Frankliniella occidentalis* (Pergrande), on chrysanthemums was similar when *B. bassiana* was applied on its own or combined with an attractant, indicating that the fungus did not affect plant colonisation by thrips (Ludwig and Oetting 2002). A final example is that of the aphid-specific fungal pathogen *P. neoaphidis* which had no effect on the colonisation of bean plants by the pea aphid, *Acyrthosiphon pisum* (Harris), in cage experiments (Baverstock et al. 2005a). This resulted in transmission of conidia during plant colonisation and, to a lesser extent, during in situ feeding. However, the ability of an insect to detect entomopathogenic fungi may not only be dependent on the species and isolate of the fungus, but also on the substrate on which the fungus is deposited. Meyling and Pell (2006) found that whilst the generalist aphid predator *Anthocoris nemorum* (L.) avoided leaf surfaces contaminated with *B. bassiana* and rapidly withdrew from contact with *B. bassiana*-sporulating cadavers, its behaviour on soil was not altered by the presence of the fungus.

In contrast, Ormond (2007) found that the seven-spot ladybird *Coccinella septempunctata* (L.) detected and avoided *B. bassiana* on leaves and in soil.

Whilst an inability to avoid entomopathogenic fungi is beneficial for control of a pest, it may be detrimental towards other natural enemies of the pest. For entomopathogenic fungi to be effective control agents, repellence by the fungus and/or a minimal loss of the other guild members to infection is required (Lord 2001). In some cases this threat may come indirectly via the target pest. For example, the hosts of hymenopteran parasitoids face a greater risk of infection by entomopathogenic fungi than the parasitoid itself. Parasitoids would therefore gain a selective advantage through detecting and avoiding hosts that are contaminated with fungus (Baverstock et al. 2005b; Powell et al. 1986). The ability of parasitoids to detect and avoid hosts infected with entomopathogenic fungi from both the Hypocreales and Entomophthorales has been assessed (Baverstock et al. 2005b; Fransen and van Lenteren 1993; Lord 2001). Larval saw-toothed grain beetles *Oryzaephilus surinamensis* (L.) are attacked by the ectoparasitoid *Cephalonomia tarsalis* (Ashmead) and are also susceptible to infection by *B. bassiana* (Lord 2001). However, *C. tarsalis* larvae are also susceptible to *B. bassiana* and died within one day of oviposition on host larvae infected with the fungal pathogen (Lord 2001). Despite this, *C. tarsalis* was unable to detect the presence of *B. bassiana* and entered grain samples containing the fungus where it subsequently oviposited on *B. bassiana*-infected larvae (Lord 2001). A second example is the interaction between the aphid parasitoid *Aphidius ervi* (Haliday) and *P. neoaphidis*. Baverstock et al. (2005b) found that *A. ervi* would enter aphid colonies containing *P. neoaphidis* and forage on plants contaminated with the fungus. On making contact with fungus-infected aphids the parasitoid would attempt to oviposit. Indeed, it was only once the aphid had succumbed to infection and was sporulating that the parasitoid appeared to detect the presence of the fungus and did not attempt to oviposit (Baverstock et al. 2005b). The apparent inability of *A. ervi* to detect and respond to *P. neoaphidis*-infected hosts may be due to spatial and temporal separation reducing the encounter rate between the two natural enemies and, therefore, the selection pressure for avoidance behaviour to evolve (Baverstock et al. 2005b).

Contrary to this there are examples of parasitoids being able to detect hosts infected with entomopathogenic fungi. *Encarsia formosa* (Gahan) is used as a control agent against the greenhouse whitefly *Trialeurodes vaporariorum* (Westwood) to protect several glasshouse crops including vegetables and ornamentals. Fransen and van Lenteren (1993) assessed the interaction between *E. formosa* and the entomopathogenic fungus *Aschersonia aleyrodis* (Webber) (Ascomycota: Hypocreales). Although the parasitoid adopted an oviposition posture on hosts showing signs of infection, these were rejected after probing, indicating that the parasitoid could detect the presence of the fungus. Further studies revealed that if fungal infection occurred within three days of parasitisation, there was a significant reduction in the number of parasitised hosts. However, if parasitisation occurred 4, 7 or 10 days after parasitisation, there was no effect on the number of parasitised hosts. This suggested that *E. formosa* and *A. aleyrodis* could be used together to control *T. vaporariorum*.

Unlike parasitoids which suffer a direct fitness cost from ovipositing in hosts infected with entomopathogenic fungi, competition with fungal pathogens for prey items is not always detrimental to predators. The seven-spot ladybird *C. septempunctata* and *P. neoaphidis* are both commonly occurring natural enemies of aphids in temperate regions. Although both of these species compete for aphids, the coccinellid does not avoid foraging on plants where the fungus is present (Baverstock 2004). This is unsurprising given that *C. septempunctata* is not susceptible to infection by the fungus. Indeed, *C. septempunctata* is an asymmetric intraguild predator of *P. neoaphidis*, predating living aphids infected with the fungus as well as dead sporulating cadavers (Pell et al. 1997; Roy and Pell 2000; Roy et al. 1998, 2001). However, some species of entomopathogenic fungi can have a direct negative effect on a predator if infected hosts are less suitable as prey, and it is surprising that avoidance of sub-optimal prey items has not been observed. Simelane et al. (2008) found that adult and larval *C. septempunctata* readily predated aphids infected with *Neozygites fresenii* (Nowakowski) (Entomophthorales: Neozygitaceae), this is despite the fungus having significant negative effects on the development of the coccinellid, even without direct infection. When consuming aphids infected with *N. fresenii*, the development time of the coccinellid was significantly longer, mortality between second and fourth instars was significantly greater, body size was significantly smaller and egg production significantly lower compared to conspecifics fed a diet of uninfected aphids (Simelane et al. 2008). Similar results were found by Roy et al. (2008) who showed that the fecundity of the harlequin ladybird, *Harmonia axyridis* (Pallas), was reduced dramatically when it was infected with *B. bassiana* at doses of 10^5–10^9 conidia ml^{-1}, although only a low number of harlequin ladybirds succumbed to the fungal disease. Whilst the reproductive success of the two-spot ladybird, *Adalia bipunctata* (L.), was also reduced when inoculated with a dose of 10^9 conidia ml^{-1}, mortality was also high (Roy et al. 2008).

Although there is a large body of research which indicates that insects are either not attracted to entomopathogenic fungi or are not able to detect their presence, there are exceptions. Dromph and Vestergaard (2002) assessed the susceptibility of three species of collembolans (*Folsomia fimet aria* (L.), *Hypogastrura assimilis* (Krausbauer) and *Proisotoma minuta* (Tullberg)) to three species of entomopathogenic fungi, *B. bassiana*, *Beauveria brongniartii* (Saccardo) Petch (Ascomycota: Hypocreales) and *M. anisopliae*. *Folsomia fimet aria* was shown to be susceptible to both *B. brongniartii* and *M. anisopliae* when fed them on inoculated sphagnum, however, it was attracted to these fungal pathogens in a choice arena. In a pair-wise comparison test, the order of attractiveness was found to be similar for all three species of collembolan, with *B. brongniartii* being the most attractive pathogen and *B. bassiana* the least attractive. Further to this, a positive relationship was found between the concentration of all three fungi and the attractiveness to collembolans.

Transmission of entomopathogenic fungi between insects

Transmission is a key parameter that determines the rate of spread of entomopathogenic fungi within host populations and, therefore, the pathogens potential for use as a microbial control agent (Steinkraus 2006). Transmission is the dispersal of infective propagules from an infected host to a new host and is the most "perilous" part of the lifecycle of a fungal pathogen (Hajek and St. Leger 1994). Transmission can occur horizontally (within a generation) and vertically (between generations) within a species, between species and from a local scale on a single plant to a landscape scale. Movement of entomopathogenic fungi by host and non-host invertebrates to susceptible hosts is one of the most important mechanisms for transmitting to new habitats (Fuxa and Tanada 1987; Roy et al. 2001).

Within species transmission

Horizontal transmission between individuals of the same species (autodissemination) can occur through direct contact between contaminated and uncontaminated individuals or indirectly via conidia that have been deposited on the substrate (Quesada-Moraga et al. 2008; Roy and Pell 2000; Vega et al. 2000). Whilst it is relatively simple to quantify fungal transmission, the determination of the underlying mechanisms is more challenging. For example, although it was observed in the 1980's that collembolans are able to transmit *B. bassiana*, *M. anisopliae* and *Paecilomyces farinosus* (Holm ex Gray) A. H. S. Brown & G. Sm. (Ascomycota: Eurotiomycetes), it was not until 2001 that the mechanisms that facilitate this dispersal were described (Dromph 2001; Visser et al. 1987; Zimmermann and Bode 1983). Dromph (2001) found that *F. fimet aria*, *H. assimilis* and *P. minuta* were able to transmit viable conidia of *B. bassiana*, *B. brongniartii* and *M. anisopliae* on either their cuticle or within their gut. Transmission of the three species of entomopathogenic fungi by each of the species of collembolan was similar. However, whilst ingestion had no affect on the viability of *B. bassiana* or *B. brongniartii*, the viability of *M. anisopliae* was reduced from 98.8% in the undigested control to 24.3% and 54% following ingestion by *F. fimet aria* and *P. minuta* respectively.

Direct transmission between contaminated and uninfected individuals is less variable and more efficient than indirect transmission via conidia that have been deposited on the substrate, and can lead to high mortality rates even when the number of contaminated individuals is low (DeKesel 1995). Kreutz et al. (2004) found that a lethal dose of *B. bassiana* could be transmitted from a single contact between treated male and untreated female adult spruce bark beetles, *Ips typographus* (L.), resulting in a 75% mortality rate when there was a 1:20 ratio of treated and untreated beetles. This mortality rate increased to 90% when the ratio was 1:1 (Kreutz et al. 2004). Direct transmission of *B. bassiana* and *M. anisopliae* between termite workers (*C. formosanus*) and other colony members occurred readily whilst conidia of *Laboulbenia slackensis* (Cepede and Picard) (Ascomycota: Laboulbeniales), which form adherent thread-like structures, enabled direct transmission between contaminated and uninfected salt marsh beetles, *Pogonus chalceus* (Marsham) (Carabidae) (DeKesel 1995; Jones et al. 1996). Direct transmission from males to females during copulation is widespread and has been reported for both ascomycetous and entomophthoralean fungi (Furlong and Pell 2001; Quesada-Moraga et al. 2008; Kaaya and Okech 1990; Toledo et al. 2007). Potentially, direct transmission between males and females could be exploited for biological control through releasing males inoculated with an entomopathogenic fungus into wild populations (Toledo et al. 2007). Male tsetse flies (Glossinidae) were able to transmit *B. bassiana* and *M. anisopliae* to females, successfully infecting 65% and 55% of females, respectively (Kaaya and Okech 1990). Transmission of *B. bassiana* to the Mexican fruit fly, *Anastrepha ludens* (Loew), occurred during mating and, to a lesser extent, through contact during courtship (Toledo et al. 2007). The efficiency of transmission during copulation varies depending on whether it is the male or the female that is contaminated. Male-to-female transmission of *M. anisopliae* within populations of the Mediterranean fruit fly, *Ceratitis capitata* (Wiedemann), was greatest when males were inoculated with the fungus (Quesada-Moraga et al. 2008). However, the efficiency of horizontal transmission could be reduced if females preferentially selected uninfected males over those contaminated with entomopathogenic fungi.

The efficiency of direct transmission between males and females can also be dependent on the species and/or dose of entomopathogenic fungi. The transmission of *B. bassiana* from inoculated male diamond-back moths, *Plutella xylostella* (L.), to male and females was greater than the transmission of *Zoophthora radicans* (Brefeld) Batko (Zygomycetes: Entomophthorales) (Furlong and Pell 2001). Similarly, transmission of *B. bassiana* from inoculated males to foraging larvae was greater than transmission of *Z. radicans* (Furlong and Pell 2001). It was hypothesised that the difference in transmission between the two species of fungi may have been due to a greater concentration of *B. bassiana* being applied to the male moths, the infective secondary conidia were then dislodged and contaminated uninfected *P. xylostella*. However, subsequent secondary transmission amongst larvae was less for *B. bassiana* than for *Z. radicans* (Furlong and Pell 2001).

Effect of insect movement on transmission of entomopathogenic fungi

A positive relationship between insect movement and transmission of entomopathogenic fungi has been observed in a number of systems. Aphids release alarm pheromone (E-β-farnesene) when threatened with attack or during predation. This induces an escape response in surrounding conspecifics in which they unplug their stylets and move to another part of the plant. Roy et al. (1999) demonstrated the effects of *P. neoaphidis* infection on the alarm response of infected pea aphids, *A. pisum*. Infected aphids produced alarm pheromone but ceased responding to it. Therefore, infected aphids would continue to elicit a response in neighbouring aphids, which could enhance transmission. Indeed, Roditakis et al. (2000) found that the number of peach potato aphids, *Myzus persicae* (Sulzer), which became infected with a *Lecanicillium* spp. (=*Verticillium lecanii* (Zimmermann) Viégas) that had been deposited on the leaf surface was increased when alarm pheromone was released. However, the authors did not believe that the addition of alarm pheromone would be a practical pest control option and explored different methods to increase aphid movement. An alternative was to use a sub-lethal dose of the chloronicotinyl insecticide, imidacloprid. This insecticide inhibited aphid settling and increased searching behaviour and, therefore, the probability of the insect making contact with conidia on the leaf surface was enhanced (Roditakis et al. 2000). The use of imidacloprid to alter behaviour and increase fungal infection has been assessed in other insect orders. Imidacloprid reduced larval mobility of the root weevil, *Diaprepes abbreviatus* (L.), and has been associated with a decrease in conidial avoidance and increased infection with *B. bassiana* and *M. anisopliae* whereas in the termite, *R. flavipes*, imidacloprid was found to affect hygiene function (e.g. grooming) resulting in increased infection with

B. bassiana (Boucias et al. 1996; Roditakis et al. 2000; Quintela and McCoy 1998). In addition to increasing contact between insects and entomopathogenic fungi, sub-lethal doses of insecticides may provide immediate protection through affecting the behaviour of the pest, which ultimately succumbs to infection by the slower acting fungus. For example, Shah et al. (2007) found that sub-lethal doses of imidacloprid or a second insecticide, fipronil, prevented feeding by black vine weevils, *Otiorhynchus sulcatus* (F.), which were subsequently killed by *M. anisopliae*.

Insect behaviour, such as foraging or predator avoidance, may also affect transmission of entomopathogenic fungi. Transmission of *P. neoaphidis* to *A. pisum* is approximately double during plant colonisation and subsequent feeding than through in situ feeding alone (Baverstock et al. 2005a). In addition to this, transmission of *P. neoaphidis* to *A. pisum* colonising bean plants is also enhanced in the presence of foraging predators and parasitoids. Roy et al. (1998) found that *C. septempunctata* increased transmission of *P. neoaphidis* despite the coccinellid predating sporulating fungal cadavers. Foraging by the hymenopteran parasitoid *A. ervi* has also been shown to increase transmission of *P. neoaphidis* to *A. pisum*, however, the increased reproductive success of the fungus was correlated with a decrease in the reproductive success of the parasitoid (Baverstock et al. 2009). The enhanced transmission of entomopathogenic fungi in the presence of foraging arthropods is not limited to interactions that occur within crops and has been observed in populations of non-pest aphids found on plants in field margins. The nettle aphid, *Microlophium carnosum* (Buckton), and the aphid predator *A. nemorum* (L.) were able to distribute *B. bassiana* from the soil to nettle leaves (Meyling et al. 2006) whilst foraging *C. septempunctata* increased the transmission of *P. neoaphidis* in populations of *Uroleucon jacea* (L.) infesting knapweed and *M. carnosum* infesting nettles (Ekesi et al. 2005). Insects that co-occur with aphids and entomopathogenic fungi but are not within the same guild have also been reported as enhancing fungal transmission. Transmission of *P. neoaphidis* to *M. carnosum* was enhanced to a similar level in the presence of the parasitoid *Aphidius microlophii* (Pennacchio & Tremblay), which is an enemy of the aphid, and the non-enemy parasitoid *Aphidius colemani* (Viereck), which feeds on the honeydew produced by the aphid (Baverstock et al. 2008). In addition to this, foraging by caterpillars of the peacock butterfly, *Inachis io* (L.), also enhanced transmission of *P. neoaphidis* to *M. carnosum* (Baverstock et al. 2008). However, it was speculated that enhanced transmission in the presence of an herbivore is dependent on the degree of herbivory, with low levels of herbivory increasing transmission through the disturbance of aphids whereas high levels of herbivory would reduce transmission due to the displacement of aphids.

Transmission of entomopathogenic fungi to hosts is also affected by abiotic conditions and the substrate on which they are deposited. Growth and topography of the host plant influenced the susceptibility of the mustard beetle, *Phaedon cochleariae* (F.), to *M. anisopliae* that has been sprayed on the plant (Inyang et al. 1998). At higher temperatures, leaf expansion diluted the inoculum density of the pathogen resulting in decreased mustard beetle mortality. In addition to this, host plant species affected transmission, with the number of larvae that acquired conidia on oilseed rape being greater than those on Chinese cabbage or turnip. Shanley and Hajek (2008) found that rainfall increased the transmission of *M. anisopliae* through aiding dispersal from fungal bands onto bark where it could infect the Asian longhorn beetle, *Anoplophora glabripennis* (Motschulsky) whilst, in contrast, Pell et al. (1998) showed that heavy rainfall was capable of knocking *P. neoaphidis*-sporulating cadavers from leaves onto soil where they were subsequently destroyed.

Vectoring of entomopathogenic fungi

Vectoring of fungal conidia occurs when the fungus is transported by a third party that is either not susceptible to the fungus or is not the target prey species. Vectoring of conidia from either the substrate or from an infected host has been reported for a variety of insect-entomopathogenic fungi associations. Collembolans interact with entomopathogenic fungi in soil and, although they have been reported as consuming pathogens, they also enhanced the dispersal of the fungi by transporting conidia that had become attached to their cuticles or in their guts (Broza et al. 2001; Dromph 2001). Three species of collembolans (*F. fimet aria*, *H. assimilis* and *P. minuta*) were all able to vector a sufficient quantity of *B. bassiana*, *B. brongniartii* or *M. anisopliae* from soil to cause mortality in the mealworm, *Tenebrio molitor* (L.) (Dromph 2003). The ability to vector fungi was primarily dependent on body size, with larger insects being able to vector more conidia (Dromph 2003). Vectoring of entomopathogenic fungi could also be exploited to control pest insects. For example, when artificially contaminated with *L. longisporum*, the common black ant, *Lasius niger* (L.), was able to vector conidia of the fungus to colonies of the rosy apple aphid *Dysaphis plantaginea* (Passerini), resulting in mortality of 68.3%, 30.8% and 3.7% of aphids when assessed under laboratory, semi-field and field conditions, respectively (Bird et al. 2004). However, *L. niger* workers were also observed removing *L. longisporum*-infected aphid cadavers, a process that would remove an inoculum source which may otherwise have infected aphids within the colony. *Lasius niger* also vectors *L. longisporum* to the black-bean aphid, *Aphis fabae* (Scopoli) (Flower 2002). The coccinellid *Hippodamia convergens* (Guerin) was able to vector conidia of *Isaria* (*Paecilomyces*) *fumosoroseus* (Wize) Brown & Smith (Ascomycota: Eurotiomycetes) to uninfected Russian wheat aphids, *Diuraphis noxia* (Kurdjumov), if it became contaminated when sprayed directly with the fungus, through predating aphids that had been sprayed with the fungus or through foraging on plants that contained sporulating *D. noxia* cadavers (Pell and Vandenberg 2002). A second coccinellid, *C. septempunctata*, was also reported as being able to vector entomopathogenic fungi. Both adult and larvae that were artificially contaminated with *P. neoaphidis* vectored the fungus directly to colonies of uninfected pea aphids, *A. pisum*, and indirectly through the deposition of infective conidia on the leaf surface (Roy et al. 2001). Further studies have revealed that *C. septempunctata* vectors *P. neoaphidis* from non-crop plants that are commonly found in field margins such as nettle, knapweed or bird's-foot trefoil to *A. pisum* feeding on bean plants, resulting in an aphid mortality rate of up to 13% (Ekesi et al. 2005). However, vectoring efficiency is affected by prey species and although *C. septempunctata* was able to vector *P. neoaphidis* to populations of *A. pisum*, it was unable to vector the fungus to the cereal aphid *Sitobion avenae* (F.) (Roy et al. 2001).

The examples described illustrate that transmission and vectoring of entomopathogenic fungi occurs at a local scale. However, long distance transmission of fungal pathogens within infected alate insects also occurs. Aphids are able to disperse up to 1,600 km through a combination of active hovering and passive flight on wind currents (Robert 1987). Various species of entomopathogenic fungi have been identified in migratory alate aphids trapped from the air (Chen and Feng 2004a; Feng et al. 2007; Huang, et al. 2008). Zhang et al. (2007) found that the dispersal ability of alate *M. persicae* that were inoculated with *Conidiobolus obscurus* (Hall & Dunn) (Zygomycota: Entomophthorales) was not different to uninfected aphids and, following dispersal, infected aphids were able to reproduce and transmit the pathogen to their progeny. Several other species of aphid have been recorded as transmitting entomopathogenic fungi when migrating as alates, including *S. avenae*, *Rhopalosiphum padi* (L.) and *Schizaphis graminum* (Rondani) (Feng et al. 2004). In a field study by Chen and Feng (2002), 760 alate *M. persicae* were trapped and observed for fungal infection. Of these, 87.6% died due to mycosis, 94.4% of which succumbed to infection by Entomophthorales with the remaining being infected with the Hypocrealean *B. bassiana*. Of those infected with Entomophthorales over two-thirds were infected with *P. neoaphidis*. This study was repeated at a larger scale, trapping 7,139 migratory alates from nine species of aphids, from which eight species of fungal pathogens were identified (Feng et al. 2004). Using a computer-monitored flight mill system, *S. avenae* that had been inoculated with *P. neoaphidis* were able to fly for several hours before initiating colonies, reproducing and transmitting the fungus to their progeny (Chen and Feng 2004b; Feng et al. 2004). Further studies have revealed that whilst the number of aphids trapped does not vary consistently with temperature or humidity, there is a positive relationship between humidity and mortality due to fungal infection, and this is most apparent with insects infected with Entomophthorales (Chen et al. 2008). It is not just entomopathogenic fungi that are dispersed within aphids, alates that are parasitised by either *Aphidius gifuensis* (Ashmead) or *Diaeretiella rapae* (McIntosh) have also been recorded (Feng et al. 2007). Whilst co-infection between two species of entomopathogenic fungi within migratory alates is rare, low numbers of alates have been recorded as being co-infected with *P. neoaphidis* and either *Zoophthora anhuiensis* (Li) Humber (Zygomycetes:

Entomophthorales), *Conidiobolus* spp. or *N. fresenii* (Chen and Feng 2004a, 2006).

Assisted autodissemination

Assisted-autodissemination utilises artificial devices to inoculate insects with entomopathogenic fungi. The insects that are inoculated can either be the pest species or non-target insects that act as vectors of the pathogen. Assisted auto-dissemination works by attracting an insect into an inoculation device where it becomes contaminated with the infective conidia before returning to the crop and disseminating the pathogen to pest insects (Vega et al. 2000; Vickers et al. 2004). Assisted-autodissemination has been widely assessed for a number of insect and fungal species and has several advantages over the mass application of entomopathogenic fungi, the benefits include: (1) it is highly efficient, especially when incorporating a target specific pheromone, (2) it can be species specific, (3) dissemination devices are simple to construct and maintain, (4) it is cost effective as the ratio of fungal inoculum to hosts is low and, (5) contaminated insects will return to their habitats, therefore dispersing the pathogen (Vega et al. 2000; Yasuda 1999). Although these inoculation devices could contain insecticides which would kill the individual that enters, entomopathogenic fungi is self-replicating and, once vectored, can be transmitted throughout entire colonies of the pest species (Grace and Zoberi 1992). The following examples illustrate the developmental procedure and principles of assisted-autodissemination.

Dusky sap beetles, *Carpophilus lugubris* (Murray), contaminated with *B. bassiana* conidia from an auto-inoculator were able to cause high levels of mortality in populations of unexposed beetles in laboratory bioassays (Vega et al. 1995). Field experiments using traps containing coloured tracer dye then showed that *C. lugubris* was able to vector the coloured dye to apple orchards and fields of corn (Vega et al. 1995). Finally, field trials using auto-inoculation devices containing *B. bassiana* showed that *C. lugubris* that are moving into overwintering sites such as tree holes could be contaminated with the fungus which may then spread throughout the overwintering population (Dowd and Vega 2003). Although this example utilised a complex inoculation device, this is not always necessary. Shimazu (2004) found that young adult pine sawyers, *Monochamus alternatus* (Hope), could be inoculated with a lethal dose of *B. bassiana* through walking on a non-woven fabric strip that had been contaminated with the fungus at a concentration of 3.5×10^8 conidia cm^{-2}.

The efficiency of attracting insects to inoculation devices can be increased by utilising pheromones. The brown winged green bug, *Plautia crossota stali* (Scott), is a pest in fruit orchards in Japan but is susceptible to *B. bassiana* (Tsutsumi et al. 2003). Through incorporating aggregation pheromone into woven sheets that were contaminated with *B. bassiana* and attached to trees, both males and females became infected with the fungus and average mortality rates of 70–75% were achieved (Tsutsumi et al. 2003). Alternatively, pheromones can be incorporated into food baits to increase their efficiency at attracting pest insects. Renn et al. (1999) combined sex pheromone with bait containing sugar solution to attract house flies, *Musca domestica* (L.) into inoculation devices where they could become inoculated with *M. anisopliae*. These inoculation devices were efficient, with fly mortality rates of between 95.2% and 100%. In addition to this, when flies entered in pairs, a greater quantity of conidia was acquired by each individual compared to insects entering alone, indicating that mutual interference occurred between the insects (Renn et al. 1999).

Care needs to be taken when using simple devices where the fungal pathogen and pheromone are incorporated onto the same substrate to ensure that the pheromone does not inhibit the infectivity of the pathogen. Smith et al. (1999) found that *B. bassiana* and an aggregation pheromone could be incorporated into fat pellets and used to contaminate the larger grain borer, *Prostephanus truncatus* (Horn). However, the pheromone caused either a slight decrease in the viability of the *B. bassiana* conidia or a delay in the germination of the pathogen. Using complex traps allows pheromones and fungal entomopathogens to be stored in separate containers and, therefore, the chemicals do not interfere with the pathogenicity of the entomopathogenic fungi. For example, aggregation pheromones have been used to lure adult spruce bark beetles, *Ips typographus* (L.), into inoculation devices where they were contaminated with *B. bassiana* resulting in a significant reduction in the number of bore holes and maternal galleries

(Kreutz et al. 2004). Sex pheromones are widely used as lures in autoinoculation devices and are usually designed to contaminate males with the fungus, which they then pass on to females during mating. Male sweet potato weevils, *Cylas formicarius* (F.) are attracted to devices containing synthetic sex pheromone where they become contaminated with *B. bassiana* (Yasuda 1999). Of those insects assessed, 57.9% of males were contaminated with the fungus whilst 31.6% of females were also found to be contaminated. It was suggested that the females became contaminated through mating with contaminated males. The potential of using either artificial sex pheromones or those released from females to lure adult male diamond back moths, *P. xylostella*, into devices where they were contaminated with *Z. radicans* has been assessed (Furlong et al. 1995; Pell et al. 1993). Whereas males only entered devices containing virgin female moths between dusk and dawn (when sex pheromone is naturally released), males entered devices containing synthetic pheromone throughout the day. Males spent a mean of 88 s within devices before leaving, in which time they were contaminated with a lethal dose of *Z. radicans* which they could transmit to conspecifics. Proof for this concept of transmission was obtained in a subsequent study where adult *P. xylostella* were inoculated with *Z. radicans* and released into field cages containing plants infested with *P. xylostella* larvae (Vickers et al. 2004). After six days 79% of the larvae were found to be infected with *Z. radicans*.

Autoinoculation devices are not restricted to inoculating the target prey and can be used to attract and contaminate non-host vectors of entomopathogenic fungi. For example, bees are able to vector pathogens to control both plant and insect pests (Kapongo et al. 2008a; Al-mazra'awi et al. 2006; Carreck et al. 2007). Bumble bees, *Bombus impatiens* (Cresson), were able to vector *B. bassiana* to sweet pepper plants in glasshouses where they induced infection in populations of the tarnished plant bug, *Lygus lineolaris* (Palisot de Beauvois) and the western flower thrip *F. occidentalis* (Al-mazra'awi et al. 2006). Carreck et al. (2007) demonstrated that honeybees, *Apis mellifera* (L.), could be used to vector *M. anisopliae* into field populations of the pollen beetle, *Meligethes aeneus* (F.). However, care needs to be taken when utilising beneficial insects as vectors of entomopathogenic fungi to prevent the vector from becoming infected with the pathogen. Kapongo et al. (2008b) found that significantly more bumble bees died when using a high concentration (2×10^{11} g^{-1}) of *B. bassiana* compared to medium and low doses of the fungus (6.24×10^{10} and 9×10^{9} g^{-1} conidia respectively). Dose, therefore, needs to be optimised to maximise infection whilst minimising the mortality of the vector.

Summary

The interactions between fungal entomopathogens and their hosts are being unravelled through eloquent research. The importance of subtle behavioural interactions in determining the success or failure of entomopathogenic fungi as biological control agents cannot be underplayed. Insect pathologists can no longer assess simple bi-trophic interactions between pathogens and their prey within the laboratory to determine the impact of entomopathogenic fungi as biological control agents. Behavioural responses of target and non-target arthropods to entomopathogenic fungi needs to be assessed at the population scale and under natural biotic and abiotic conditions to fully determine the impact of entomopathogens on both the target prey and the communities in which they occur. To increase the efficacy of biocontrol programmes incorporating entomopathogenic fungi, future research focusing on multitrophic interactions (Cory and Ericsson 2009), including above and below ground signalling, is required along with the development of technologies to enhance the efficacy of pathogen transmission through the manipulation of host behaviour.

Acknowledgments Jason Baverstock and Judtih K Pell are supported by the Department for Environment, Food and Rural Affairs of the United Kingdom (Defra). Rothamsted Research is an Institute of the Biotechnology and Biological Sciences Research Council of the United Kingdom. The Centre for Ecology & Hydrology is an institute of the Natural Environment Research Council of the United Kingdom.

References

Al-mazra'awi MS, Shipp L, Broadbent B, Kevan P (2006) Biological control of *Lygus lineolaris* (Hemiptera: Miridae) and *Frankliniella occidentalis* (Thysanoptera: Thripidae) by *Bombus impatiens* (Hymenoptera: Apidae)

vectored *Beauveria bassiana* in greenhouse sweet pepper. Biol Control 37:89–97
Baverstock J. (2004). Interactions between aphids, their insect and fungal natural enemies and the host plant. Ph.D. thesis, University of Nottingham, Loughborough, UK
Baverstock J, Alderson PG, Pell JK (2005a) *Pandora neoaphidis* transmission and aphid foraging behaviour. J Invertebr Pathol 90:73–76
Baverstock J, Alderson PG, Pell JK (2005b) Influence of the aphid pathogen *Pandora neoaphidis* on the foraging behaviour of the aphid parasitoid *Aphidius ervi*. Ecol Entomol 30:665–672
Baverstock J, Baverstock KE, Clark SJ, Pell JK (2008) Transmission of *Pandora neoaphidis* in the presence of co-occurring arthropods. J Invertebr Pathol 98:356–359
Baverstock J, Clark SJ, Alderson PG, Pell JK (2009) Intraguild interactions between the entomopathogenic fungus *Pandora neoaphidis* and an aphid predator and parasitoid at the population scale. J Invertebr Pathol 102:167–172
Belmain SR, Simmonds MSJ, Blaney WM (2002) Influence of odor from wood-decaying fungi on host selection behavior of deathwatch beetle, *Xestobium rufovillosum*. J Chem Ecol 28:741–754
Bird AE, Hesketh H, Cross JV, Copland M (2004) The common black ant, *Lasius niger* (Hymenoptera: Formicidae), as a vector of the entomopathogen *Lecanicillium longisporum* to rosy apple aphid, *Dysaphis plantaginea* (Homoptera: Aphididae). Biocontrol Sci Tech 14:757–767
Boucias DG, Stokes C, Storey G, Pendland JC (1996) The effects of imidacloprid on the termite *Reticulitermes flavipes* and its interaction with the mycopathogen *Beauveria bassiana*. Pflanzenschutz-Nachrichten Bayer (English ed.) 49:103–144
Broza M, Pereira RM, Stimac JL (2001) The non-susceptibility of soil Collembola to insect pathogens and their potential as scavengers of microbial pesticides. Pedobiologia 45:523–534
Bush AO, Fernandez JC, Esch GW, Seed JR (2001) Parasitism: the diversity and ecology of animal parasites. Cambridge University Press, Cambridge
Carreck NL, Butt TM, Clark SJ, Ibrahim L, Isger EA, Pell JK, Williams IH (2007) Honey bees can disseminate a microbial control agent to more than one inflorescence pest of oilseed rape. Biocontrol Sci Tech 17:179–191
Chen C, Feng MG (2002) Evidence for transmission of aphid-pathogenic fungi by migratory flight of *Myzus persicae* alates. Chin Sci Bull 47:1987–1989
Chen C, Feng MG (2004a) Observation on the initial inoculum source and dissemination of Entomophthorales-caused epizootics in populations of cereal aphids. Sci China C Life Sci 47:38–43
Chen C, Feng MG (2004b) *Sitobion avenae* alatae infected by *Pandora neoaphidis*: their flight ability, post-flight colonization, and mycosis transmission to progeny colonies. J Invertebr Pathol 86:117–123
Chen C, Feng MG (2006) Experimental simulation of transmission of an obligate aphid pathogen with aphid flight dispersal. Environ Microbiol 8(1):69–76
Chen B, Li ZY, Feng MG (2008) Occurrence of entomopathogenic fungi in migratory alate aphids in Yunnan Province of China. BioControl 53:317–326
Chouvenc T, Su NY, Elliott ML (2008) Interaction between the subterranean termite *Reticulitermes flavipes* (Isoptera: Rhinotermitidae) and the entomopathogenic fungus *Metarhizium anisopliae* in foraging arenas. J Econ Entomol 101:885–893
Cory JS, Ericsson JD (2009) Fungal entomopathogens in a tritrophic context. BioControl. doi:10.1007/s10526-009-9247-4 (this SI)
Cory JS, Hoover K (2006) Plant-mediated effects in insect-pathogen interactions. Trends Ecol Evol 21:278–286
DeKesel A (1995) Relative importance of direct and indirect infection in the transmission of *Laboulbenia slackensis* (Ascomycota, Laboulbeniales). Belg J Bot 128:124–130
Dowd PF, Vega FE (2003) Autodissemination of *Beauveria bassiana* by sap beetles (Coleoptera: Nitidulidae) to overwintering sites. Biocontrol Sci Tech 13:65–75
Dromph KM (2001) Dispersal of entomopathogenic fungi by collembolans. Soil Biol Biochem 33:2047–2051
Dromph KM (2003) Collembolans as vectors of entomopathogenic fungi. Pedobiologia 47:245–256
Dromph KM, Vestergaard S (2002) Pathogenicity and attractiveness of entomopathogenic hyphomycete fungi to collembolans. Appl Soil Ecol 21:197–210
Ekesi S, Shah PA, Clark SJ, Pell JK (2005) Conservation biological control with the fungal pathogen *Pandora neoaphidis*: implications of aphid species, host plant and predator foraging. Agric For Entomol 7:21–30
Fäldt J, Jonsell M, Nordlander G, Borg-Karlson AK (1999) Volatiles of bracket fungi *Fomitopsis pinicola* and *Fomes fomentarius* and their functions as insect attractants. J Chem Ecol 25:567–590
Feng MG, Chen C, Chen B (2004) Wide dispersal of aphid-pathogenic Entomophthorales among aphids relies upon migratory alates. Environ Microbiol 6:510–516
Feng MG, Chen C, Shang SW, Ying SH, Shen ZC, Chen XX (2007) Aphid dispersal flight disseminates fungal pathogens and parasitoids as natural control agents of aphids. Ecol Entomol 32:97–104
Flower E (2002) The Trojan ant: infecting aphids with the fungus *V. lecanii* using the ant *Lasius niger*. BSc thesis, Imperial College of Science, Technology and Medicine, Wye, UK
Fransen JJ, van Lenteren JC (1993) Host selection and survival of the parasitoid *Encarsia formosa* on greenhouse whitefly, *Trialeurodes vaporariorum*, in the presence of hosts infected with the fungus *Aschersonia aleyrodis*. Entomol Exp Appl 69:239–249
Furlong MJ, Pell JK (2001) Horizontal transmission of entomopathogenic fungi by the diamondback moth. Biol Control 22:288–299
Furlong MJ, Pell JK, Choo OP, Rahman SA (1995) Field and laboratory evaluation of a sex-pheromone trap for the autodissemination of the fungal entomopathogen *Zoophthora radicans* (Entomophthorales) by the diamond-back moth, *Plutella xylostella* (Lepidoptera, Yponomeutidae). Bull Entomol Res 85:331–337
Fuxa JR, Tanada Y (1987) Epidemiological concepts applied to epizootiology. In: Fuxa JR, Tanada Y (eds) Epizootiology of insect diseases. Wiley, New York, pp 3–21
Grace JK, Zoberi MH (1992) Experimental evidence for transmission of *Beauveria bassiana* by *Reticulitermes*

flavipes workers (Isoptera, Rhinotermitidae). Sociobiology 20:23–28

Hajek AE, St Leger RJ (1994) Interactions between fungal pathogens and insect hosts. Annu Rev Entomol 39:293–322

Huang ZH, Feng MG, Chen XX, Liu SS (2008) Pathogenic fungi and parasitoids of aphids present in air captures of migratory alates in the low-latitude plateau of Yunnan, China. Environ Entomol 37:1264–1271

Inyang EN, Butt TM, Ibrahim L, Clark SJ, Pye BT, Beckett A, Archer S (1998) The effect of plant growth and topography on the acquisition of conidia of the insect pathogen *Metarhizium anisopliae* by larvae of *Phaedon cochleariae*. Mycol Res 102:1365–1374

Jones WE, Grace JK, Tamashiro M (1996) Virulence of seven isolates of *Beauveria bassiana* and *Metarhizium anisopliae* to *Coptotermes formosanus* (Isoptera: Rhinotermitidae). Environ Entomol 25:481–487

Kaaya GP, Okech MA (1990) Horizontal transmission of mycotic infection in adult tsetse, *Glossina morsitans-morsitans*. Entomophaga 35:589–600

Kapongo JP, Shipp L, Kevan P, Broadbent B (2008a) Optimal concentration of *Beauveria bassiana* vectored by bumble bees in relation to pest and bee mortality in greenhouse tomato and sweet pepper. BioControl 53:797–812

Kapongo JP, Shipp L, Kevan P, Sutton JC (2008b) Co-vectoring of *Beauveria bassiana* and *Clonostachys rosea* by bumble bees (*Bombus impatiens*) for control of insect pests and suppression of grey mould in greenhouse tomato and sweet pepper. Biol Control 46:508–514

Klinger E, Groden E, Drummond F (2006) *Beauveria bassiana* horizontal infection between cadavers and adults of the Colorado potato beetle, *Leptinotarsa decemlineata* (Say). Environ Entomol 35:992–1000

Kreutz J, Zimmermann G, Vaupel O (2004) Horizontal transmission of the entomopathogenic fungus *Beauveria bassiana* among the spruce bark beetle, *Ips typographus* (Col., Scolytidae) in the laboratory and under field conditions. Biocontrol Sci Technol 14:837–848

Long DW, Groden E, Drummond FA (2000) Horizontal transmission of *Beauveria bassiana* (Bals.) Vuill. Agric For Entomol 2:11–17

Lord JC (2001) Response of the wasp *Cephalonomia tarsalis* (Hymenoptera: Bethylidae) to *Beauveria bassiana* (Hyphomycetes: Moniliales) as free conidia or infection in its host, the sawtoothed grain beetle, *Oryzaephilus surinamensis* (Coleoptera: Silvanidae). Biol Control 21:300–304

Ludwig SW, Oetting RD (2002) Efficacy of *Beauveria bassiana* plus insect attractants for enhanced control of *Frankliniella occidentalis* (Thysanoptera: Thripidae). Fla Entomol 85:270–272

Meyling NV, Pell JK (2006) Detection and avoidance of an entomopathogenic fungus by a generalist insect predator. Ecol Entomol 31:162–171

Meyling NV, Pell JK, Eilenberg J (2006) Dispersal of *Beauveria bassiana* by the activity of nettle insects. J Invertebr Pathol 93:121–126

Milner RJ (1997) Prospects for biopesticides for aphid control. Entomophaga 42:227–239

Myles TG (2002) Alarm, aggregation, and defense by *Reticulitermes flavipes* in response to a naturally occurring isolate of *Metarhizium anisopliae*. Sociobiology 40:243–255

Ormond E (2007) The overwintering interactions of the seven spot ladybird (*Coccinella septempunctata*) and the entomopathogenic fungus *Beauveria bassiana*. Ph.D. thesis, Anglia Ruskin University, Cambridge

Pell JK, Vandenberg JD (2002) Interactions among the aphid *Diuraphis noxia*, the entomopathogenic fungus *Isaria fumosoroseus* and the coccinellid *Hippodamia convergens*. Biocontrol Sci Tech 12:217–224

Pell JK, Wilding N, Player AL, Clark SJ (1993) Selection of an isolate of *Zoopthora radicans* (Zygomycetes, Entomophthorales) for biocontrol of the diamondback moth *Plutella xylostella* (Lepidoptera, Yponomeutidae). J Invertebr Pathol 61:75–80

Pell JK, Pluke R, Clark SJ, Kenward MG, Alderson PG (1997) Interactions between two aphid natural enemies, the entomopathogenic fungus *Erynia neoaphidis* Remaudiere & Hennebert (Zygomycetes: Entomophthorales) and the predatory beetle *Coccinella septempunctata* L. (Coleoptera: Coccinellidae). J Invertebr Pathol 69:261–268

Pell JK, Tydeman C, MaCartney A (1998) Impact of rainfall on the persistence and transmission of *Erynia neoaphidis*. IOBC/WPRS Bull 21:49

Powell W, Wilding N, Brobyn PJ, Clark SJ (1986) Interference between parasitoids (hym, Aphididae) and fungi (Entomophthorales) attacking cereal aphids. Entomophaga 31:293–302

Quesada-Moraga E, Martin-Carballo I, Garrido-Jurado I, Santiago-Alvarez C (2008) Horizontal transmission of *Metarhizium anisopliae* among laboratory populations of *Ceratitis capitata* (Wiedemann) (Diptera: Tephritidae). Biol Control 47:115–124

Quintela ED, McCoy CW (1998) Conidial attachment of *Metarhizium anisopliae* and *Beauveria bassiana* to the larval cuticle of *Diaprepes abbreviatus* (Coleoptera: Curculionidae) treated with imidacloprid. J Invertebr Pathol 72:220–230

Rath AC (2000) The use of entomopathogenic fungi for control of termites. Biocontrol Sci Tech 10:563–581

Renn N, Bywater AF, Barson G (1999) A bait formulated with *Metarhizium anisopliae* for the control of *Musca domestica* L-(Dipt., Muscidae) assessed in large-scale laboratory enclosures. J Appl Entomol 123:309–314

Robert Y (1987) Dispersion and migration. In: Minks AK, Harrewijn P (eds) Aphids, their biology, natural enemies and control, vol A. Elsevier, Amsterdam, pp 299–313

Roditakis E, Couzin ID, Balrow K, Franks NR, Charnley AK (2000) Improving secondary pick up of insect fungal pathogen conidia by manipulating host behaviour. Ann Appl Biol 137:329–335

Roy HE, Pell JK (2000) Interactions between entomopathogenic fungi and other natural enemies: implications for biological control. Biocontrol Sci Tech 10:737–752

Roy HE, Pell JK, Clark SJ, Alderson PG (1998) Implications of predator foraging on aphid pathogen dynamics. J Invertebr Pathol 71:236–247

Roy HE, Pell JK, Alderson PG (1999) Effects of fungal infection on the alarm response of pea aphids. J Invertebr Pathol 74:69–75

Roy HE, Pell JK, Alderson PG (2001) Targeted dispersal of the aphid pathogenic fungus *Erynia neoaphidis* by the aphid predator *Coccinella septempunctata*. Biocontrol Sci Tech 11:99–110

Roy HE, Steinkraus DC, Eilenberg J, Hajek AE, Pell JK (2006) Bizarre interactions and endgames: entomopathogenic fungi and their arthropod hosts. Annu Rev Entomol 51:331–357

Roy HE, Brown PMJ, Rothery P, Ware RL, Majerus MEN (2008) Interactions between the fungal pathogen *Beauveria bassiana* and three species of coccinellid: *Harmonia axyridis*, *Coccinella septempunctata* and *Adalia bipunctata*. BioControl 53:265–276

Shah PA, Pell JK (2003) Entomopathogenic fungi as biological control agents. Appl Microbiol Biotechnol 61:413–423

Shah FA, Ansari MA, Prasad M, Butt TM (2007) Evaluation of black vine weevil (*Otiorhynchus sulcatus*) control strategies using *Metarhizium anisopliae* with sublethal doses of insecticides is disparate horticultural growing media. Biol Control 40:246–252

Shanley RP, Hajek AE (2008) Environmental contamination with *Metarhizium anisopliae* from fungal bands for control of the Asian longhorned beetle, *Anoplophora glabripennis* (Coleoptera: Cerambycidae). Biocontrol Sci Tech 18:109–120

Shimazu M (2004) A novel technique to inoculate conidia of entomopathogenic fungi and its application for investigation of susceptibility of the Japanese pine sawyer, *Monochamus alternatus*, to *Beauveria bassiana*. Appl Entomol Zool 39:485–490

Shimizu S, Yamaji M (2003) Effect of density of the termite, *Reticulitermes speratus* Kolbe (Isoptera: Rhinotermitidae), on the susceptibilities to *Metarhizium anisopliae*. Appl Entomol Zool 38:125–130

Simelane DO, Steinkraus DC, Kring TJ (2008) Predation rate and development of *Coccinella septempunctata* L. influenced by *Neozygites fresenii*-infected cotton aphid prey. Biol Control 44:128–135

Smith SM, Moore D, Karanja LW, Chandi EA (1999) Formulation of vegetable fat pellets with pheromone and *Beauveria bassiana* to control the larger grain borer, *Prostephanus truncatus* (Horn). Pest Sci 55:711–718

Staples JA, Milner RJ (2000) A laboratory evaluation of the repellency of *Metarhizium anisopliae* conidia to *Coptotermes lacteus* (Isoptera: Rhinotermitidae). Sociobiology 36:133–148

Steinkraus DC (2006) Factors affecting transmission of fungal pathogens of aphids. J Invertebr Pathol 92:125–131

Sullivan BT, Berisford CW (2004) Semiochemicals from fungal associates of bark beetles may mediate host location behavior of parasitoids. J Chem Ecol 30:703–717

Sun JZ, Fuxa JR, Richter A, Ring D (2008) Interactions of *Metarhizium anisopliae* and tree-based mulches in repellence and mycoses against *Coptotermes formosanus* (Isoptera: Rhinotermitidae). Environ Entomol 37:755–763

Thompson SR, Brandenburg RL (2005) Tunnelling responses of mole crickets (Orthoptera: Gryllotalpidae) to the entomopathogenic fungus, *Beauveria bassiana*. Environ Entomol 34:140–147

Toledo J, Campos SE, Flores S, Liedo P, Barrera JF, Villasenor A, Montoya P (2007) Horizontal transmission of *Beauveria bassiana* in *Anastrepha ludens* (Diptera: Tephritidae) under laboratory and field cage conditions. J Econ Entomol 100:291–297

Tsutsumi T, Teshiba M, Yamanaka M, Ohira Y, Higuchi T (2003) An autodissemination system for the control of brown winged green bug, *Plautia crossota stali* Scott (Heteroptera: Pentatomidae) by an entomopathogenic fungus, *Beauveria bassiana* E-9102 combined with aggregation pheromone. Jpn J Appl Entomol Zool 47:159–163

Vega FE, Dowd PF, Bartelt RJ (1995) Dissemination of microbial agents using an autoinoculating device and several insect species as vectors. Biol Control 5:545–552

Vega FE, Dowd PF, Lacey LA, Pell JK, Jackson DM, Klein MG (2000) Dissemination of beneficial microbial agents by insects. In: Lacey LA, Kaya HK (eds) Field manual of techniques in invertebrate pathology. Kluwer, London, pp 153–177

Vickers RA, Furlong MJ, White A, Pell JK (2004) Initiation of fungal epizootics in diamondback moth populations within a large field cage: proof of concept for auto-dissemination. Entomol Exp Appl 111:7–17

Villani MG, Krueger SR, Schroeder PC, Consolie F, Consolie NH, Preston-Wilsey LM, Roberts DW (1994) Soil application effects of *Metarhizium anisopliae* on Japanese-beetle (Coleoptera, Scarabaeidae) behaviour and survival in turfgrass microcosms. Environ Entomol 23:502–513

Villani MG, Allee LL, Preston-Wilsey L, Consolie N, Xia Y, Brandenburg RL (2002) Use of radiography and tunnel castings for observing mole cricket (Orthoptera: Gryllotalpidae) behavior in soil. Am Entomol 48:42–50

Visser S, Parkinson D, Hassall M (1987) Fungi associated with *Onychiurus subtenuis* (Collembola) in an Aspen woodland. Can J Bot 65:635–642

Wang CL, Powell JE (2004) Cellulose bait improves the effectiveness of *Metarhizium anisopliae* as a microbial control of termites (Isoptera: Rhinotermitidae). Biol Control 30:523–529

Yasuda K (1999) Auto-infection system for the sweet potato weevil, *Cylas formicarius* (Fabricius) (Coleoptera: Curculionidae) with entomopathogenic fungi, *Beauveria bassiana* using a modified sex pheromone trap in the field. Appl Entomol Zool 34:501–505

Zhang GZ, Feng MG, Chen C, Ying SH (2007) Opportunism of *Conidiobolus obscurus* stems from depression of infection in situ to progeny colonies of host alatae as disseminators of the aphid-pathogenic fungus. Environ Microbiol 9: 859–868

Zimmermann G, Bode E (1983) Investigations on the dispersal of the entomopathogenic fungus *Metarhizium-anisopliae* (Fungi Imperfecti, Moniliales) by soil arthropods. Pedobiologia 25:65–71

Author Biographies

Jason Baverstock works in Dr Judith K. Pell's group in the Department for Plant and Invertebrate Ecology at Rothamsted Research. The group's research focuses on the ecology of entomopathogenic fungi, to elucidate their role in population regulation and community structure and to inform biological control strategies. Specifically: intraguild interactions; the relationships between guild diversity, habitat diversity and ecosystem function; pathogen-induced host behavioural change.

Helen E. Roy leads zoological research in the Biological Records Centre at the NERC Centre for Ecology & Hydrology. The focus of her research is insect community interactions with particular emphasis on the effects of environmental change.

Judith K. Pell is the head of the group in the Department for Plant and Invertebrate Ecology at Rothamsted Research.

BioControl (2010) 55:103–112
DOI 10.1007/s10526-009-9236-7

ORIGINAL PAPER

Fungal entomopathogens in the rhizosphere

Denny J. Bruck

Received: 29 June 2009 / Accepted: 28 September 2009 / Published online: 24 October 2009

Abstract The ecology of fungal entomopathogens in the rhizosphere is an understudied area of insect pathology. The rhizosphere is the region of soil in which the release of root exudates influences the soil microbiota, and may provide a favorable environment for fungal entomopathogens. The objective of this review is to bring together the relatively scant data available to date on the subject of fungal entomopathogens colonizing the rhizosphere and to highlight the importance of these findings. Gaining a better understanding of the ecology of fungal entomopathogens in the rhizosphere will help in the development of successful microbial control strategies against root-feeding insect pests.

Keywords *Metarhizium anisopliae* · *Beauveria bassiana* · Fungal ecology · Rhizosphere competent

Introduction

The rhizosphere encompasses a few millimeters of soil surrounding the plant root, an area where multifaceted ecological and biological processes take place. It is in the rhizosphere that complex interactions between roots, root exudates, beneficial and pathogenic micro-organisms, and invertebrates take place. Hiltner (1904) was the first to define the "rhizosphere effect" by observing that the number and activity of micro-organisms increased in the vicinity of plant roots. A large array of microbes can inhabit the rhizosphere and it is widely accepted that members from all microbial groups perform important functions in the rhizosphere (Giri et al. 2005). However, most studies of rhizosphere microbiology have focused on bacteria and fungi (Bowen and Rovira 1999). Two types of microbial interactions are recognized in the rhizosphere, those based on dead plant material (detritus-based) affecting nutrient and energy flows, and those based on living plant roots (Barea et al. 2005). Root exudates fall into two main classes of compounds: (1) low-molecular weight compounds such as amino acids, organic acids, sugars, phenolics, and other secondary metabolites, and (2) high-molecular weight compounds such as polysaccharides and proteins (Marschner 1995). Bais et al. (2006) published a comprehensive review on the role of root exudates on interactions between plant roots and other plants, microbes, and nematodes present in the rhizosphere.

There are three separate, but interacting, regions that make up the rhizosphere: the outer rhizosphere, the rhizoplane and the root (Kennedy 1998; Bowen and Rovira 1999). The outer rhizosphere contains the soil that is loosely adhered to the roots and is the

Handling Editor: Helen Roy

D. J. Bruck (✉)
USDA-ARS, Horticultural Crops Research Unit, 3420 N.W. Orchard Avenue, Corvallis, OR 97330, USA
e-mail: Denny.Bruck@ars.usda.gov

region where the root exudates influence the soil microbiota. The rhizoplane is the portion of the rhizosphere directly in contact with the root surface resulting in the soil being tightly adhered to the roots. The roots themselves are also an important component of the rhizosphere, particularly for endophytic microorganisms (Kennedy 1998; Bowen and Rovira 1999). Because of the secluded nature of the rhizosphere it is an under-studied area of science. However, even in light of this fact, there have been significant discoveries particularly in the areas of the biological control of root pathogens (Whipps 1997, 2001) and phytoremediation (Pilon-Smits 2005).

Entomopathogenicity is a lifestyle that has arisen and been lost multiple times in many fungal lines (Roberts and Humber 1981; Rehner and Buckley 2005; Humber 2008). Hypocreales contains the largest number of fungal entomopathogens including two of the most widely studied, *Beauveria bassiana* (Balsamo) Vuillemin (Ascomycete: Hypocreales) and *Metarhizium anisopliae* (Metchnikoff) Sorokin (Ascomycete: Hypocreales) both of which have been used for the microbial control of a wide array of foliar and soil-borne invertebrate pests (Lacey and Kaya 2007). Most studies have focused on the use of these fungi as replacements for chemical insecticides with little consideration of their ecological niche in the environment. The successful use of these fungal entomopathogens as microbial insecticides has been sporadic, due in large part to our incomplete understanding of their ecology. While commercial microbial control products based on *B. bassiana* and *M. anisopliae* have been registered around the world, they are used primarily in small niche markets and not large acreage crops. Several factors have limited the adoption of microbial control agents in the industrialized world including: regulatory constraints, activist resistance, benign and efficacious chemicals and limited research funding (Lord 2005). Other factors include inconsistent control, poor persistence, erratic product quality, poor shelf life and elevated costs relative to chemicals. To be effective, biological control agents must proliferate in the environment; a fundamental difference with chemical agents (Nelson et al. 1994). As a discipline, insect pathology must attain a better understanding of the ecology of fungal entomopathogens in order to improve the chances of success in agricultural production systems (Jaronski 2007; Vega et al. 2009). Entomopathogenic fungal isolates have traditionally been selected for development as microbial control agents based on laboratory bioassay results. Little emphasis has been placed on understanding the ecology of individual isolates. A preoccupation with killing insect pests has blinded us to the importance of fungal ecology when screening, selecting and releasing fungal entomopathogens in the field.

The soil has long been considered the natural reservoir for fungal many entomopathogens (Harrison and Gardner 1991; Bing and Lewis 1993; Chandler et al. 1997; Bidochka et al. 1998, 2001; Klingen et al. 2002; Shapiro Ilan et al. 2003; Bruck 2004). Isolating fungal entomopathogens from soil offers insight into their biodiversity and provides a pool of potential microbial control agents. Traditionally, isolation is followed by bioassays against target pests in the laboratory to identify the isolate with the lowest LC_{50} and LT_{50} values. A much needed third step, following isolation and laboratory bioassays, should involve the characterization of the ecological constraints of the candidate isolates relative to the environment in which pests are being targeted. Understanding the dynamic interactions between the insect pests, the fungi and the host plant should be important considerations in the development and understanding of fungal entomopathogens as microbial control agents. History provides us with clear examples of the benefits of understanding fungal ecology for enhanced microbial control of insects. Lewis and colleagues (Bing and Lewis 1991, 1992) observed that *B. bassiana* grew endophytically within the green tissues of *Zea mays* L. (Cyperales: Poaceae). They also demonstrated that endophytic isolates of *B. bassiana* effectively controlled European corn borer, *Ostrinia nubilalis* (Lepidoptera: Crambidae; Lewis et al. 2002) while being non-pathogenic to *Z. mays* (Lewis et al. 2001). This pioneer research has in recent years led to investigation of *B. bassiana* as an endophyte of a wide variety of plants (Vega 2008). Insect pathology is not the only discipline to benefit from an enhanced understanding of microbial ecology. In the field of plant pathology, the "disease triangle" is a central concept based on the principle that disease is the result of an interaction between a host, a pathogen, and the environment (McNew 1960; Agrios 2005; Jones 1998). Insect pathologists developing microbial control programs would benefit greatly by integrating the disease triangle concept into their studies.

The objective of this review is to bring together the relatively scant data available to date on the ecology of fungal entomopathogens. Because this chapter is focused on fungal entompathogens in the rhizosphere, I will limit the discussion to the control of root-feeding insects.

The rhizosphere as a key microenvironment for fungal entomopathogens

Rhizosphere competent microorganisms are those that show enhanced growth in response to developing roots (Schmidt 1979). The discovery of *M. anisopliae* as rhizosphere competent was serendipitous (Hu and St. Leger 2002). Field trials by Hu and St. Leger (2002) were designed to determine the fate of fungal clones of *M. anisopliae* in the field. This was accomplished by employing a *gfp* gene driven by a constitutive promoter which strongly labeled the fungus with no impact on fungal growth or pathogenicity. Samples were collected from a variety of locations in and around the field to monitor for fungal distribution and persistence. Soil samples were collected 4–5 cm from, as well as adjacent to the cabbage taproot. During the six months following fungal application, the fungal titer in the bulk soil decreased from 10^5 propagules g^{-1} in the top 3 cm of soil to 10^3 propagules g^{-1}. However, fungal titers in the rhizosphere remained at 10^5 propagules g^{-1} six months after fungal application resulting in a 100:1 ratio in fungal densities between the rhizosphere and bulk soil (Hu and St. Leger 2002). The rhizosphere effect was most pronounced in the top 3 cm of soil and may be explained by a combination of two factors: (1) roots were most numerous in the top 3 cm of soil, and (2) fungal spores applied to the field were concentrated in the upper soil profile.

At the time that the Hu and St. Leger (2002) manuscript was published, we were performing experiments to determine the persistence of *M. anisopliae* (F52, Novozymes Biologicals Inc., Salem VA, USA) in bark and peat-based soilless potting media. Subsequent to learning that at least one isolate of *M. anisopliae* was rhizosphere competent, we sought to determine if *M. anisopliae* (F52) colonized the rhizosphere of *Picea abies* (L.) Karst. 'Nidiformis' (Pinales: Pinaceae). On each of the subsequent sample dates the difference between the fungal population in the rhizosphere and surrounding bulk media was significantly greater than zero, indicating that not only did *M. anisopliae* colonize the rhizosphere of *P. abies*, but the fungal population responded favorably to the rhizosphere microenvironment. The mean difference in *M. anisopliae* population levels between the rhizosphere and bulk soil ranged from 0.65 to 1.28 $\log_{10}$ CFU g^{-1} media. Data analysis of the mean difference between the rhizosphere and bulk media fungal population of each plant sampled showed that potting media type was the only parameter measured that had any significant effect on the size of the difference observed. The difference in *M. anisopliae* population levels between the rhizosphere and bulk media was greatest in the peat-based potting media on three of the five sample dates (Bruck 2005). Positive response to root exudates by *M. anisopliae* in the field was also suggested by Klingen et al. (2002), although the fungal population in the rhizosphere was not quantified. Studies of *M. anisopliae* population dynamics in the rhizosphere and surrounding bulk soil help describe the density as well as the temporal and spatial dynamics of the inoculum in soil. A more complete understanding of the relationship between the density of fungal entomopathogen inoculum and insect disease incidence is critical to understanding the outcome of microbial control efforts.

Subsequent studies demonstrated isolate variability in rhizosphere competence between plants. Studies were performed to determine the "rhizosphere host range" of F52 as well as *M. anisopliae* isolates collected from nursery soils in Oregon, USA (Bruck 2004). Bare root cuttings of *P. abies*, *Picea glauca* (Moench) Voss (Pinales: Pinaceae) and *Taxus baccata* L. (Taxales: Taxaceae) were planted into soilless potting media (Sunshine Mix #3, Sun Gro Horticulture, Bellevue, WA, USA) incorporated with one of three *M. anisopliae* isolates (F52, IP99, IP285). Four plants from each treatment were randomly selected at 6, 10 and 14 weeks after planting and the fungal population in the bulk and rhizosphere soil determined as described by Bruck (2005). The bulk soil populations of all isolates remained relatively steady or declined over the 14 week period (Figs. 1, 2, 3). The rhizosphere population response of each isolate to the various plant species was distinctive. The isolates F52 and IP99 were rhizosphere competent on the roots of *P. abies* with their populations increasing nearly 10-fold over a 14 week period. However, the rhizosphere population of IP285 on the roots of *P. abies* remained flat (Fig. 1). These data confirm our earlier work demonstrating a significant population increase of F52 in the rhizosphere of *P. abies* (Bruck 2005). All of the isolates tested colonized the rhizosphere of *P. glauca* with a nearly 10-fold increase in their populations over the 14 week period (Fig. 2). None of the isolates tested responded favorably to the rhizosphere of *T. baccata* over the course of 14 weeks (Fig. 3).

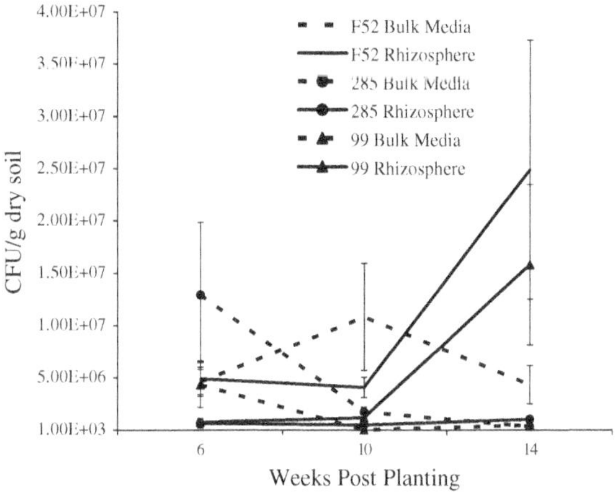

Fig. 1 Fungal population ± SE (cfu g^{-1} dry soil) of three *M. anisopliae* isolates in the bulk soil and the rhizosphere soil of *Picea abies* 6, 10 and 14 weeks after planting in fungal inoculated soil

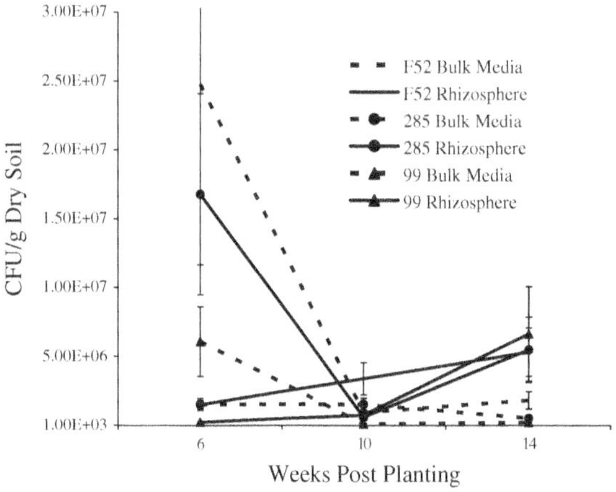

Fig. 2 Fungal population ± SE (cfu g^{-1} dry soil) of three *M. anisopliae* isolates in the bulk soil and the rhizosphere soil of *Picea glauca* 6, 10 and 14 weeks after planting in fungal inoculated soil

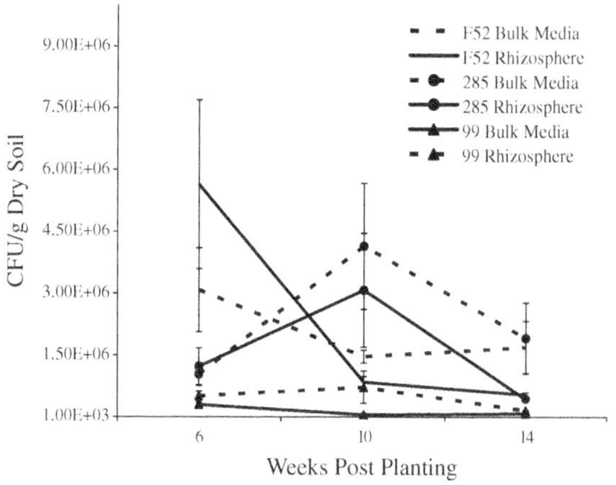

Fig. 3 Fungal population ± SE (cfu g^{-1} dry soil) of three *M. anisopliae* isolates in the bulk soil and the rhizosphere soil of *Taxus baccata* 6, 10 and 14 weeks after planting in fungal inoculated soil

Tritrophic interactions

Tritrophic interactions are well described for terrestrial systems (Sabelis and van de Baan 1983; Dicke et al. 1990; Turlings et al. 1990; Dicke et al. 1993; Turlings et al. 1995; Kessler and Baldwin 2001). In above-ground systems, herbivore feeding elicits systemic production of secondary metabolites by plants that serve as attractants to predators and parasitoids (Turlings and Tumlinson 1992; Dicke et al. 1993). Tritrophic interactions may also involve entomopathogens, plants, and insects (Cory and Ericsson 2009). Currently, it is unclear if plants manipulate 'bodyguard' entomopathogens similarly to their manipulation of predators and parasitoids (Sabelis et al. 1999; Elliot et al. 2000). While bodyguard traits are yet to be demonstrated with microbial entomopathogens, these microorganisms are clearly involved in tritrophic interactions and that multitrophic relationships also exist (Cory and Hoover 2006). One example of the complex interactions occurs between secondary plant metabolites and the fungal entomopathogen *Neozygites tanajoae* Delalibera Jr., Humber & Hajek (Zygomycetes: Entomophthorales) used in the control of cassava green mites *Mononychellus tanajoa* (Bondar) (Acari: Tetranychidae). Cassava green plant volatiles suppress the germination of *N. tanajoae* in the absence of mite feeding (Hountondji et al. 2005). However, plant volatiles released in response to green mite feeding on leaves trigger conidiation, allowing the fungus to release infective spores when mites are present (Hountondji et al. 2005).

Tritrophic interactions have been found to operate below ground as well. One case involves the entomopathogenic nematode, *Heterorhabditis megidis* Poinar Jackson & Klein (Rhabditidae: Heterorhabditidae) and its orientation to black vine weevil *Otiorhynchus sulcatus* (F.) (Coleoptera: Curculionidae) larvae. Boff et al. (2001) observed *H. megidis* attraction towards strawberry plants fed upon by black vine weevil larvae. However, they were unable to determine if the orientation was due to chemical cues emitted from the plant. The attraction of *H. megidis* to chemical cues released by the conifer *Thuja occidentalis* L. (Pinales: Cupressaceae) fed upon by black vine weevil larvae feeding was confirmed by van Tol et al. (2001). Since these initial findings, the production of natural enemy attractants in response to root herbivory has been identified in turnips (Neveu et al. 2002), tulips (Aratchige et al. 2004) and corn (Rasmann et al. 2005).

There is contradictory evidence in the literature concerning the ability of fungal entomopathogens in the soil to influence insect behavior. Villani et al. (1994) observed that Japanese beetle, *Popillia japonica* Newman (Coleoptera: Scarabaeidae) oviposited preferentially on bare soil treated with *M. anisopliae* mycelia over non-inoculated bare soil, possibly in response to CO_2 released during mycelial growth. However, Japanese beetle grubs avoided regions of sod treated with *M. anisopliae* (Villani et al. 1994). Rath (2000) found that isolates of *M. anisopliae* vary in their repellency in the laboratory and field against termites. The termites *Reticulitermes flavipes* (Kollar) and *Coptotermes formosanus* Shiraki (Isoptera: Rhinotermitidae) were attracted to *M. anisopliae* mycelial preparations and volatile extracts (Engler and Gold 2004). Mole crickets (Orthoptera: Gryllotalpidae) modified their behavior in response to *M. anisopliae* and *B. bassiana* incorporated into soil so as to reduce their exposure to these fungal entomopathogens (Villani et al. 2002; Thompson and Brandenburg 2005). Rath (2000) as well as Thompson and Brandenburg (2005) demonstrated that termite and cricket avoidance behavior, respectively, was dependent on the fungal isolate, which may partially account for the behavioral differences observed among fungal isolates and insect species. More recently, wireworms *Agriotes obscurus* L. (Coleoptera: Elateridae) were repelled by *M. anisopliae*-contaminated soil at a rate that increased with conidia concentration in the soil. However, the rate of emigration was reduced when food was present (Kabaluk and Ericsson 2007a).

St. Leger (2008) speculates that *M. anisopliae* could provide a "repellent barrier" around plant roots which would provide more effective protection to the plant than direct fungal infection of the herbivore, primarily due to the time lag between infection and cessation of feeding. This may well be the case with some fungal entomopathogen isolates, however. The opposite phenomenon in which insects are attracted to plants when their rhizosphere is colonized may also occur (Kepler and Bruck 2006). When placed in a two-choice soil olfactometer, black vine weevil larvae were significantly more attracted to *P. abies* roots growing in *M. anisopliae* inoculated potting media than plants grown in uninoculated media, revealing a tritrophic interaction that differs significantly from previous reports (Kepler and Bruck 2006). In our studies, it was not a natural enemy whose behavior was altered in response to secondary plant metabolites (Turlings and Tumlinson 1992; Dicke et al. 1993; Boff et al. 2001; van Tol et al. 2001; Rasmann et al. 2005), but rather the behavior of the pest itself. From an evolutionary standpoint of the fungal entomopathogen this makes sense as *M. anisopliae* spores in the soil are not able to actively seek out insect hosts. If *M. anisopliae* is in fact utilizing the rhizosphere as a bridge between insect hosts, preferentially attracting hosts to the fungus in the rhizosphere may substantially shorten the length of the bridge. Unfortunately, we can only speculate on whether the fungus or plant is the source of the attractive compound(s). The evidence seems to indicate that the plant in association with the fungus produces compounds attractive to black vine weevil larvae. However, it is also plausible that when colonizing the rhizosphere, the fungus produces attractive compounds that are not produced in the absence of plant roots. There may be an evolutionary benefit to the plant in having root-feeding insects attracted to fungal colonized plants in a community in which there is not 100% fungal colonization. In such a scenario, root-feeding insects preferentially feed on roots colonized with fungal entomopathogens subsequently becoming infected which results in a net reduction in root-feeding in the plant community.

A bodyguard interaction between host plant and the herbivore via an entomopathogen is by definition an indirect one (Elliot et al. 2000). Plants may have an indirect impact on entomopathogens by: (1) maintaining a population of the entomopathogen, (2) increasing contact rate between the insect and the entomopathogen and, (3) by increasing the susceptibility of the insect to the entomopathogen (Elliot et al. 2000). In the case of *M. anisopliae*, fungal propagules in the rhizosphere increase in response to root exudates (Hu and St. Leger 2002; Bruck 2005) and the presence of the fungus in the rhizosphere, at least in some cases, results in increased exposure of insects to the fungus (Kepler and Bruck 2006). An increase in insect susceptibility to fungal entomopathogens in the rhizosphere has yet to be demonstrated, but all three of the above scenarios outlined by Elliot et al. (2000) need not occur for the bodyguard interaction to be successful. In addition, the employment of a fungal entomopathogen as a bodyguard by a plant must result in a net positive return on investment, must complement the plants other defenses, and the investment must be secure (Elliot et al. 2000). Preliminary data suggest that at least in the case of *M. anisopliae* colonizing the rhizosphere of *P. abies*, there is no measurable cost to plant fitness (Kepler and Bruck, unpublished data). Cooperation between host plants and microorganisms should benefit both partners, given their differing resource needs and metabolic capabilities (Hoeksema and Schwartz 2003). However, these mutual benefits do not guarantee that the cooperation is evolutionarily stable (Kiers and Denison 2008).

Role of fungal entomopathogens in the rhizosphere for controlling root-feeding insects

We have demonstrated the pest management potential of rhizosphere-competent fungal entomopathogens (Bruck 2005). Colonization of the rhizosphere of *P. abies* by a rhizosphere competent isolate of *M. anisopliae* provided nearly 80% control of black vine weevil larvae within two weeks of exposure to inoculated roots (Bruck 2005). This work was the first to demonstrate that roots colonized with a fungal entomopathogen resulted in high levels of insect infection through root feeding. Hu and St. Leger (2002) also noted that the carrying capacity of *M. anisopliae* (2575-GFP) in the cabbage rhizosphere (10^5 propagules g^{-1}) was higher than the LC_{50} value of the isolate against a number of insect pests. While our understanding of the ecology and significance of *M. anisopliae* in the rhizosphere is in its infancy, it is clear that an increased understanding of this relationship is likely to be an important aspect in the microbial control of root-feeding insects. Currently, data on the pest management potential of rhizosphere competent fungal entomopathogens are scant. However, the prospective ramifications of this relationship are tremendous. A simple calculation of the economic benefits that can be realized by utilizing rhizosphere competent fungal entomopathogens yields savings significant enough to warrant further investigation. For example, a grower of container-grown ornamentals utilizes approximately 10× the amount of potting media annually to grow production plants as is used in the propagation of new plant material at their operation. The use of a rhizosphere competent fungal entomopathogen incorporated into soil during plant propagation would result in a 10-fold reduction in the amount of fungal inoculum required. The use of rhizosphere competent fungal entomopathogens could result in effective control of root-feeding insect pests without the added cost of treating the surrounding bulk soil with large numbers of fungal propagules. Great numbers of fungal entomopathogen propagules are applied or incorporated into soil for the control of root-feeding insects, most of which are not involved in control.

Soil adapting traits

Habitat and proximity to potential insect hosts are important driving forces in the population structure of *M. anisopliae* and *B. bassiana* (Bidochka et al. 1998, 2001, 2002; Humber 2008). Bidochka et al. (1998) found *M. anisopliae* occurred more frequently in agricultural habitats while *B. bassiana* was predominately isolated from forested habitats. Genomic analysis of *M. anisopliae* revealed two non-recombining lineages of *M. anisopliae* var. *anisopliae* in southern Ontario, Canada; one lineage typically occurred in agricultural soils while the other was more common in forest soils (Bidochka et al. 2001).

Recent analyses have determined that the two mutually exclusive groups reported by Bidochka et al. (2001) are *M. robertsii* and *M. brunneum* (Bischoff et al. 2009). Conversely Inglis et al. (2008), observed that two closely-related cosmopolitan genotypes of *M. anisopliae* var. *anisopliae* predominated urban, agricultural, and forest soils in southwestern British Columbia, Canada. The discrepancy between these studies may be the result of the geographic isolation which restricted emigration of *M. anisopliae* into southwest British Columbia (Inglis et al. 2008) or cryptic species (Bischoff et al. 2009). Within any particular habitat, it is not unreasonable to assume that rhizosphere colonization may play a key role in which fungal entomopathogens are present. Plants growing in soil containing fungal entomopathogens would result in long-term exposure of fungi to certain plant communities putting a tremendous amount of selection pressure on the fungi to select for those that can "bridge" the gap between insect hosts by persisting in the rhizosphere of plants in that particular habitat. As stated by Humber (2008) "Natural selection may also lead a fungus to an increasing or decreasing level of nutritional and biological adjustment to its food source; such adjustments could move a fungus in any direction along the nutritional continuum from beneficial to commensal to saprobic to parasitic to pathogenic associations with the source of its nutrients". Two differing sets of selection pressure appear to be at play on fungal entomopathogens: survival in soil and virulence towards insects (Prior 1992). A review by St Leger (2008) outlines the adaptations of *M. anisopliae* to life in the soil. *M. anisopliae* expresses a different subset of genes to persist and colonize insect and plant tissues suggesting that the ability to adapt to life in the soil and as an insect pathogen requires different subsets of genes (Wang et al. 2005). *M. anisopliae* produces two different proteins (MAD1 and MAD2) used for adhesion to insect and plant surfaces. MAD1 and MAD2 are differentially produced in response to insect hemolymph and plant root exudates, respectively. Expression of MAD1 and MAD2 in yeast cells allowed them to adhere to insect cuticle and a plant surface, respectively. *M. anisopliae* is able to adapt its adhesive properties to insects or plant roots through regulation, localization, and specificity control in the functional distinction between MAD1 and MAD2 (Wang and St. Leger 2007).

Conclusions

Jaronski (2007) considered the ecology of fungal entomopathogens in soil and stated "If a generalization can be made, it is that one simply cannot generalize." The result of any one study of the ecology of fungal entomopathogens in soil cannot be used to make broad generalizations on their ecological role. The soil habitat and all of the complex biotic and abiotic interactions that occur in the soil are extremely complex and it is evident that not all fungal entomopathogens are performing the same role. Our current knowledge serves as the foundation for future research to advance our understanding of the ecological niche of soil-borne fungal entomopathogens. Studies of fungal ecology in the rhizosphere to date have focused on *M. anisopliae*. However, natural rhizosphere colonization by *M. anisopliae* and *B. bassiana* readily occurs on a variety of plants (Bruck unpublished data). It is plausible that as research continues, other fungal entomopathogens will be isolated from the rhizosphere as well. Natural rhizosphere colonization indicates that this phenomenon is not an artifact of the relatively short duration or the inundative release of fungal spores into the environment that took place in studies to date. The employment of molecular approaches will provide better insight into the genotypic diversity and aid in our understanding of the ecology of naturally-occurring fungal entomopathogens in soil and the rhizosphere. Bischoff et al. (2009) recognized nine distinct phylogenic species with the *M. anisopliae* lineage. The ability to objectively differentiate cryptic species using molecular tools allows for systematic efforts to differentiate physiological and ecological features that may further differentiate these phylogenic species (Bischoff et al. 2009).

Much is left to be done to fully understand the role that rhizosphere competent fungal entomopathogens play in regulating pest populations and how we can use that knowledge to design and implement more effective microbial control programs. Questions of particular importance to consider are highlighted by Vega et al. (2009) and include the following: (1) Do plants benefit from a rhizosphere association with fungal entomopathogens? (2) Is the 'bodyguard' concept relevant in soil? If so, what is the signaling mechanism between trophic levels? (3) Do different phylogenetic groups of fungal entomopathogens display different strategies in their association with plants? (4) How do soil-borne fungal entomopathgens interact between above and below ground ecosystems? (5) What is the mechanism of yield increases in *Z. mays* reported by Kabaluk and Ericsson (2007b)? *M. anisopliae* increased the stand density and fresh weight of *Z. mays* when conidia were applied to seeds prior to planting. Unfortunately, the mechanism for the yield increase is unknown. (6) Does plant diversity impact fungal entompathogen diversity at the landscape or local level, and what is its impact on natural pest control? In addition to the basic scientific questions posed above, there are a number of applied questions that require further investigation as well: (1) What is the most effective approach for inoculating roots with rhizosphere competent isolates? Approaches will need to be identified for plants propagated via seed, cuttings and tissue culture. (2) How long do rhizosphere competent isolates persist on the root system of annual and perennial plants? (3) Will the use of rhizosphere competent isolates provide consistent and acceptable levels of pest control?

Future prospects

Clearly, further investigation is necessary before we have even an elementary understanding of the ecology of fungal entomopathogens in soil. Early indications are that the rhizosphere, up until recently, has been an under appreciated niche for soil-borne fungal entomopathogens. A more complete understanding of fungal ecology is likely to aid in not only the development of the next generation of microbial control programs but may also lead to other benefits including increased yields (Kabaluk and Ericsson 2007b), direct disease antagonism, compatibility with other beneficial microorganisms in the rhizosphere (Jaronski et al. 2006), and plant growth promotion.

Acknowledgments I would like to thank Leslie Lewis and David Shapiro-Ilan for helpful suggestions which improved the manuscript. I would also like to thank Helen Roy, Fernando Vega, Mark Goettel, Judith Pell, Eric Wajnberg and David Chandler for the invitation to prepare this review. Mention of trade names or commercial products in this publication is solely for the purpose of providing specific information and does not imply recommendation or endorsement by the U.S. Department of Agriculture.

References

Agrios G (2005) Plant pathology, 5th edn. Elsevier Academic Press, San Diego

Aratchige NS, Lesna I, Sabelis MW (2004) Below-ground plant parts emit herbivore-induced volatiles: olfactory responses of a predatory mite to tulip bulbs infested by rust mites. Exp Appl Acarol 33:21–30

Bais HP, Weir TL, Perry LG, Gilroy S, Vivanco JM (2006) The role of root exudates in rhizosphere interactions with plants and other organisms. Annu Rev Plant Biol 57:233–266

Barea J-M, Pozo MJ, Azcón R, Azcón-Aguilar C (2005) Microbial co-operation in the rhizosphere. J Exp Bot 56:1761–1778

Bidochka MJ, Kasperski JE, Wild GAM (1998) Occurrence of the entomopathogenic fungi *Metarhizium anisopliae* and *Beauveria bassiana* in soils from temperate and near-northern habitats. Can J Bot 76:1198–1204

Bidochka MJ, Kamp AM, Lavender TM, Dekoning J, Amritha de Croos JN (2001) Habitat association in two genetic groups of the insect-pathogenic fungus *Metarhizium anisopliae*: uncovering cryptic species? Appl Environ Microbiol 67:1335–1342

Bidochka MJ, Menzies FV, Kamp AM (2002) Genetic groups of the insect-pathogenic fungus *Beauveria bassiana* are associated with habitat and thermal growth preferences. Arch Microbiol 178:531–537

Bing LA, Lewis LC (1991) Suppression of *Ostrinia nubilalis* (Hübner) (Lepidoptera: Pyralidae) by endophytic *Beauveria bassiana* (Balsamo) Vuillemin. Environ Entomol 20:1207–1211

Bing LA, Lewis LC (1992) Endophytic *Beauveria bassiana* (Balsamo) Vuillemin in corn: the influence of the plant growth stage and *Ostrinia nubilalis* (Hübner). Biocontrol Sci Technol 2:39–47

Bing LA, Lewis LC (1993) Occurrence of the entomopathogen *Beauveria bassiana* (Balsamo) Vuillemin in different tillage regimes and in *Zea mays* L. and virulence towards *Ostrinia nubilalis* (Hübner). Agric Ecosystems Environ 45:147–156

Bischoff JF, Rehner SA, Humber RA (2009) A multilocus phylogeny of the *Metarhizium anisopliae* lineage. Mycologia 101:512–530

Boff MIC, Zoon FC, Smits PH (2001) Orientation of *Heterorhabditis megidis* to insect hosts and plant roots in a Y-tube sand olfactometer. Entomol Exp Appl 98:329–337

Bowen GD, Rovira AD (1999) The rhizosphere and its management to improve plant growth. Advanc Agron 66:1–102

Bruck DJ (2004) Natural occurrence of entomopathogens in Pacific Northwest nursery soils and their virulence to the black vine weevil, *Otiorhynchus sulcatus* (F.) (Coleoptera: Curculionidae). Environ Entomol 33:1335–1343

Bruck DJ (2005) Ecology of *Metarhizium anisopliae* in soilless potting media and the rhizosphere: implications for pest management. Biol Control 32:155–163

Chandler D, Hay D, Reid AP (1997) Sampling and occurrence of entomopathogenic fungi and nematodes in UK soils. Appl Soil Ecol 5:133–141

Cory J, Ericsson JD (2009) Fungal entomopathogens in a tritrophic context. BioControl. doi:10.1007/s10526-009-9247-4 (this SI)
Cory JS, Hoover K (2006) Plant-mediated effects in insect-pathogen interactions. Trends in Ecol Evol 21:278–286
Dicke M, Sabelis MW, Takabayashi J, Bruin J, Posthummus MA (1990) Plant strategies of manipulating predator–prey interactions through allelochemicals: prospects for application in pest control. J Chem Ecol 16:3091–3118
Dicke M, van Baarlen P, Wessels R, Dijkman H (1993) Herbivory induces systemic production of plant volatiles that attract predators of the herbivore: extraction of endogenous elicitor. J Chem Ecol 19:581–599
Elliot SL, Sabelis MW, Janssen A, van der Geest LPS, Beerling EAM, Fransen J (2000) Can plants use entompathogens as bodyguards? Ecol Lett 3:228–235
Engler KM, Gold RE (2004) Effects of multiple generations of *Metarhizium anisopliae* on subterranean termite feeding and mortality (Isoptera: Rhinotermitidae). Sociobiology 44:211–240
Giri B, Giang PH, Kumari R, Prasad R, Varma A (2005) Microbial diversity in soils. In: Buscot F, Varma S (eds) Microorganisms in soils: roles in genesis and functions. Springer-Verlag, Heidelberg, pp 19–55
Harrison RD, Gardner WA (1991) Occurrence of entomogenous fungus *Beauveria bassiana* in pecan orchard soils in Georgia. J Entomol Sci 26:360–366
Hiltner L (1904) Uber neure Erfahrungen und probleme auf dem gebeit der bodenenbackteriologie und unter besonderer berucksichtigung der grundungung und brache. Arb Deut Landwirsch Ges 98:59–78
Hoeksema JD, Schwartz MW (2003) Expanding comparative-advantage biological market models: contingency of mutualism on partners' resource requirements and acquisition trade-offs. Proc R Soc London Ser B 270:913–919
Hountondji FCC, Sabelis MW, Hanna R, Janssen A (2005) Herbivore-induced plant volatiles trigger sporulation in entomopathogenic fungi: the case of *Neozygites tanajoae* infecting the cassava green mite. J Chem Ecol 31:1003–1021
Hu G, Leger RJ (2002) Field trials using a recombinant mycoinsecticide (*Metarhizium anisopliae*) reveal that it is rhizosphere competent. Appl Environ Microbiol 68:6383–6387
Humber RA (2008) Evolution of entomopathogenicity in fungi. J Invertebr Pathol 98:262–266
Inglis GD, Duke GM, Goettel MS, Kabaluk JT (2008) Genetic diversity of *Metarhizium anisopliae* var. *anisopliae* in southwestern British Columbia. J Invertbr Pathol 98:101–113
Jaronski ST (2007) Soil ecology of the entomopathogenic ascomycetes: a critical examination of what we (think) we know. In: Ekesi S, Maniania NK (eds) Use of entomopathogenic fungi in biological pest management. Research Signpost, Kerala India, pp 91–144
Jaronski ST, Larson B, Fuller-Schaefer C (2006) Effect of three bacterial disease control agents on the entomopathogenic fungi, *Metarhizium ansisopliae* and *Beauveria bassiana*. The American Phytopathological Society, North Central Division Meeting, June 13–15, 2006. Fargo, North Dakota
Jones DG (1998) The epidemiology of plant diseases. Kluwer AcademicPublishers, Dordrecht, The Netherlands
Kabaluk JT, Ericsson JD (2007a) Environmental and behavioral constraints on the infection of wireworms by *Metarhizium anisopliae*. Environ Entomol 36:1415–1420
Kabaluk JT, Ericsson JD (2007b) *Metarhizium anisopliae* seed treatment increases yield of field corn when applied for wireworm control. Agron J 99:1377–1381
Kennedy AC (1998) The rhizosphere and spermosphere. In: Sylvia DM, Fuhrmann JJ, Hartel PG, Zuberer DA (eds) Principles and applications of soil microbiology. Prentice Hall, New Jersey, pp 389–407
Kepler RM, Bruck DJ (2006) Examination of the interaction between the black vine weevil (Coleoptera: Curculionidae) and an entompathogenic fungus reveals a new tritrophic interaction. Environ Entomol 35:1021–1029
Kessler A, Baldwin IT (2001) Defensive function of herbivore-induced volatile emissions in nature. Science 291:2141–2144
Kiers ET, Denison RF (2008) Sanctions, cooperation, and the stability of plant-rhizosphere mutualisms. Annu Rev Ecol Evol 39:215–236
Klingen I, Hajek A, Meadow R, Renwick JAA (2002) Effect of brassicaceous plants on the survival and infectivity of insect pathogenic fungi. BioControl 47:411–425
Lacey LA, Kaya HK (2007) Field manual of techniques in invertebrate pathology. Springer, Dordrecht, The Netherlands
Leger RJ (2008) Studies on adaptations of *Metarhizium anisopliae* to life in soil. J Invertebr Pathol 98:271–276
Lewis LC, Bruck DJ, Gunnarson RD, Bidne KG (2001) Assessment of plant pathogenicity of endophytic *Beauveria bassiana* in Bt transgenic and non-transgenic corn. Crop Sci 41:1395–1400
Lewis LC, Bruck DJ, Gunnarson RD (2002) On-farm evaluation of *Beauveria bassiana* for control of *Ostrinia nubilalis* in Iowa, USA. BioControl 47:167–176
Lord JC (2005) From Metchnikoff to Monsanto and beyond: the path of microbial control. J Invertebr Pathol 89:19–29
Marschner H (1995) Mineral nutrition of higher plants, 2nd edn. Academic Press, London
McNew GL (1960) The nature, origin, and evolution of parasitism. In: Horsfall JG, Dimond AE (eds) Plant pathology: an advanced treatise, vol 2. Academic Press, New York, pp 19–69
Nelson EB, Burpee LL, Lawton MB (1994) Biological control of turfgrass diseases. In: Leslie A (ed) Handbook of integrated pest management for turf and ornamentals. CRC Press, Boca Raton, FL, pp 409–421
Neveu N, Grandgirard J, Nenon JP, Cortesero AM (2002) Systemic release of herbivore-induced plant volatiles by turnips infested by concealed root-feeding larvae *Delia radicum* L. J Chem Ecol 28:1717–1732
Pilon-Smits E (2005) Phytoremediation. Annu Rev Plant Biol 56:15–39
Prior C (1992) Discovery and characterization of fungal pathogens for locust and grasshopper control. In: Lomer CJ, Prior C (eds) Biological control of locusts and grasshoppers. CAB International, Wallingford, pp 159–180
Rasmann S, Köllner TG, Degenhardt J, Hiltpold I, Toepfer S, Kuhlmann U, Gershenzon J, Turlings TCJ (2005) Recruitment of entomopathogenic nematodes by insect-damaged maize roots. Nature 434:732–737

Rath AC (2000) The use of entomopathogenic fungi for control of termites. Biocontrol Sci Technol 10:563–581

Rehner SA, Buckley E (2005) A *Beauveria* phylogeny inferred from nuclear ITS and EF1-α sequences: evidence for cryptic diversification and links to *Cordyceps* teleomorphs. Mycologia 97:84–98

Roberts DW, Humber RA (1981) Entomogenous fungi. In: Cole GT, Kendrick WB (eds) Biology of conidal fungi, vol 2. Academic Press, New York, pp 201–236

Sabelis MW, van de Baan HE (1983) Location of distant spidermite colonies by phyoseiid predators: demonstration of specific kairomones emitted by *Tetranychus urticae* and *Panonychus ulmi*. Entomol Exp Appl 33:303–314

Sabelis MW, Janssen A, Pallini A, Venzon M, Bruin J, Drukker B, Scutareanu P (1999) Behavioral responses of predatory and herbivorous arthropods to induced plant volatiles: from evolutionary ecology to agricultural applications. In: Agrawal AA, Tuzun S, Bent E (eds) Induced plant defenses against pathogens and herbivores: biochemistry, ecology, and agriculture. American Phytopathological Society, St. Paul, Minnesota, pp 269–296

Schmidt EL (1979) Initiation of plant root-microbe interactions. Ann Rev Microbiol 33:355–376

Shapiro-Ilan DI, Gardner WA, Fuxa JR, Wood BW, Nguyen KB, Adams BJ, Humber RA, Hall MJ (2003) Survey of entomopathogenic nematodes and fungi endemic to pecan orchards of the Southeastern United States and their virulence to the pecan weevil (Coleoptera: Curculionidae). Environ Entomol 32:187–195

Thompson SR, Brandenburg RL (2005) Tunneling responses of mole crickets (Orthoptera: Gryllotalpidae) to the entomopathogenic fungus, *Beauveria bassiana*. Environ Entomol 34:140–147

Turlings TCJ, Tumlinson JH (1992) Systemic release of chemical signals by herbivore-injured corn. Proc Natl Acad Sci USA 89:8399–8402

Turlings TCJ, Tumlinson JH, Lewis WJ (1990) Exploitation of herbivore-induced plant odors by host-seeking parasitic wasps. Science 250:1251–1253

Turlings TCJ, Loughrin JH, McCall PJ, Rose USR, Lewis WJ, Tumlinson JH (1995) How caterpillar-damaged plants protect themselves by attracting parasitic wasps. Proc Natl Acad Sci USA 92:4169–4174

van Tol RWHM, van der Sommen ATC, Boff MIC, van Bezooijen J, Sabelis MW, Smits PH (2001) Plants protect their roots by alerting the enemies of grubs. Ecol Lett 4:292–294

Vega FE (2008) Insect pathology and fungal endophytes. J Invertebr Pathol 98:277–279

Vega FE, Goettel MS, Blackwell M, Chandler D, Jackson MA, Keller S, Koike M, Maniania NK, Monzón A, Ownley BH, Pell JK, Rangel DEN, Roy HE (2009) Fungal entomopathogens: new insights on their ecology. Fungal Ecol 2:149–159

Villani MG, Krueger SR, Schroeder PC, Consolie R, Consolie NH, Preston-Wilsey LM, Roberts DW (1994) Soil application effects of *Metarhizium anisopliae* on Japanese beetle (Coleoptera: Scarabaediae) behavior and survival in turfgrass microcosms. Environ Entomol 23:502–513

Villani MG, Allee LL, Preston-Wilsey L, Consolie N, Xia Y, Brandenburg RL (2002) Use of radiography and tunnel castings for observing mole cricket (Orthoptera: Gryllotalpidae) behavior in soil. Am Entomol 48:42–50

Wang C, Leger RJ (2007) The MAD1 adhesion of *Metarhizium anisopliae* links adhesion with blastospore production and virulence to insects, and the MAD2 adhesion enables attachment to plants. Eukary Cell 6:808–816

Wang C, Hu G, Leger RJ (2005) Differential gene expression by *Metarhizium anisopliae* growing in root exudate and host (*Manduca sexta*) cuticle or hemolymph reveals mechanisms of physiological adaptation. Fungal Genet Biol 42:7.4–718

Whipps JM (1997) Developments in the biological control of soil-borne plant pathogens. Advan Botanic Res 26:1–134

Whipps JM (2001) Microbial interactions and biocontrol in the rhizosphere. J Exper Bot 52:487–511

Author Biography

Dr. Denny J. Bruck develops biologically-based management programs for pests of small fruit and ornamental nursery plants at the USDA-ARS Horticultural Crops Research Unit. The focus of his research is developing new concepts fundamental to understanding the ecology of entomopathogens.

BioControl (2010) 55:113–128
DOI 10.1007/s10526-009-9241-x

Endophytic fungal entomopathogens with activity against plant pathogens: ecology and evolution

Bonnie H. Ownley · Kimberly D. Gwinn · Fernando E. Vega

Received: 17 September 2009 / Accepted: 12 October 2009 / Published online: 28 October 2009

Abstract Dual biological control, of both insect pests and plant pathogens, has been reported for the fungal entomopathogens, *Beauveria bassiana* (Bals.-Criv.) Vuill. (Ascomycota: Hypocreales) and *Lecanicillium* spp. (Ascomycota: Hypocreales). However, the primary mechanisms of plant disease suppression are different for these fungi. *Beauveria* spp. produce an array of bioactive metabolites, and have been reported to limit growth of fungal plant pathogens in vitro. In plant assays, *B. bassiana* has been reported to reduce diseases caused by soilborne plant pathogens, such as *Pythium, Rhizoctonia*, and *Fusarium*. Evidence has accumulated that *B. bassiana* can endophytically colonize a wide array of plant species, both monocots and dicots. *B. bassiana* also induced systemic resistance when endophytically colonized cotton seedlings were challenged with a bacterial plant pathogen on foliage. Species of *Lecanicillium* are known to reduce disease caused by powdery mildew as well as various rust fungi. Endophytic colonization has been reported for *Lecanicillium* spp., and it has been suggested that induced systemic resistance may be active against powdery mildew. However, mycoparasitism is the primary mechanism employed by *Lecanicillium* spp. against plant pathogens. Comparisons of *Beauveria* and *Lecanicillium* are made with *Trichoderma*, a fungus used for biological control of plant pathogens and insects. For *T. harzianum* Rifai (Ascomycota: Hypocreales), it has been shown that some fungal traits that are important for insect pathogenicity are also involved in biocontrol of phytopathogens.

Keywords *Beauveria bassiana* · Fungal endophyte · Hypocreales · Induced systemic resistance · *Lecanicillium* · Mycoparasite · *Trichoderma*

Handling Editor: Helen Roy.

B. H. Ownley (✉) · K. D. Gwinn
Department of Entomology and Plant Pathology, The University of Tennessee, 2431 Joe Johnson Drive, 205 Ellington Plant Sciences Bldg, Knoxville, TN 37996-4560, USA
e-mail: bownley@utk.edu

K. D. Gwinn
e-mail: kgwinn@utk.edu

F. E. Vega
Sustainable Perennial Crops Laboratory, United States Department of Agriculture, Agricultural Research Service, Building 001, BARC-West, Beltsville, MD 20705, USA
e-mail: fernando.vega@ars.usda.gov

Introduction

Resource availability can trigger shifts in functionality within a fungal species, thereby changing the ecological role of the organism (Termorshuizen and Jeger 2009). Shifts from one resource to another may necessitate significant adaptations in metabolism, particularly if the resources are dissimilar (Leger et al. 1997). Among members of the Hypocreales, animal, fungal, and plant resources are exploited. These fungi gain nutrition in a variety of ways, including: saprotrophs that colonize the rhizosphere and phyllosphere, endophytic saprotrophs, hemibiotrophs and necrotrophs of plants, entomopathogens, and mycoparasites. Some of these fungi function in more than one econutritional mode. Fungi traditionally known for their entomopathogenic characteristics, such as *Beauveria bassiana* (Bals.-Criv.) Vuill. (Ascomycota: Hypocreales) and *Lecanicillium* spp. (Ascomycota: Hypocreales), have recently been shown to engage in plant-fungus interactions (Vega 2008; Vega et al. 2008), and both have been reported to effectively suppress plant disease (Goettel et al. 2008; Ownley et al. 2008).

Mechanisms of plant disease suppression by biocontrol fungi

Biological control of plant pathogens usually refers to the use of microorganisms that reduce the disease-causing activity or survival of plant pathogens. Several different biological control mechanisms against plant pathogens have been identified. With some mechanisms, such as antibiosis, competition, and parasitism, the biocontrol organism is directly involved. With other modes of biological control, such as induced systemic resistance and increased growth response, endophytic colonization by the biocontrol organism triggers responses in the plant that reduce or alleviate plant disease.

Antibiosis, competition, and mycoparasitism

The mechanism of antibiosis includes production of antibiotics, bioactive volatile organic compounds (VOCs), and enzymes. Volatile bioactive compounds include acids, alcohols, alkyl pyrones, ammonia, esters, hydrogen cyanide, ketones, and lipids (Ownley and Windham 2007). The fungal endophyte *Muscodor albus* Worapong, Strobel & W.M. Hess (Ascomycota: Xylariales) produces a mixture of VOCs that are lethal to a variety of microorganisms (Strobel et al. 2001; Mercier and Jiménez 2004; Mercier and Smilanick 2005; Strobel 2006), as well as to insects (Riga et al. 2008; Lacey et al. 2009). In the first report of VOCs released by a fungal entomopathogen, carbon source played a major role in VOC production by *B. bassiana*. When cultured on glucose-based media, the VOCs identified were diisopropyl naphthalenes (<50%), ethanol (ca. 10%) and sesquiterpenes (6%), but in media with n-octacosane (an insect-like alkane), the primary VOCs were *n*-decane (84%) and sesquiterpenes (15%) (Crespo et al. 2008).

Enzymes involved in antibiosis are distinctly different from those involved in mycoparasitism of plant pathogens. For example, the biocontrol fungus *Talaromyces flavus* Tf1 (Klöcker) Stolk & Samson (Ascomycota: Eurotiales) produces the enzyme glucose oxidase, whose reaction product, hydrogen peroxide, kills microsclerotia of phytopathogenic *Verticillium* (Fravel 1988).

Fungal biocontrol organisms actively compete against plant pathogens for niche or infection site, carbon, nitrogen, and various microelements. The site of competition is often the rhizosphere, phyllosphere, or intercellularly within the plant. Successful competition is often a matter of timing as resources are likely to go to the initial colonizer.

Mycoparasitism is the parasitism of one fungus by another. Varying degrees of host specificity are displayed by mycoparasites. Within a given species of mycoparasite, some isolates may infect a large number of taxonomically diverse fungi, while others demonstrate a high level of specificity (Askary et al. 1998). As reviewed in Harmon et al. (2004), parasitism by the biocontrol fungus *Trichoderma* (Ascomycota: Hypocreales) begins with detection of the fungal host before contact is made. *Trichoderma* produces low levels of an extracellular exochitinase, which diffuse and catalyze the release of cell-wall oligomers from the target host fungus. This activity induces *Trichoderma* to release fungitoxic endochitinases, which also degrade the fungal host cell wall. Attachment of the mycoparasite to the host fungus is mediated by binding of carbohydrates in the *Trichoderma* cell wall to lectins in the cell wall of the fungal host. Upon contact, hyphae of *Trichoderma* coil around the host fungus and form appressoria. Several lytic enzymes are involved in degradation of the cell walls of fungal and oomycetous plant pathogens, including chitinases, ß-1,3 gluconases, proteases, and lipases.

In many cases, mechanisms of biocontrol are not mutually exclusive, i.e. multiple mechanisms may be operating against a specific plant pathogen, or a given biocontrol fungus may employ different mechanisms against different phytopathogens. For example, control of *Botrytis cinerea* Pers. (Ascomycota: Helotiales) on grapes (*Vitis*) with *Trichoderma* involves competition for nutrients and mycoparasitism of sclerotia, the overwintering, long-term survival structure of *Botrytis*. Both mechanisms contribute to suppression of the pathogen's capability to cause and perpetuate disease (Dubos 1987). Following application to leaves as a preventative, *Trichoderma* induced resistance to downy mildew, *Plasmopara viticola* (Berk. & M.A. Curtis) Berl. & De Toni (Oomycota: Peronosporales), in grape (Perazzolli et al. 2008). Therefore, it is possible that induced systemic resistance may also play a role in biocontrol of *Botrytis*. Induced resistance to *Botrytis*, following application of *T. harzianum* T39 Rifai (Ascomycota: Hypocreales) to roots and leaves of several ecotypes of *Arabidopsis thaliana* (L.) Heynh. has been reported (Korolev et al. 2008).

Induced systemic resistance

Plants are sessile organisms that must develop a complex chemical arsenal in order to withstand biotic and abiotic attack. Colonization of plants with nonpathogenic fungi and bacteria can lead to induced systemic resistance (ISR) in the host plant. Induced resistance is a plant-mediated biocontrol mechanism whereby the biocontrol agent and the phytopathogen do not make physical contact with one another. Plants react to the presence of a pathogen with a rapid expression of defense-related genes. For example, dramatic cellular changes, characterized by rapid necrotization of lemon (*Citrus* × *limon* (L.) Burm. f.) fruit exocarp cells were observed in fruit treated with *Lecanicillium muscarium* DAOM 198499 (Petch) Zare & W. Gams (formerly *Cephalosporium muscarium* Petch). Phenolic compounds and phenol oxidase were both present in reactive cells (Benhamou 2004).

In contrast, gene expression changes in plants infected with beneficial fungi tend to be mild, and the relationship is allowed to develop resulting in an infected or colonized plant. The signaling mechanisms for this induced resistance are based on jasmonic acid (JA) and ethylene (Van Loon et al. 1998; Van Wees et al. 2008; Gutjahr and Paszkowski 2009). Induction of systemic resistance via the JA/ethylene signaling pathway has been reported primarily for plant growth-promoting bacteria, however, it is also operative for many mycorrhizal fungi (Gutjahr and Paszkowski 2009) and biocontrol fungi (Harmon et al. 2004; Vinale et al. 2008).

Endophytism by fungal entomopathogens

Even though the term "endophyte" has several definitions (Hyde and Soytong 2008), it is widely accepted that endophytes are microorganisms present in plant tissues without causing any apparent symptoms. Fungal endophytes are widespread and quite diverse in nature (Arnold et al. 2000; Arnold 2007). For example, Vega et al. (2009b) reported 257 unique ITS genotypes for fungal endophytes isolated from coffee plants in Hawaii, Mexico, Colombia, and Puerto Rico. Infection by fungal endophytes can be localized (i.e., not systemic; see Saikkonen et al. 1998 and references therein), and establishing a long-term systemic infection with endophytic fungal entomopathogens that can act against plant pathogens will remain a challenge, and should be the focus of intensive study.

Isolation of *B. bassiana* as a fungal endophyte has been reported for many plants under natural conditions, as well as in plants inoculated using various methods (Vega 2008; Vega et al 2008). In contrast to the several studies dealing with endophytic *Beauveria* spp., only a handful of studies have been conducted on endophytic *Lecanicillium* spp. For example, *Lecanicillium dimorphum* (J.D. Chen) Zare & W. Gams and *L cf. psalliotae* (Treschew) Zare & W. Gams have been introduced as endophytes in date palms (*Phoenix dactylifera* L.) (Gómez-Vidal et al. 2006), and *L. muscarium* strain DAOM 198499 (=*Verticillium lecanii* (Zimm.) Viégas) and *L. muscarium* strain B-2 have been introduced as endophytes in cucumber (*Cucumis sativus* L.) roots (Benhamou and Brodeur 2001; Hirano et al. 2008). In cytological investigations of cucumber roots, the entomopathogen grew actively at the root surface and colonized a small number of epidermal and cortical cells, without inducing extensive host cell damage. Ingress into the root tissue was primarily intercellular and cell wall penetration was seldom observed (Benhamou and Brodeur 2001). *Verticillium*

(=*Lecanicillium*) *lecanii* has been reported as a natural endophyte in an Araceae (Petrini 1981), in *Arctostaphylos uva-ursi* (L.) (Widler and Müller 1984), and in *Carpinus caroliniana* Walter (Bills and Polishook 1991).

Although traditionally categorized as a soil saprophyte, *Beauveria* spp. are considered to be poor competitors for organic resources against other ubiquitous saprophytic soil fungi (Keller and Zimmermann 1989; Hajek 1997). The endophytic habit of *B. bassiana* may provide benefits to both plant and fungus. It is well known that plant species has a significant impact on shaping plant-associated microbial communities (Berg et al. 2005; reviewed in Berg and Smalla 2009). As suggested by the bodyguard hypothesis, the plant gains through reduction of damage against herbivorous insects (Elliot et al. 2000; White et al. 2002) or plant diseases; the fungus benefits through protection from environmental stress, acquisition of limited nutrients from endophytic colonization as well as exudates on the plant surface, and use of the plant surface as a staging platform for insect parasitism. On tomato (*Solanum lycopersicum* L.) and other dicots, as well as monocots, colonization by *B. bassiana* is not restricted to growth as an endophyte (Ownley et al. 2008; Powell et al. 2009; authors, unpublished data). From initial establishment as a seed treatment, the fungus can be found on the outer surfaces as the plant ages, particularly in areas where new leaves or shoots have emerged. The fungus also gains from nutrients acquired during saprophytic colonization of the plant when it, or parts of it senesce. Similar epiphytic growth was observed by Posada and Vega (2005) with cocoa (*Theobroma cacao* L.) seedlings.

Beauveria bassiana: Potential for biological control of plant pathogens

Beauveria bassiana is known to occur naturally in more than 700 species of insect hosts (Inglis et al. 2001). Infection of host insects results in the production of large numbers of conidia, thereby serving to increase the population size of the fungus (Meyling and Eilenberg 2007). There is now substantial evidence that *B. bassiana* can provide protection against some soilborne plant pathogens (Ownley et al. 2004; Ownley et al. 2008; Vega et al. 2009a, b). It is likely that more than one mode of action is operative in suppression of plant disease by *B. bassiana*. Isolates of the fungus are known to produce numerous secondary metabolites (e.g. beauvericin, beauverolides, bassianolides, oosporein, cyclosporin A, and oxalic acid) with antibacterial, antifungal, cytotoxic, and insecticidal activities (Grove and Pople 1980; Genthner et al. 1994; Gupta et al. 1995; Boucias and Pendland 1998; Copping and Menn 2000). Effects of these compounds on microorganisms and insects have been reported (Kanaoka et al 1978; Taniguchi et al. 1984; Eyal et al. 1994; Boucias et al. 1995). Recently, another antimicrobial compound, bassianolone, from *B. bassiana* fermentation culture under low nitrogen conditions, was characterized (Oller-López et al. 2005). Bassianolone has activity against fungi and Gram-positive cocci. Antibiosis assays with *B. bassiana* against various plant pathogens in vitro have been reported (Table 1). However, the antimicrobial compounds were not identified.

Beauveria bassiana strain 11-98 suppresses plant disease caused by the soilborne plant pathogens *Rhizoctonia solani* Kühn (Basidiomycota: Cantharellales) (Ownley et al. 2004) and *Pythium myriotylum* Drechsler (Oomycota: Pythiales) (Clark et al. 2006). This isolate produces beauvericin (Leckie et al. 2008) and oosporein (authors, unpublished data), but it is not known if these compounds play a role in suppression of plant disease. Biological control of plant pathogens with *B. bassiana* 11-98 is likely to involve competition for resources (Ownley et al. 2004), since the fungus is a plant colonist. Application of *B. bassiana* 11-98 to tomato seed resulted in endophytic and epiphytic colonization of seedlings and subsequent protection against damping-off. Similarly, seed treatment of cotton (*Gossypium hirsutum* L.) reduced severity of *R. solani* damping-off in seedlings (Griffin 2007; Ownley et al. 2008). In both tomato and cotton, the degree of disease control achieved with *Beauveria bassiana* was correlated with the population density of conidia established on seed (Ownley et al. 2008; authors, unpublished data). Smaller seeds, such as tomato were protected more effectively with rates of 1×10^6–10^7 CFU/seed, while higher rates (1×10^7–10^9 CFU/seed) gave the greatest protection against seedling disease in cotton.

Parasitism of *Pythium myriotylum* by *B. bassiana* may be involved in suppression of Pythium damping-

Table 1 Studies reporting activity of *Beauveria* spp. against plant pathogens

Strain or species of *Beauveria*	Type of study	Plant pathogen	Activity against plant pathogen	Reference
Beauveria bassiana, isolated from wheat rhizosphere	In vitro bioassay	*Gaeumannomyces graminis* var. *tritici* J. Walker (Ascomycota: Sordariomycetidae)	Inhibited growth; produced chitinase and β-gluconases	Renwick et al. (1991)
	In planta (wheat), pot assays		Suppressed take-all disease	
Beauveria bassiana (Bals.-Criv.) Vuill. (Ascomycota: Hypocreales), five different isolates	In vitro bioassay	*Fusarium oxysporum* E.F. Smith & Swingle (Ascomycota: Hypocreales)	All *Beauveria* isolates inhibited mycelial growth of the pathogens tested	Reisenzein and Tiefenbrunner (1997)
		Armillaria mellea (Vahl) P. Kumm (Basidiomycota: Agaricales)		
		Rosellinia necatrix Berl. ex Prill. (Ascomycota: Xylariales)		
Culture filtrate of *B. bassiana*	In vitro bioassay	*Fusarium oxysporum* f. sp. *lycopersici* (Sacc.) W.C. Snyder & H.N. Hansen (Ascomycota Hypocreales)	Inhibited mycelial growth;	Bark et al. (1996)
		Botrytis cinerea Pers. (Ascomycota: Helotiales)	Inhibited and delayed conidial germination	
B. bassiana	In vitro bioassay	*Pythium ultimum* Trow (Oomycota: Pythiales), *Pythium debaryanum* R. Hesse (Oomycota: Pythiales), *Septoria nodorum* (=*Phaeosphaeria nodorum* (E. Müll.) Hedjar. (Ascomycota: Pleosporales)	Caused cell lysis; inhibited mycelial growth	Vesely and Koubova (1994)
B. brongniartii (Sacc.) Petch (Ascomycota: Hypocreales)		*Rhizoctonia solani* Kühn (Basidiomycota: Canthareliales), *Pythium irregular* Buisman (Oomycota: Pythiales), *Phoma betae* (=*Pleospora betae* Björl. (Ascomycota: Pleosporales)), *Phoma exigua* var. *foveata* Malc. & E.G. Gray (Ascomycota: Pleosporales)	Did not inhibit mycelial growth of these pathogens	
Culture filtrates of *Beauveria* sp.	In vitro bioassay	*Rhizoctonia solani*	Inhibited mycelial growth; stimulated growth of cucumber	Lee et al. (1999)
B. bassiana 142, applied to onion bulbs	*In planta* (onion), field and greenhouse	*Fusarium oxysporum* f. sp. *cepae* (Hanzawa) W.C. Snyder & H.N. Hansen (Ascomycota: Hypocreales)	Increased bulb germination; reduced plant infection	Flori and Roberti (1993)
B. bassiana 11-98, applied as a seed treatment	*In planta* (tomato), greenhouse	*Rhizoctonia solani*	Reduced damping off of seedlings; increased plant growth	Ownley et al. (2000) and Ownley et al. (2004)
B. bassiana 11-98, applied as a seed treatment	*In planta* (tomato), growth chamber	*Pythium myriotylum* Drechsler (Oomycota: Pythiales)	Reduced damping off of seedlings	Clark et al. (2006)
B. bassiana 11-98	In vitro bioassay	*Rhizoctonia solani*	Did not inhibit mycelial growth of *R. solani*; but hyphae of 11-98 coiled around hyphae of *P. myriotylum*, which suggested parasitism	Griffin (2007) and Ownley et al. (2008)
B. bassiana 11-98, applied as a seed treatment	*In planta* (cotton), growth chamber	*Pythium myriotylum*	Reduced damping-off of seedlings	

off in tomato seedlings. In dual culture, hyphae of isolate 11-98 were observed coiling around the larger coenocytic hyphae of *P. myriotylum* (Griffin 2007).

The extent of endophytic colonization of tomato by *B. bassiana* 11-98 was also correlated with the rate of conidia applied to seed. Rates that were most effective in disease control also resulted in the greatest degree of plant colonization. *Beauveria bassiana* was detected in root, stem, and leaf sections of surface-sterilized tomato seedlings with standard dilution plating procedures onto semi-selective medium (Ownley et al. 2008). In addition to seedlings, *B. bassiana* 11-98 has been recovered from foliage, stem, and root tissues of surface-sterilized 18-week-old tomato plants produced from treated seed (Powell et al. 2009). *Beauveria bassiana* has also been recovered as an endophyte of eastern purple coneflower (*Echinacea purpurea* L. Moench), cotton, snap bean (*Phaseolus vulgaris* L.), soybean (*Glycines max* L.), and switchgrass (*Panicum virgatum* L.) following application of conidia to seed (Griffin 2007; Ownley et al. 2008; authors, unpublished data).

Endophytic *B. bassiana* 11-98 has been observed with scanning electron microscopy (SEM), and detected with polymerase chain reaction (PCR) in cotton seedlings (Griffin 2007). Using SEM on seedlings maintained in a sterile system, conidial germination and hyphal growth were observed in association with areas of leaf exudation. Penetration points through epithelial cells were observed, without formation of a specialized structure. Hyphae ramified through the palisade parenchyma and mesophyll layers of leaf tissues. *Beauveria bassiana* 11-98 was also detected with PCR in a mixed DNA sample of 1 part *B. bassiana* DNA to 1,000 parts cotton DNA, and from surface-sterilized tissues of cotton seedlings grown from *B. bassiana*-treated seed (Griffin 2007; Ownley et al 2008; authors, unpublished data).

The results of a study with cotton seedlings suggested that induced systemic resistance is also a probable mechanism of biological control for *B. bassiana* 11-98 (Griffin 2007; Ownley et al. 2008; authors, unpublished data). Isolate 11-98 was evaluated for its ability to induce systemic resistance in cotton against *Xanthomonas axonopodis* pathovar *malvacearum* (causes bacterial blight). Conidia of *B. bassiana* were applied as a root drench to 5-day old seedlings, 13 days prior to pathogen challenge. Treatment with *B. bassiana* (at 10^7 CFU/seedling root) resulted in significantly lower foliar disease ratings for bacterial blight than the untreated control and was as effective as 2,6-dichloro-isonicotinic acid, which has been shown to induce systemic resistance against plant pathogens.

Lecanicillium spp. and biological control of plant pathogens

Lecanicillium spp. (formerly classified in the single species *Verticillium lecanii*) are well known as entomopathogens of aphids and scale insects (Hall 1981; Goettel et al. 2008). These fungi are also known as mycoparasites of species of plant pathogenic, biotrophic powdery mildew (Hall 1980; Verhaar et al. 1996) and rust fungi (Spencer and Atkey 1981; Allen 1982; Whipps 1993) on various vegetable, fruit, and ornamental crops, and as pathogens of plant parasitic nematodes (Meyer et al. 1990; Shinya et al. 2008). Activity of *Lecanicillium* spp. against both plant pathogens and insects has been demonstrated in bioassays (Askary et al. 1998; Askary and Yarmand 2007; Kim et al. 2007) and greenhouse studies (Kim et al. 2008) (Table 2).

Commercial products containing *Lecanicillium* spp. have not been developed for plant disease control. However, a formulation of *L. longisporum* (Petch) Zare & W. Gams, known as Vertalec®, is available for control of insect pests. *Lecanicillium longisporum* (applied as Vertalec®), *Lecanicillium attenuatum* Zare & W. Gams CS625, and *Lecanicillium* sp. DAOM 198499 suppressed development of powdery mildew, *Podosphaera fuliginea* (Schltdl.) U. Braun & S. Takam. (Ascomycota: Erysiphales) (=synonym *Sphaerotheca fuliginea*) on cucumber leaf discs when applied one or eight days after powdery mildew inoculation. When applied to highly infected leaf discs 11–15 days after pathogen inoculation, *Lecanicillium* treatments significantly suppressed subsequent production of powdery mildew spores, compared to controls (Kim et al. 2007). In greenhouse experiments, *L. longisporum* (applied as Vertalec®) suppressed spore production of powdery mildew on potted cucumber plants under conditions of low and high infection levels (Kim et al. 2008).

Askary et al. (1997) provided ultrastructural and cytochemical evidence for the process of parasitism of *P. fuliginea* by *Lecanicillium* sp. DAOM 198499

Table 2 Studies on *Lecanicillium* spp. as dual biological controls for plant pathogens and insect pests

Species or strain of *Lecanicillium*[a]	Type of study	Plant pathogen	Mode of action against plant pathogen	Insect	Reference
V. lecanii Vertalec DAOM 216596 (see below) DAOM 198499 (see below)	Laboratory bioassay	*Podosphaera fuliginea* (Schltdl.) U. Braun & S. Takam. (Ascomycota: Erysiphales) (syn. *Sphaerotheca fuliginea*) Powdery mildew	Parasitism/ antibiosis	*Macrosiphum euphorbiae* (Hemiptera: Aphididae)	Askary et al. (1998)
L. muscarium (Petch) Zare & W. Gams (Ascomycota: Hypocreales) strain DAOM 198499	Laboratory bioassay	*P. fuliginea* (syn. *S. fuliginea*)	Parasitism	*M. euphorbiae* *Aphidius nigripes* (Hymenoptera: Braconidae)	Askary and Yarmand (2007)
L. longisporum (Petch) Zare & W. Gams (Ascomycota: Hypocreales) (Vertalec) *L. attenuatum* Zare & W. Gams (Ascomycota: Hypocreales) strain CS625 *Lecanicillium* sp. strain DAOM 198499	Laboratory bioassay	*P. fuliginea* (syn. *S. fuliginea*)	Not reported	*Myzus persicae* (Hemiptera: Aphididae) *M. euphorbiae* *Aulacorthum solani* (Hemiptera: Aphididae)	Kim et al. (2007)
L. longisporum (Vertalec)	Greenhouse	*P. fuliginea* (syn. *S. fuliginea*)	Not reported	*Aphis gossypii* (Hemiptera: Aphididae)	Kim et al. (2008)
L. lecanii (Zimm.) Zare & W. Gams (Ascomycota: Hypocreales)	Field (survey)	*Hemileia vastatrix* Berk. & Broome (Basidiomycota: Pucciniales) Coffee leaf rust	Parasitism	*Coccus viridis* (Hemiptera: Coccidae)	Vandermeer et al. (2009)

[a] Name listed is the same as was given in the reference

(formerly *V. lecanii* DAOM 198499), including production of cell-wall degrading enzymes such as chitinases. They suggested that prior to invasion of *P. fuliginea*, the powdery mildew fungus was weakened by antibiotics produced by *Lecanicillium* (Askary et al. 1997). Subsequently, Benhamou and Brodeur (2000) showed that this strain does produce anti-fungal compounds in culture that are effective against *Penicillium digitatum* (Pers.) Sacc. (Ascomycota: Eurotiales), which causes postharvest green mold of citrus. It has been suggested that production of antimicrobial compounds that weaken or kill the target host cells prior to parasitism is a form of specialized saprophytism, rather than parasitism (Bélanger and Labbé 2002).

In most of the studies with *Lecanicillium* as a biological control against plant pathogens, activity has been attributed to parasitism. Indeed, an array of extracellular lytic enzymes have been reported for isolates of *Lecanicillium*, including cellulases, proteases, β-1,3-glucanases, chitinases (Bidochka et al. 1999; Saksirirat and Hoppe 1991) and more recently, pectinases (Benhamou and Brodeur 2001). However, induction of plant host defense reactions against *P. digitatum* (Benhamou and Brodeur 2000; Benhamou 2004), *Pythium ultimum* Trow (Oomycota: Pythiales) (Benhamou and Brodeur 2001), and powdery mildew (Hirano et al. 2008) have been reported. In studies on biological control of *P. ultimum*, *Lecanicillium* sp. DAOM 198499 grew intercellularly among epidermal and cortical cells on cucumber roots treated with the fungus (Benhamou and Brodeur 2001). Endophytic colonization of cucumber roots was also observed when blastospores of *L. muscarium* B-2 were applied to roots. Subsequently induced resistance to powdery mildew on the cucumber leaf surface was reported (Hirano et al. 2008). Koike et al. (2004) demonstrated that *L. muscarium* B-2 is also a very successful epiphytic colonist of cucumber leaf surfaces, suggesting that competition for nutrients and space may also be operative against powdery mildew.

Fungal endophytism and induced systemic resistance

Recently, protcomic analysis of *P. dactylifera* infected with endophytic *B. bassiana* or two *Lecanicillium* spp. was reported by Gómez-Vidal et al. (2009). Colonization by *B. bassiana*, *L. dimorphum*, or *L. cf. psalliotae* resulted in induction of proteins related to plant defense or stress response, and proteins involved in energy metabolism and photosynthesis were also affected. As additional studies on molecular analysis of plants infected with endophytic fungal entomopathogens are conducted, it will become evident that endophytism is inducing important changes in plant metabolism, even though the plant does not present any symptoms of endophyte infection. It will be important to take into consideration that endophytes may cause plants to enter a "primed state" (*sensu* Conrath et al. 2006; see also Schulz and Boyle 2005), which could be contributing to the antagonistic effects of *B. bassiana* and *Lecanicillium* on plant pathogenic fungi. It is also possible that endophyte infection might result in positive effects such as enhanced plant growth (Ernst et al. 2003; Schulz and Boyle 2005). Plant growth-related variables should be measured in all studies dealing with the introduction of fungal entomopathogens as possible endophytes, as was recently done by Tefera and Vidal (2009) for sorghum plants inoculated with *B. bassiana*, although it will be difficult to elucidate the role of a specific endophyte if others are already present in the plant.

When endophytism results in "primed" plants, subsequent biotic challenge leads to a transitory period of strongly potentiated gene expression that is associated with accelerated defense responses. These responses confer broad-spectrum resistance to pathogens and insects (Van Wees et al. 2008). In this respect, plants colonized by fungal entomopathogens resemble plants colonized with plant growth-promoting rhizobacteria (Harmon et al. 2004). Much of the research on systemic resistance of plants infected with endophytic beneficial fungi has focused on mycorrhizal fungi (reviewed in Gutjahr and Paszkowski 2009). These obligate fungi live on plant roots and stimulate plant growth and development by increasing nutrient uptake and decreasing disease and insect problems. While plants infected with hypocrealean fungi do not have the complex structures associated with mycorrhizal infection, they can occupy a nutritional niche in or on the plant and develop an active cross talk with their plant hosts that results in induced resistance (Vinale et al. 2008). Induction of plant resistance has been reported for several species of *Trichoderma* (Harmon et al. 2004; Jeger et al. 2009), and mechanisms for induced resistance are beginning to emerge (Segarra et al. 2007; Vinale et al 2008). Mechanisms for induced resistance by other hypocrealean fungi are scant, but much information on mechanisms of induced resistance obtained from studies with *Trichoderma* can be applied to other fungal entomopathogens.

Many species of *Trichoderma* have been commercially developed for biological control of plant diseases and insects (Harmon et al. 2004; Shakeri and Foster 2007). Some of these isolates induce resistance to plant pathogens (Table 3). Typically, *Trichoderma* is applied to soil or to plant roots grown in co-culture with the fungus. However, some species induce systemic resistance when leaves are treated with *Trichoderma* conidia (Perazzolli et al. 2008; Korolev et al. 2008). Plant hosts in which resistance is induced are taxonomically diverse and include both monocots and dicots. Several recent studies support jasmonate/ethylene signaling as the mechanism for induced systemic resistance (Table 3), further suggesting that the response is similar to that induced by rhizobacteria (reviewed in Harmon et al. 2004). Induced resistance is broad spectrum, and subsequent challenges of the primed plant by taxonomically diverse pathogens (e.g., bacteria, necrotrophic fungi, biotrophic fungi) induce a rapid and intense activation of cellular defense mechanisms somewhat reminiscent of hypersensitive responses.

Species in the genus *Trichoderma* (Ascomycota: Hypocreales) are well known for the production of bioactive metabolites that play a role in the mycoparasitic or entomopathogenic lifestyles of the

Table 3 Recent evidence for involvement of the jasmonate/ethylene pathway in systemic resistance induced by *Trichoderma* species

Species and strain or extract	Plant	Pathogen	Evidence of effects	Efficacy	References
T. asperellum Samuels, Lieckf. & Nirenberg (Ascomycota: Hypocreales) strain T34, (10^7 spores)	*Cucumis sativus* L. (cucumber)	*Pseudomonas syringae* pv *lachrymans*	Significant increase of jasmonic acid (JA), but not salicylic acid (SA) at 1 h, both peaked at 3 h; JA levels not above untreated control after 6 h, SA decreased until 24 h; Significant increase of peroxidase by 6 h	Reduced bacterial colony forming units by ca. 50%	Segarra et al. (2007)
T. harzianum Rifai (Ascomycota: Hypocreales) strain T39	*Arabidopsis thaliana* (L.) Heynh.	*Botrytis cinerea* Pers. (Ascomycota: Helotiales)	Col-0 ecotype, and auxin-resistant and SA acid mutants were ISR-inducible; Mutants impaired in ABA, gibberillic acid, or ethylene/JA were not ISR-inducible	Disease severity reduced in Col-0 following either root or leaf application	Korolev et al. (2008)
T. harzianum strain T39	*Vitis vinifera* L. cv. Pinot Noir (grape)	*Plasmopara viticola* (Berk. & M.A. Curtis) Berl. & De Toni (Oomycota: Peronosporales)	Timing and persistence differed from BTH which is SA-dependent	Leaf treatment decreased disease severity; Root treatment did not	Perazzolli et al. (2008)
T. virens (J.H. Mill., Giddens & A.A. Foster) Arx (Ascomycota: Hypocreales) strain Gv29-8	*Zea mays* L. (corn)	*Colletotrichum graminicola* (= *Glomerella graminicola* D.J. Politis (Ascomycota: Sordariomycetidae)	Induction of JA and green leaf volatile biosynthetic genes	Reduced lesion area in leaves from endophytic plants	Djonović et al. (2007)

fungus, as well as in the induction of resistance in plant hosts. Elicitors or resistance inducers can be divided into three broad categories: proteins with enzymatic activity, avirulence-like gene products, and low molecular weight compounds released from cell walls (either fungal or plant) as a result of hydrolytic enzymes (e.g., chitinase, glucanase) (Vinale et al. 2008). In several recent studies, various proteins and peptides from *Trichoderma* have been shown to induce host defense responses (Table 4). Volatiles released after treatment with alamethicin, a 20-amino acid polypeptide isolated from *T. viride* Pers., affect the behavior of the parasitoid *Cotesia glomerata* (L.) (Hymenoptera: Braconidae) (Bruinsma et al. 2009). Wasps chose alamethicin-treated plants over nontreated plants, but chose plants on which *Pieris brassicae* (L.) (Lepidoptera: Pieridae) had fed over alamethicin-treated plants.

Sm1, a hydrophobin-like small protein secreted by *Trichoderma virens* (J.H. Mill., Giddens & A.A. Foster) Arx, was the first non-enzymatic proteinaceous elicitor determined to be involved in induced resistance responses in rice (*Oryza sativa* L.), cotton, and maize (*Zea mays* L.) (Djonović et al. 2006, 2007). Recently a second small hydrophobin-like protein (Epl1) was isolated from *Hypocrea atroviride* (=*Hypocrea atroviridis* Dodd, Lieckf. & Samuels (Ascomycota: Hypocreales)) (teleomorph of *T. atroviride* P. Karst.) (Vargas et al. 2008). Epl1 was produced as a dimer. Sm1 can also be a dimer, but upon dimerization, the glycosyl moiety and activity are lost. Both hydrophobins are active as resistance inducers when configured as a monomer. Vargas et al. (2008) have proposed that aggregation of the elicitor disrupts the molecular cross-talk between the beneficial fungal colonizer and plant.

Recent proteomic studies provide a glimpse into the complexity of the *Trichoderma*-plant interaction. In cucumber, 51 proteins were different in treatments with *T. asperellum* Samuels, Lieckf. & Nirenberg and untreated controls; 17 proteins were up-regulated, and 11 were down-regulated. Proteins were divided

Table 4 Effects of selected *Trichoderma*-derived peptides and proteins on host defense responses

Peptide/protein	Plant	Effects and efficacy	Reference	Similar compounds described for *Beauveria* or *Lecanicillium* spp.
Alamethicin: Ion channel-forming peptide mixture	*Brassica oleracea* L. var. *gemmifera* DC. 'Cyrus' (brussel sprouts)	20-fold more potent inducer of ISR than JA; volatile emissions; increased preference for parasitoid wasps (*Cotesia glomerata* (L.) (Hymenoptera: Braconidae))	Bruinsma et al. (2009)	
	Suspension cells of *Arabidopsis thaliana* (L.) Heynh. (Col-1) and *Nicotiana tabacum* L. 'BY-2' (tobacco)	Activation of callose synthase; callose deposition	Aidemark et al. (2009)	
Mitogen-activated protein kinase TMK1: Serine-threonine kinases	*Phaseolus vulgaris* L. (var. nanus L.) (bean)	Deletion tmk1 mutants had reduced mycoparasitism and host-specific regulation of *ech42* gene transcription; deletion mutants had an increased ability to protect plants against *Rhizoctonia solani* Kühn (Basidiomycota: Cantharellales)	Reithner et al. (2007)	Zhang et al. (2009)—*Beauveria*—regulation of environmental stress and virulence to insects
Sm1: Cerato-platanin protein that is hydrophobin-like	*Zea mays* L. (corn)	Deletion or over-expression of Sm1 in mutants did not affect normal growth and development of *Trichoderma virens* (J.H. Mill., Giddens & A.A. Foster) Arx (Ascomycota: Hypocreales); Root colonization was not affected in mutants, but ability to induce resistance to a foliar pathogen was reduced in deletion mutants and increased in some over-expression mutants	Djonović et al. (2007)	Ying and Feng (2004) *Beauveria*—relationship between hydrophobins and thermotolerance Kamp (2002) *Lecanicillium*—Hydrophobins abundant in sporulating cultures, but not in mycelial cultures
	Oryza sativa L. 'M-202' (rice); *Gossypium hirsutum* L. 'Paymaster 2326BG/RR' and 'DeltaPine 50' (cotton)	Induced expression of defense genes (glucanase, chitinase) locally and systemically; H_2O_2 produced in Sm1-treated levels, but no resulting necrosis	Djonović et al. (2006)	
Ethylene-inducing xylanase: 18 Kd protein similar to serine protease	*Gossypium hirsutum* 'DeltaPine 50' (cotton)	The 18 Kd protein increased terpenoid production and peroxidase activity	Hanson and Howell (2004)	
ThPG1 endopolygalacturonase: Cell-wall degrading enzyme associated with pectin degradation	*Lycopersicon esculentum* (=*Solanum lycopersicon* L. var. *lycopersicon*) 'Marmande' (tomato)	ThPG1-silenced mutants had lower polygalacturonase activity and less growth on pectin medium; protection against *Botrytis cinerea* Pers. (Ascomycota: Helotiales) was the same for ThPG1-silenced mutants and wild type, even though root colonization by mutants was lower	Morán-Diez et al. (2009)	Fenice et al. (1997) *Lecanicillium*—Antarctic strains of *V.* (=*Lecanicillium*) *lecanii* had wide enzymatic competence, including polygalacturonase activity
ABC transporter membrane pump: ATP-binding cassette with transmembrane domain	*L. esculentum*	Gene up-regulated in fungus by pathogen-secreted metabolites and some fungicides; deletion mutants were sensitive to fungicides and lost ability to protect against *Pythium ultimum* Trow (Oomycota: Pythiales) and *R. solani*	Ruocco et al. (2009)	

into four categories: stress and defense, energy and metabolism, secondary metabolism, and protein synthesis/folding (Segarra et al. 2007). In maize, 114 proteins were up-regulated and 50 were down-regulated in response to treatment with *T. harzianum*. Most of the upregulated genes were for proteins involved in carbohydrate metabolism, defense, and photosynthesis (Shoresh and Harman 2008).

There are several parallels between *Trichoderma* and *Beauveria* and/or *Lecanicillium* spp. that suggest similar mechanisms of induced resistance:

1. These fungi can live endophytically between plant cells without causing negative effects on plant growth and development. Genes with similar function (e.g., plant defense/stress response, energy metabolism, and photosynthesis) are up-regulated in plants colonized by *Beauveria* and *Lecanicillium* (Gómez-Vidal et al. 2009) and those colonized by *Trichoderma* spp. (Segarra et al. 2007; Shoresh and Harman 2008).
2. Plant colonization can be established horizontally by application of spores to seed, roots, or leaves. Even though the relationship between the fungi and their hosts is intimate, plants can easily be infected. This is similar to mycorrhizae but contrasts markedly with the grass endophytes in the genus *Neotyphodium* (Ascomycota: Hypocreales), which are transmitted vertically via seed (Giménez et al. 2007; Hartley and Gange 2009).
3. *Beauveria* and *Trichoderma* spp. are natural and introduced colonists of a wide variety of plants that include both dicots and monocots. Although there is less information available on the plant host range of *Lecanicillium* spp., it has also been recovered as a natural and introduced endophyte of monocots and dicots.
4. All three fungi produce a wide array of enzymes and avirulence-like products. Hydrolytic enzymes that can attack substrates as diverse as plant cell walls, insect cuticle, and oomycetous and fungal plant pathogens are important for the varied nutritional niches occupied by these fungi.
5. *Beauveria bassiana* and many species of *Trichoderma* produce hydrophobins or hydrophobin-like molecules. It has been suggested that the functions of hydrophobins in the life cycle of fungi include: formation of protective layers, attachment, structural components of cell walls, and reduction of surface tension to allow aerial growth (Linder 2009). Hydrophobins produced by *B. bassiana* have been shown to be important in conidial thermotolerance (Ying and Feng 2004) and attachment to substrates (Holder and Keyhani 2005). Hydrophobins of *T. asperellum* were proposed to protect hyphae from defense compounds during the early stages of infection (Viterbo and Chet 2006). Therefore, it is possible that they play a similar role in *B. bassiana*. Hydrophobins have been detected in *Lecanicillium* (Kamp 2002), but little is known on their role in the fungal life cycle.
6. Mitogen-associated protein kinases (MAP kinases) in the subfamily HOG-1 (High osmolarity glycerol (1) are associated with host infection and with protection from osmotic stress in *Beauveria* and *Trichoderma* spp. The MAP kinases interfere with the ability of *T. atroviride* to induce resistance to the soilborne plant pathogen, *R. solani*, in bean plants. Deletion mutants had a greater ability than wild type to protect the plants. In *B. bassiana*, MAP kinases regulated response of the fungus to stress. Deletion mutants were more sensitive to hyper-osmotic stress, high temperature, and oxidative stress than the wild type (Zhang et al. 2009). When transcript levels of hydrophobin-encoding genes in the deletion mutants were low, conidial attachment to cicada hind wings was severely impaired (Zhang et al. 2009).
7. Both *Beauveria* and *Trichoderma* spp. can induce systemic resistance to bacterial pathogens. In cucumber, plants infected by *T. asperellum* (10^7 conidia ml^{-1}) supported less than 50% the number of colony-forming units (CFU) after challenge with *Pseudomonas syringae* pathovar *lachrymans* (Segarra et al. 2007). Treatment of cotton with 1×10^7 CFU *B. bassiana* 11-98 per root induced systemic resistance against bacterial blight (*Xanthomonas axonopodis* pathovar *malvacearum*) on cotton foliage. Although bacterial populations were not assessed, foliar disease ratings were significantly lower for *Beauveria*-treated plants than the untreated control (Griffin 2007).
8. Both *Lecanicillium* and *Trichoderma* spp. can induce systemic resistance to oomycetous plant pathogens. Host plant signaling and subsequent

intense defense responses have been proposed for *Lecanicillium*-treated cucumber. Ingress of *P. ultimum* into roots resulted in the deposition of an electron-opaque material that frequently encircled pathogen hyphae and accumulated in uninfected xylem vessels (Benhamou and Brodeur 2001). Inoculation of roots with *L. muscarium* resulted in root colonization and endophytic growth. Plant leaves were protected from powdery mildew, but defense enzymes were not different in colonized and non-colonized plants (Hirano et al. 2008). *Trichoderma harzianum* induced systemic resistance in pepper plants grown from seed treated with *T. harzianum* spores (Ahmed et al. 2000). Stem lesions, caused by inoculation with *Phytophthora capsici* Leonian (Oomycota: Peronosporales), were 40% shorter than lesions in inoculated plants grown from non-treated seed. *P. capsici* was isolated from zones immediately contiguous with the necrotic tissue, but *T. harzianum* was not, suggesting that there was no direct contact between them. The percentage of *P. capsici* isolated nine days after inoculation was greater in non-treated inoculated plants than in *Trichoderma*-treated plants inoculated with *P. capsici*. In addition to induced resistance against *P. capsici* in the upper part of the plant, concentration of the phytoalexin capsidiol was more than 7-fold greater than in non-treated plants inoculated with *P. capsici*, six days after inoculation (Ahmed et al. 2000).

Conclusions

The ability of many hypocrealean entomopathogens to occupy nutritional niches as diverse as insects, fungi, and plants provides unique opportunities for biological control of multiple plant pathogens and insect pests. Use of these fungi may overcome some of the challenges faced in plant disease control. For example, many foliar phytopathogens have a very high sporulation rate and are well-suited for widespread dissemination as air-borne propagules. If genetic resistance is not available in the crop, fungicide applications are often the primary means of disease control. The rapid reproduction rate of foliar pathogens coupled with frequent applications of systemic fungicides, many of which are narrow spectrum, increases the chances of developing fungicide resistance in these pathogens (Fry 1982). The ability of the hypocrealean fungi to use several strategies reduces the probability of development of resistance. For example, treatment of roots or seeds with *Beauveria* or *Lecanicillium* spp. conidia potentially produces endophyte-infected plants that reduce initial establishment of the disease through induced resistance. Studies have shown that both *Beauveria* and *Lecanicillium* spp. can become established as epiphytes, which provides opportunities for plant disease suppression through antibiosis, competition, or mycoparasitism. Endophytic and epiphytic populations of these fungi could also reduce insect damage to the plant.

Plant diseases caused by soilborne fungi are notoriously difficult to control since these fungi generally have wide host ranges and can survive in soil for long periods of time as saprophytes or as specialized survival structures (e.g., sclerotia, chlamydospores). Resistant cultivars are available for a limited number of host-pathogen combinations. Soilborne pathogens often cause disease at multiple life stages of the plant (i.e., seed rot, damping-off of seedlings, and root rots), but typically, the greatest impact is on the seed or newly emerged seedling. Use of hypocrealean fungi as plant, seed, or soil treatments facilitates rapid colonization of plant hosts and creates potential for subsequent induced resistance. Older plants may be protected from root rots by induced systemic resistance, although this has not been documented. Seed treatment may also create a potential 'antibiotic' spermosphere that inhibits populations of seed rot pathogens. Mycoparasitism by hypocrealean fungi can be directed against survival structures of soilborne plant pathogens, thus reducing their inoculum potential.

Although much has been accomplished in the commercial development of *Beauveria* and *Lecanicillium* spp. as fungal entomopathogens in plant production, more work is needed to understand the roles of these fungi as epiphytes and endophytes involved in suppression of plant diseases. Some strains of these fungi have been approved for use as bioinsecticides. Use in plant disease control extends development of these products. Future studies should focus on the ecology of these fungi (Vega et al. 2009a, b), their role in plant-microbe interactions, and their antagonism against pathogenic and nontarget microorganisms.

References

Ahmed AS, Sánchez CP, Candela ME (2000) Evaluation of induction of systemic resistance in pepper plants (*Capsicum annuum*) to *Phytophthora capsici* using *Trichoderma harzianum* and its relation with capsidiol accumulation. Eur J Plant Pathol 106:817–824

Aidemark M, Andersson C-J, Rasmusson AG, Widell S (2009) Regulation of callose synthase activity *in situ* in alamethicin-permeabilized *Arabidopsis* and tobacco suspension cells. BMC Plant Biol 9:27 (in press)

Allen DJ (1982) *Verticillium lecanii* on the bean rust fungus, *Uromyces appendiculatus*. Trans Br Mycol Soc 79: 362–364

Arnold AE (2007) Understanding the diversity of foliar endophytic fungi: progress, challenges, and frontiers. Fungal Biol Rev 21:51–66

Arnold AE, Maynard Z, Gilbert GS, Coley PD, Kursar TD (2000) Are tropical fungal endophytes hyperdiverse? Ecol Lett 3:267–274

Askary H, Yarmand H (2007) Development of the entomopathogenic hyphomycete *Lecanicillium muscarium* (Hyphomycetes: Moniliales) on various hosts. Eur J Entomol 104:67–72

Askary H, Benhamou N, Brodeur J (1997) Ultrastructural and cytochemical investigations of the antagonistic effect of *Verticillium lecanii* on cucumber powdery mildew. Phytopathology 87:359–368

Askary H, Carrière Y, Bélanger RR, Brodeur J (1998) Pathogenicity of the fungus *Verticillium lecanii* to aphids and powdery mildew. Biocontrol Sci Technol 8:23–32

Bark YG, Lee DG, Kim YH, Kang SC (1996) Antibiotic properties of an entomopathogenic fungus, *Beauveria bassiana*, on *Fusarium oxysporum* and *Botrytis cinerea*. Korean J Plant Pathol 12:245–250

Bélanger RR, Labbé C (2002) Control of powdery mildews without chemicals: prophylactic and biological alternatives for horticultural crops. In: Bélanger RR, Bushnell WR, Dik AJ, Carver TLW (eds) The powdery mildews: a comprehensive treatise. American Phytopathological Society, St. Paul, pp 256–267

Benhamou N (2004) Potential of the mycoparasite, *Verticillium lecanii*, to protect citrus fruit against *Penicillium digitatum*, the causal agent of green mold: a comparison with the effect of chitosan. Phytopathology 94:693–705

Benhamou N, Brodeur J (2000) Evidence of antibiosis and induced host defense reaction in the interaction between *Verticillium lecanii* and *Penicillium digitatum*, the causal agent of green mold. Phytopathology 90:932–943

Benhamou N, Brodeur J (2001) Pre-inoculation of Ri T-DNA transformed cucumber roots with the mycoparasite, *Verticillium lecanii*, induces host defense reactions against *Pythium ultimum* infection. Physiol Mol Plant Pathol 58:133–146

Berg G, Smalla K (2009) Plant species and soil type cooperatively shape the structure and function of microbial communities in the rhizosphere. FEMS Microbiol Ecol 68:1–13

Berg G, Zachow C, Lottmann J, Götz M, Costa R, Smalla K (2005) Impact of plant species and site on rhizosphere-associated fungi antagonistic to *Verticillium dahlia* Kleb. Appl Environ Microbiol 71:4203–4213

Bidochka MJ, Burke S, Ng L (1999) Extracellular hydrolytic enzymes in the fungal genus *Verticillium*: adaptations for pathogenesis. Can J Microbiol 45:856–864

Bills GF, Polishook JD (1991) Microfungi from *Carpinus caroliniana*. Can J Bot 69:1477–1482

Boucias DG, Pendland JC (1998) Principles of insect pathology. Kluwer Acad Pub, Boston

Boucias DG, Mazet I, Pendland J, Hung SY (1995) Comparative analysis of the *in vivo* and *in vitro* metabolites produced by the entomopathogen *Beauveria bassiana*. Can J Bot 73(Suppl 1):S1092–S1099

Bruinsma M, Pang B, Mumm R, van Loon JJA, Dicke M (2009) Comparing induction at an early and late step in signal transduction mediating indirect defence in *Brassica oleracea*. J Exp Bot 60:2589–2599

Clark MM, Gwinn KD, Ownley BH (2006) Biological control of *Pythium myriotylum*. Phytopathology 96:S25

Conrath U, Beckers GJM, Flors B, García-Agustín P, Jakab G, Mauch F, Newman M-A, Pieterse CMJ, Poinssot B, Pozo MJ, Pugin A, Schaffrath Ton J, Wendehenne D, Zimmerli L, Mauch-Mani B (2006) Priming: getting ready for battle. Mol Plant Microbe Interact 19:1062–1071

Copping LG, Menn JJ (2000) Biopesticides: a review of their action, applications and efficacy. Pest Manag Sci 56: 651–676

Crespo R, Pedrini N, Juárez MP, Dal Bello GM (2008) Volatile compounds released by the entomopathogenic fungus *Beauveria bassiana*. Microbiol Res 163:148–151

Djonović S, Pozo MJ, Dangott LJ, Howell CR, Kenerley CM (2006) Sm1, a proteinaceous elicitor secreted by the biocontrol fungus *Trichoderma virens* induces plant defense responses and systemic resistance. Mol Plant Microbe Interact 19:838–853

Djonović S, Vargas WA, Kolomiets MV, Horndeski M, Wiest A, Kenerley CM (2007) A proteinaceous elicitor Sm1 from the beneficial fungus *Trichoderma virens* is required for induced systemic resistance in maize. Plant Physiol 145:875–889

Dubos B (1987) Fungal antagonism in aerial agrobiocenoses. In: Chet I (ed) Innovative approaches to plant disease control. Wiley-Liss, New York, pp 107–135

Elliot SL, Sabelis MW, Janssen A, van der Geest LPS, Beerling EAM, Fransen J (2000) Can plants use entomopathogens as bodyguards? Ecol Lett 3:228–235

Ernst M, Medgen KW, Wirsel SGR (2003) Endophytic fungal mutualists: seed borne *Stagonospora* spp. enhance reed biomass production in axenic microcosms. Mol Plant Microbe Interact 16:580–587

Eyal J, Mabud MDA, Fischbein KL, Walter JF, Osborne LS, Landa Z (1994) Assessment of *Beauveria bassiana* Nov. EO-1 strain, which produces a red pigment for microbial control. Appl Biochem Biotechnol 44:65–80

Fenice M, Selbmann L, Zucconi L, Onofri S (1997) Production of extracellular enzymes by Antarctic fungal strains. Polar Biol 17:275–280

Flori P, Roberti R (1993) Treatment of onion bulbs with antagonistic fungi for the control of *Fusarium oxysporum* f. sp. *cepae*. Difesa delle Piante 16:5–12

Fravel DR (1988) Role of antibiosis in the biocontrol of plant diseases. Annu Rev Phytopathology 26:75–91
Fry WE (1982) Principles of plant disease management. Academic Press, New York
Genthner FJ, Cripe GM, Crosby DJ (1994) Effect of *Beauveria bassiana* and its toxins on *Mysidopsis bahia* (Mysidacea). Arch Environ Contam Toxicol 26:90–94
Giménez C, Cabrera R, Reina M, González-Coloma M (2007) Fungal endophytes and their role in plant protection. Curr Org Chem 11:707–720
Goettel MS, Koike M, Kim JJ, Aiuchi D, Shinya R, Brodeur J (2008) Potential of *Lecanicillium* spp. for management of insects, nematodes and plant diseases. J Invertebr Pathol 98:256–261
Gómez-Vidal S, Lopez-Llorca LV, Jansson H-B, Salinas J (2006) Endophytic colonization of date palm (*Phoenix dactylifera* L.) leaves by entomopathogenic fungi. Micron 37:624–632
Gómez-Vidal S, Salinas J, Tena M, Lopez-Llorca LV (2009) Proteomic analysis of date palm (*Phoenix dactylifera* L.) responses to endophytic colonization by entomopathogenic fungi. Electrophoresis 30:2996–3005
Griffin MR (2007) *Beauveria bassiana*, a cotton endophyte with biocontrol activity against seedling disease. Ph.D. Dissertation, The University of Tennessee, Knoxville, TN, USA
Grove J, Pople M (1980) The insecticidal activity of beauvericin and the enniatin complex. Mycopathologia 70:103–105
Gupta S, Montllor C, Hwang Y-S (1995) Isolation of novel beauvericin analogues from the fungus *Beauveria bassiana*. J Nat Prod 58:733–738
Gutjahr C, Paszkowski U (2009) Weights in the balance: jasmonic acid and salicylic acid signaling in root-biotroph interactions. Mol Plant Microbe Interact 22:763–772
Hajek AE (1997) Ecology of terrestrial fungal entomopathogens. Adv Microb Ecol 15:193–249
Hall RA (1980) Laboratory infection of insects by *Verticillium lecanii* strains isolated from phytopathogenic fungi. Trans Br Mycol Soc 74:445–446
Hall RA (1981) The fungus *Verticillium lecanii* as a microbial insecticide against aphids and scales. In: Burges HD (ed) Microbial control of pests and plant diseases 1970–1980. Academic Press, London, pp 483–498
Hanson LE, Howell CR (2004) Elicitors of plant defense responses from biocontrol strains of *Trichoderma virens*. Phytopathology 94:171–176
Harmon GE, Howell CR, Viterbo A, Chet I, Lorito M (2004) *Trichoderma* species–opportunistic, avirulent plant symbionts. Nature Rev Microbiol 2:43–56
Hartley SE, Gange AC (2009) Impacts on plant symbiotic fungi on insect herbivores: mutualism in a multitrophic context. Annu Rev Entomol 54:323–342
Hirano E, Koike M, Aiuchi D, Tani M (2008) Pre-inoculation of cucumber roots with *Verticillium lecanii* (*Lecanicillium muscarium*) induces resistance to powdery mildew. Res Bull Obihiro Univ 29:82–94
Holder DJ, Keyhani NO (2005) Adhesion of the entomopathogenic fungus *Beauveria* (*Cordyceps*) *bassiana* to substrata. Appl Environ Microbiol 71:5260–5266
Hyde KD, Soytong K (2008) The fungal endophyte dilemma. Fungal Divers 33:163–173
Inglis GD, Goettel MS, Butt TM, Strasser H (2001) Use of hyphomycetous fungi for managing insect pests. In: Butt TM, Jackson C, Magan N (eds) Fungi as biocontrol agents. Progress, problems and potential. CABI Publishing, Oxfordshire, pp 23–69
Jeger MJ, Jeffries P, Elad Y, Xu X-M (2009) A generic theoretical model for biological control of foliar plant diseases. J Theor Biol 256:201–214
Kamp AM (2002) Pleiomorphic deterioration in entomopathogenic fungi. MS thesis, Brock University, St. Catherines, Ontario, Canada
Kanaoka M, Isogai A, Murakoshi S, Ichinoe M, Suzuki A, Tamura S (1978) Bassianolide, a new insecticidal cyclodepsipeptide from *Beauveria bassiana* and *Verticillium lecanii*. Agr Biol Chem 42:629–635
Keller S, Zimmermann G (1989) Mycopathogens of soil insects. In: Wilding N, Collins NM, Hammond PM, Webber JF (eds) Insect-fungus interactions. Acad Press, London, pp 239–270
Kim JJ, Goettel MS, Gillespie DR (2007) Potential of *Lecanicillium* species for dual microbial control of aphids and the cucumber powdery mildew fungus, *Sphaerotheca fuliginea*. Biol Control 40:327–332
Kim JJ, Goettel MS, Gillespie DR (2008) Evaluation of *Lecanicillium longisporum*, Vertalec® for simultaneous suppression of cotton aphid, *Sphaerotheca fuliginea*, on potted cucumbers. Biol Control 45:404–409
Koike M, Higashio T, Komori A, Akiyama K, Kishimoto N, Masuda E, Sasaki M, Yoshida S, Tani M, Kuramoti K, Sugimoto M, Nagao H (2004) *Verticillium lecanii* (*Lecanicillium* spp.) as epiphyte and its application to biological control of arthropod pests and diseases. IOBC/WPRS Bull 27:41–44
Korolev N, David DR, Elad Y (2008) The role of phytohormones in basal resistance and *Trichoderma*-induced systemic resistance to *Botrytis cinerea* in *Arabidopsis thaliana*. BioControl 53:667–683
Lacey LA, Horton DR, Jones DC, Headrick HL, Neven LG (2009) Efficacy of biofumigant fungus *Muscodor albus* (Ascomycota: Xylariales) for control of codling moth (Lepidoptera: Tortricidae) in stimulated storage conditions. J Econ Entomol 102:43–49
Leckie BM, Ownley BH, Pereira RM, Klingeman WE, Jones CJ, Gwinn KD (2008) Mycelia and spent fermentation broth of *Beauveria bassiana* incorporated into synthetic diets affect mortality, growth and development of larval *Helicoverpa zea* (Lepidoptera: Noctuidae). Biocontrol Sci Technol 18:697–710
Lee S-M, Yeo W-H, Jee H-J, Shin S-C, Moon Y-S (1999) Effect of entomopathogenic fungi on growth of cucumber and *Rhizoctonia solani*. FRI J For Sci 62:118–125
Linder MB (2009) Hydrophobins: proteins that self assemble at interfaces. Curr Opin Colloid Interface Sci 14:356–363
Mercier J, Jiménez JI (2004) Control of fungal decay of apples and peaches by the biofumigant fungus *Muscodor albus*. Postharvest Biol Technol 31:1–8
Mercier J, Smilanick JL (2005) Control of green mold and sour rot of stored lemon by biofumigation with *Muscodor albus*. Biol Control 32:401–407

Meyer SLF, Huettel RN, Sayre RM (1990) Isolation of fungi from *Heterodera glycines* and *in vitro* bioassays for their antagonism to eggs. J Nematol 22:532–537

Meyling N, Eilenberg J (2007) Ecology of the entomopathogenic fungi *Beauveria bassiana* and *Metarhizium anisopliae* in temperate agroecosystems: potential for conservation biological control. Biol Control 43:145–155

Morán-Diez E, Hermosa R, Ambrosino P, Cardoza RE, Gutiérrez S, Lorito M, Monte E (2009) The ThPG1 endopolygalacturonase is required for the *Trichoderma harzianum*-plant beneficial interaction. Mol Plant Microbe Interact 22:1021–1031

Oller-López JL, Iranzo M, Mormeneo S, Oliver E, Cuerva JM, Oltra JE (2005) Bassianolone: an antimicrobial precursor of cephalosporolides E and F from the entomoparasitic fungus *Beauveria bassiana*. Org Biomol Chem 3:1172–1173

Ownley BH, Windham MT (2007) Biological control of plant pathogens. In: Trigiano RN, Windham MT, Windham AS (eds) Plant pathology concepts and laboratory exercises, 2nd edn. CRC Press, New York, pp 423–436

Ownley BH, Bishop DG, Pereira RM (2000) Biocontrol of Rhizoctonia damping-off of tomato with *Beauveria bassiana*. Phytopathology 90:S58

Ownley BH, Pereira RM, Klingeman WE, Quigley NB, Leckie BM (2004) *Beauveria bassiana*, a dual purpose biocontrol organism, with activity against insect pests and plant pathogens. In: Lartey RT, Caesar A (eds) Emerging concepts in plant health management. Research Signpost, Kerala, pp 256–269

Ownley BH, Griffin MR, Klingeman WE, Gwinn KD, Moulton JK, Pereira RM (2008) *Beauveria bassiana*: endophytic colonization and plant disease control. J Invertebr Pathol 3:267–270

Perazzolli M, Dagostin S, Ferrar A, Elad Y, Pertot I (2008) Induction of systemic resistance against *Plasmopara viticola* in grapevine by *Trichoderma harzianum* T39 and benzothiadiazole. Biol Control 47:228–234

Petrini O (1981) Endophytische pilze in epiphytischen Araceae, Bromeliaceae und Orchidiaceae. Sydowia 34: 135–148

Posada F, Vega FE (2005) Establishment of the fungal entomopathogen *Beauveria bassiana* (Ascomycota: Hypocreales) as an endophyte in cocoa seedlings (*Theobroma cacao*). Mycologia 97:1195–1200

Powell WA, Klingeman WE, Ownley BH, Gwinn KD (2009) Evidence of endophytic *Beauveria bassiana* in seed-treated tomato plants acting as a systemic entomopathogen to larval *Helicoverpa zea* (Lepidoptera: Noctuidae). J Entomol Sci 44:391–396

Reisenzein H, Tiefenbrunner W (1997) Growth inhibiting effect of different isolates of the entomopathogenic fungus *Beauveria bassiana* (Bals.) Vuill. to the plant parasitic fungi of the genera *Fusarium*, *Armillaria* and *Rosellinia*. Pflanzenschutz Berichte 57:15–24

Reithner B, Schuhmacher R, Stoppacher N, Pucher M, Brunner K, Zeilinger S (2007) Signaling via the *Trichoderma atroviride* mitogen-activated protein kinase Tmk1 differentially affects mycoparasitism and plant protection. Fungal Genet Biol 44:1123–1133

Renwick A, Campbell R, Coe S (1991) Assessment of *in vivo* screening systems for potential biocontrol agents of *Gaeumannomyces graminis*. Plant Pathol 40:524–532

Riga K, Lacey LA, Guerra N (2008) The potential of the endophytic fungus, *Muscodor albus*, as a biocontrol agent against economically important plant parasitic nematodes of vegetable crops in Washington State. Biol Control 45:380–385

Ruocco M, Lanzuise S, Vinale F, Marra R, Turrà D, Woo SL, Lorito M (2009) Identification of a new biocontrol gene in *Trichoderma atroviride*: the role of an ABC transporter membrane pump in the interaction with different plant pathogenic fungi. Mol Plant Microbe Interact 22:291–301

Saikkonen K, Faeth SH, Helander M, Sullivan TJ (1998) Fungal endophytes: a continuum of interactions with host plants. Annu Rev Ecol Systemat 29:319–343

Saksirirat W, Hoppe H-H (1991) Degradation of uredospores of the soybean rust fungus (*Phakopsora pachyrhizi* Syd.) by cell-free culture filtrates of the mycoparasite *Verticillium psalliotae* Treschow. J Phytopathology 132:33–45

Schulz B, Boyle C (2005) The endophytic continuum. Mycol Res 109:661–686

Segarra G, Casanova E, Bellido D, Odena MA, Oliveira E, Trillas I (2007) Proteome, salicylic acid, and jasmonic acid changes in cucumber plants inoculated with *Trichoderma asperellum* strain T34. Proteomics 7:3943–3952

Shakeri J, Foster HA (2007) Proteolytic activity and antibiotic production by *Trichoderma harzianum* in relation to pathogenicity to insects. Enzym Microb Tech 40:961–968

Shinya R, Aiuchi D, Kushida A, Tani M, Kuramochi K, Koike M (2008) Effects of fungal culture filtrates of *Verticillium lecanii* (*Lecanicillium* spp.) hybrid strains on *Heterodera glycines* eggs and juveniles. J Invertebr Pathol 97:291–297

Shoresh M, Harman GE (2008) The molecular basis of shoot responses of maize seedlings to *Trichoderma harzianum* T22 inoculation of the root: a proteomic approach. Plant Physiol 147:2147–2163

Spencer DM, Atkey PT (1981) Parasitic effects of *Verticillium lecanii* on two rust fungi. Trans Br Mycol Soc 77:535–542

St. Leger RJ, Joshi L, Roberts DW (1997) Adaptation of proteases and carbohydrases of saprophytic, phytopathogenic, and entomopathogenic fungi to the requirements of their ecological niches. Microbiology 143:1983–1992

Strobel GA (2006) *Muscodor albus* and its biological promise. J Ind Microbiol Biotechnol 33:514–522

Strobel GA, Dirkse E, Sears J, Markworth C (2001) Volatile antimicrobials from *Muscodor albus*, a novel endophytic fungus. Microbiology 147:2943–2950

Taniguchi M, Kawaguchi T, Tanaka T, Oi S (1984) Antimicrobial and respiration inhibitory activities of oosporein. Agr Biol Chem 48:1065–1067

Tefera T, Vidal S (2009) Effect of inoculation method and plant growth medium on endophytic colonization of sorghum by the entomopathogenic fungus *Beauveria bassiana*. BioControl 54:663–669

Termorshuizen AJ, Jeger MJ (2009) Strategies of soilborne plant pathogenic fungi in relation to disease suppression. Fung Ecol 1:108–114

Van Loon LC, Bakker PAHM, Pieterse CMJ (1998) Systemic resistance induced by rhizosphere bacteria. Annu Rev Phytopathology 36:453–483

Van Wees SCM, Van der Ent S, Pieterse CMJ (2008) Plant immune responses triggered by beneficial microbes. Curr Opin Plant Biol 11:443–448

Vandermeer J, Perfecto I, Liere H (2009) Evidence for hyperparasitism of coffee rust (*Hemileia vastatrix*) by the entomogenous fungus, *Lecanicillium lecanii*, through a complex ecological web. Plant Pathol 58:636–641

Vargas WA, Djonović S, Sukno SA, Kenerley CM (2008) Dimerization controls the activity of fungal elicitors that trigger systemic resistance in plants. J Biol Chem 283:19804–19815

Vega FE (2008) Insect pathology and fungal endophytes. J Invertebr Pathol 98:277–279

Vega FE, Posada F, Aime MC, Pava-Ripoll M, Infante F, Rehner SA (2008) Entomopathogenic fungal endophytes. Biol Control 46:72–82

Vega FE, Goettel MS, Blackwell M, Chandler D, Jackson MA, Keller S, Koike M, Maniania M, Monzón A, Ownley BH, Pell JK, Rangel DEN, Roy HE (2009a) Fungal entomopathogens: new insights on their ecology. Fungal Ecol 2:149–159

Vega FE, Simpkins A, Aime MC, Posada F, Peterson SW, Rehner SA, Infante F, Castillo A, Arnold AE (2009b) Fungal endophyte diversity in coffee plants from Colombia, Hawai'i, Mexico, and Puerto Rico. Fungal Ecol (in press)

Verhaar MA, Hijwegen T, Zadoks JC (1996) Glasshouse experiments on biocontrol of cucumber powdery mildew (*Sphaerotheca fuliginea*) by the mycoparasites *Verticillium lecanii* and *Sporothrix rugulosa*. Biol Control 6: 353–360

Vesely D, Koubova D (1994) In vitro effect of the entomopathogenic fungi *Beauveria bassiana* (Bals.-Criv.) Vuill. and *B. brongniartii* (Sacc.) Petch on phytopathogenic fungi. Ochr Rostl 30:113–120

Vinale F, Sivasithamparam K, Ghisalberti DL, Marra R, Woo SL, Lorito M (2008) *Trichoderma*–plant–pathogen interactions. Soil Biol Biochem 40:1–10

Viterbo A, Chet I (2006) TasHyd1, a new hydrophobin gene from the biocontrol agent *Trichoderma asperellum*, is involved in plant root colonization. Mol Plant Pathol 7:249–258

Whipps JM (1993) A review of white rust (*Puccinia horiana* Henn.) disease on chrysanthemum and the potential for its biological control with *Verticillium lecanii* (Zimm.) Viégas. Ann Appl Biol 122:173–187

White JF, Belanger F, Meyer W, Sullivan RF, Bischoff JF, Lewis EA (2002) Clavicipitalean fungal epibionts and endophytes-development of symbiotic interactions with plants. Symbiosis 33:201–213

Widler B, Müller E (1984) Untursuchungen über endophytische Pilze von *Arctostaphylos uva-ursi* (L.) Sprenger (Ericaceae). Bot Helv 94:307–337

Ying S-H, Feng M-G (2004) Relationship between thermotolerance and hydrophobin-like proteins in aerial conidia of *Beauveria bassiana* and *Paecilomyces fumosoroseus* as fungal biocontrol agents. J Appl Microbiol 97:323–331

Zhang Y, Zhao J, Fang W, Zhang J, Luo Z, Zhang M, Fan Y, Pei Y (2009) Mitogen-activated protein kinase hog1 in the entomopathogenic fungus *Beauveria bassiana* regulates environmental stress responses and virulence to insects. Appl Environ Microbiol 75:3787–3795

Author Biographies

Dr. Bonnie H. Ownley conducts research on the biology, biological and cultural control, and epidemiology of soilborne plant pathogens, and on the ecology and population dynamics of beneficial plant-associated microorganisms.

Dr. Kimberly D. Gwinn focuses her research on natural products for control of plant disease, the roles of natural products in host-symbiont relationships, and secondary metabolism of fungi.

Dr. Fernando E. Vega conducts research on biological methods to control the coffee berry borer, the most important insect pest of coffee throughout the world.

BioControl (2010) 55:129–145
DOI 10.1007/s10526-009-9240-y

Ecological considerations in producing and formulating fungal entomopathogens for use in insect biocontrol

Mark A. Jackson · Christopher A. Dunlap · Stefan T. Jaronski

Received: 26 July 2009 / Accepted: 9 October 2009 / Published online: 3 November 2009

Abstract Insect pests persist in a wide-variety of agricultural, arboreal and urban environments. Effective control with fungal entomopathogens using inundation biocontrol requires an understanding of the ecology of the target insect, fungal pathogen, and the insect-pathogen interaction. Historically, the development of production and formulation processes for biocontrol fungi has primarily focused on reducing costs by maximizing the yield of infective propagules, increasing storage stability, and improving product form for ease of application. These goals are critical for commercialization but are often in conflict with environmental and ecological considerations. Critical parameters for selecting a fungal pathogen for use in inundation biocontrol include the cost-effective production of a stable, infective propagule that is suited for use in the environment where the insect must be controlled. Production processes can be manipulated nutritionally and environmentally to produce efficacious propagules or to direct fungal differentiation to propagule forms that may be better suited for use in specific environments. Formulation development must also consider ecological and environmental factors to maximize biocontrol efficacy. A basic understanding of the surface chemistries of the fungal propagule and insect, the interactions between a fungal propagule and the insect cuticle that lead to infection, and the impact of the environment on this interaction can aid in the development of effective formulations.

Keywords Biocontrol · Fungi · Fermentation · Formulation · Conidia · Blastospores · Sclerotia · Mycoinsecticides

Handling Editor: Helen Roy

M. A. Jackson (✉) · C. A. Dunlap
United States Department of Agriculture, National Center for Agricultural Utilization Research, Crop Bioprotection Research Unit, 1815 N. University Street, Peoria, IL 61604, USA
e-mail: mark.jackson@ars.usda.gov

S. T. Jaronski
United States Department of Agriculture, Northern Plains Agricultural Research Laboratory, Pest Management Research Unit, 1500 N Central Avenue, Sydney, MT 59270, USA

Introduction

Over the past 50 years, the control of insects, weeds, and plant diseases with fungal pathogens has been a very active area of research and has resulted in a large number of commercially-available products (Butt et al. 2001; Charudattan 2001; Wraight et al. 2001; Fravel 2005; Faria and Wraight 2007). The commercial use of fungal entomopathogens to control insects is generally practiced using the inundation biocontrol approach where the environment harbouring the insect pest is inundated with high concentrations of infective fungal propagules (Eilenberg et al. 2001).

Products developed for use in the inundative approach are often termed "mycoinsecticides" or "biopesticides" in reference to their similar usage pattern compared to chemical insecticides.

Fungi are unique candidates for use in "inundation" biocontrol because of their ability to actively infect and kill the target pest. The high number of fungal propagules used in this approach requires a cost-effective production and stabilization process that delivers viable, infective fungal propagules (Goettel and Roberts 1992; Wraight et al. 2001; Jackson 2007). Production and formulation are critical to the commercial development of a fungal biocontrol agent. The biocontrol agent must provide a cost—benefit to the end user (be low cost), have a reasonably long shelf-life (remain viable and infective during storage), and provide consistent insect control under field conditions (function well in the environment of use). Shortcomings in any of these qualities can prevent the agent from becoming a commercial product. In general, product cost and storage stability have driven the development of production and formulation processes. Often, these production goals are in conflict with ecological and environmental requirements for consistent infectivity and control of the insect target. A more thorough understanding of the relationship between the insect pest and the agricultural system being managed is needed to assure success in using mycoinsecticides as stand-alone products or as part of an integrated approach to pest management (Thomas 1999; Shah and Pell 2003; Lacey and Shapiro-Ilan 2008). Many of the papers in this special issue deal with specific issues related to these "insect–pathogen–environment" interactions.

More than 150 insect biocontrol products based on fungal entomopathogens have been commercialized with over 75% of these products based on the hypocrealean fungi *Metarhizium anisopliae*, *Beauveria bassiana*, *Isaria fumosorosea*, and *B. brongniartii* (Faria and Wraight 2007). Two-thirds of these commercialized products are comprised of conidial preparations of *B. bassiana* or *M. anisopliae*, presumably using solid-substrate fermentation production processes. Both *B. bassiana* and *M. anisopliae* have a very broad insect host range with many isolates producing high concentrations of aerial conidia when grown on nutrient-rich, solid substrates (Jaronski 1997). The use of conidia as mycoinsecticides is warranted as they are a naturally-infective propagule. There are, however, ecological and environmental conditions in which the use of conidia may not be the best choice for insect biocontrol in agricultural or urban settings. For example, conidia are more shelf stable at room temperature when compared to blastospore preparations of *I. fumosorosea* but, following rehydration, can take up to 24 h to germinate compared to a 6 h germination time for rehydrated blastospores (Vega et al. 1999). Studies with subterranean termites have shown that fungal conidia readily attach to the cuticle but are effectively removed by mutual grooming (Yanagawa et al. 2008). The delivery of a blastospore preparation or other fungal propagule that germinates more rapidly may be a more effective propagule choice for control of social insects that groom nest mates or for insects that frequently moult.

Production and formulation strategies for potential mycoinsecticides must consider the environmental and ecological requirements and limitations (Vega et al. 2009). From a biotechnology standpoint, a variety of fungal propagules can be produced using solid-substrate and deep-tank fermentation by altering nutritional and environmental conditions. Likewise, formulations can be employed that alter the chemical and physical attributes of a fungal propagule for improved insecticidal activity under varied environmental conditions. Formulations developed with living, fungal entomopathogens for use in inundation biocontrol must take into account the environmental and ecological life histories of the target insect while maintaining propagule viability and efficacy (Jaronski 1997). This chapter will describe production and formulation strategies that focus not only on economic factors but also on developing fungal propagules designed for insect control in specific environments.

Selecting fungal propagules for use in inundation biocontrol

The identification of the appropriate fungal pathogen for development as a mycoinsecticide can be complex. The selection process must evaluate the potential of the fungal isolate to form a stable propagule that can be economically mass-produced, that is amenable to available application technologies, and, most importantly, is capable of consistently infecting and killing the target insect under the environmental and ecological conditions where it is a pest (Wraight et al. 2001; Jackson and Schisler 2002; Jaronski 2007). Few fungal entomopathogens are capable of meeting all these requirements. The environmental conditions present during insect control must be considered and appropriate fungi and fungal propagules selected for use in inundation biocontrol. Critical environmental factors, such as temperature, can have a profound influence on the growth and pathogenicity of a fungal entomopathogen against the target insect (Inglis et al. 1996; Faria and Wraight 2001; Yeo et al. 2003). For example, the conidia of fungal isolates collected from environments differing in climatic conditions showed dramatic variation in temperature tolerance. At 48°C, the LT_{50} was 14.3–150.3 min for conidia of various isolates of *Metarhizium* species, 10.1–61.9 min for *B. bassiana* isolates, and 2.8–6.2 min for isolates of *I. fumosorosea* (Li and Feng 2009). This variation in thermal tolerance would be a significant factor in selecting an appropriate entomopathogen for development as a mycoinsecticide

Insects inhabit diverse environments and are pest problems in agricultural, urban, forest, freshwater, and natural ecosystems. Their life histories coupled with environmental conditions conspire to make consistent insect control under field conditions difficult to achieve using fungal entomopathogens. A key consideration in the selection of a fungal entomopathogen is the fungus' ability to produce a suitable propagule for control of the insect. The efficacy of a fungal propagule is dependent on the requirements for use as a mycoinsecticide and may include enhanced virulence, desiccation tolerance, thermal tolerance, speed of germination and infection, environmental stability and reproduction, and UV tolerance (Jackson and Schisler 1992; Jackson et al. 1997; Vega et al. 1999). Numerous studies have shown that nutritional and environmental conditions during fungal growth using solid-substrate and liquid-culture fermentation influence the form and efficacy of the fungal propagule (Hallsworth and Magan 1994, 1995, 1996; Jackson et al. 1996; Jackson 1997; Magan 2001; Ying and Feng 2006). Formulation of the fungal propagule or the use of adjuvants during application can also influence efficacy (Jaronski 1997; Costa et al. 2008; Friesen et al. 2006). The selection of a fungal entomopathogen that economically produces a stable propagule which provides consistent insect control under field conditions is the ultimate goal of the selection process. An excellent description of the requirements for germination, infection and reproduction by fungal entomopathogens on the insect cuticle has been presented by Boucias and Pendland (1991) and Castrillo et al. (2005).

The life history of the insect pest and the environment in which it will be controlled dictate the fungal propagule needed for use as a mycoinsecticide. If the mycoinsecticide is to be applied as a spray (i.e., "contact" biopesticide), the production method must yield high numbers of discrete, infective propagules. Granular mycoinsecticide formulations for use in soil require the production of a persistent fungal propagule that is capable of delivering an infective inoculum to the insect host when required. Many spore forms used in spray applications are not amenable to use in granular applications. Recently, it was shown that some isolates of the entomopathogenic fungus *M. anisopliae* differentiated to form sclerotial propagules when grown in liquid culture fermentation (Jackson and Jaronski 2009). These sclerotial propagules were desiccation tolerant and germinated sporogenically in soil to produce conidia in situ that infected and killed susceptible soil-dwelling insects. The sclerotia-containing granules were more efficacious when compared to granules made from conidia of *M. anisopliae* bound to a solid nutritive carrier (Jaronski and Jackson 2008).

Production of fungal propagules for use in inundation biocontrol

Conidia production using solid substrate fermentation

Inundation biocontrol for foliar insect pests is generally practiced by spraying high concentrations of infective fungal spores. Because they can be produced in high concentration, either aerial conidia or "yeast-like" blastospores are the fungal spore forms commercially-produced for use in the spray application of mycoinsecticides. Application rates for insect control using fungal entomopathogens can approach 2.5–5 × 10^{13} spores ha^{-1} in inundation biocontrol (Faria and Wraight 2001), although lower rates have been observed to be efficacious, e.g., 2.5×10^{12} conidia ha^{-1} in the case of *M. anisopliae var. acridum* against African Orthoptera (van der Valk 2007).

The primary infective form of most fungal entomopathogens is the conidium and, in fact, the solid substrate production of aerial conidia is the most widely used production method for the mycoinsecticides *Metarhizium* and *Beauveria* (Bartlett and Jaronski 1988; Faria and Wraight 2007). Solid substrate production processes for aerial conidia can be very simple but labour intensive (autoclaved bags of moistened grain inoculated with an entomopathogen) or involve a more automated tray production system requiring higher capital costs with reduced manpower requirements (Bartlett and Jaronski 1988). Other fungi are not suited for solid substrate conidia production. Isolates of the fungal entomopathogen *I. fumosorosea* require light for significant conidia production, a characteristic that has limited its production using solid substrate fermentation (Sanchez-Murillo et al. 2004; Zimmermann 2008). Fortunately, *I. fumosorosea* and other fungal entomopathogens are dimorphic fungi and are capable of growing "yeast-like" in liquid culture to produce blastospores, which can be utilized in spray application after proper drying and formulation (Jackson 1999; Kassa et al. 2004; Jackson et al. 2006).

Submerged conidia production

Both *B. bassiana* and *M. anisopliae var acridum*, but not *M. anisopliae*, will produce submerged or microcycle conidia under certain liquid fermentation conditions (Thomas et al. 1986; Jenkins and Prior 1993; Kassa et al. 2004). These submerged conidia are not hydrophobic, unlike aerial conidia, and thus present different challenges in formulation and use. The microcycle conidia of *B. bassiana* are produced after 96 h of fermentation only in the presence of inorganic nitrogen, as nitrate, and with very high levels of carbohydrate. Submerged conidia are morphologically different from aerial conidia on an ultrastructural level, lacking one layer to their cell walls (Hegedus et al. 1990). Germination speed for submerged conidia is intermediate between aerial conidia and blastospores. Submerged conidia of *M. anisopliae var acridum* were produced on structures very similar to aerial phialides and were morphologically indistinguishable from aerial conidia, although they possessed different physical properties (Leland et al. 2005). Nitrogen, in the form of brewer's yeast, in the presence of excess sucrose was found to be essential for the production of submerged conidia by *M. anisopliae var acridum* cultures. At the present time, submerged conidia have not been commercially developed as an insect biocontrol propagule.

Blastospore production using liquid culture fermentation

Blastospores are vegetative fungal propagules that are the preferred mode of growth for many entomopathogens in the haemocoel of infected insects (Shimuzu et al. 1993; Sieglaff et al. 1997; Vestergaard et al. 1999; Askary et al. 1999). Yeast-like growth allows the fungus better access to the nutrients within the insect. Numerous entomopathogens of the genera *Isaria*, *Beauveria*, *Lecanicillium*, and *Metarhizium* can be induced to grow in a "yeast-like" fashion in submerged liquid culture. Blastospore-based mycoinsecticides are currently produced commercially by *L. lecanii* (Ascomycota: Hypocreales) and *I. fumosorosea* (Faria and Wraight 2007). Our studies with the fungal mycoinsecticide *I. fumosorosea* have demonstrated that desiccation tolerant blastospores can be rapidly produced in high concentrations if an appropriate source and concentration of nitrogen are provided (Jackson et al. 2003). Blastospores of *I. fumosorosea* are highly infective against a number of insect pests and often have a lower LD_{50} when compared to conidial preparations (Poprawski and Jackson 1999; Behle et al. 2006; Shapiro-Ilan et al. 2008).

The rapid germination rate of *I. fumosorosea* blastospores (>90% germination in 6 h) make these propagules ideal candidates for use as a contact mycoinsecticide (Vega et al. 1999). Considering environmental and ecological factors, the rapid germination rate of blastospores reduces the time required for available free-moisture and mitigates the adverse effects of extended exposure in the environment. Furthermore, the rapid germination rate of blastospores increases their chance of infecting moulting insects or social insects that groom nest mates. Blastospores of *I. fumosorosea* have also been shown to be less repellent to the Formosan subterranean termite, *Coptotermes formosanus*, when compared to conidial preparations of *I. fumosorosea* produced on solid-substrate, rice cultures (Wright et al. 2003). These differences suggest that the properties of aerial conidia, submerged conidia, and blastospores can be exploited for improved insect biocontrol, particularly if the insect target is susceptible to multiple entomopathogens capable of forming these propagules using commercial production methods. The insect's life histories and environment will dictate the appropriate fungal propagule for use as an inundative biocontrol agent.

Sclerotia production using liquid culture fermentation

Sclerotia are compact hyphal aggregates that often become melanized as they develop (Coley-Smith and Cooke 1971). These fungal structures have been reported as the overwintering propagule for many plant pathogenic fungi and for a limited number of fungal entomopathogens (Speare 1920; Evans and Samson 1982). Like many plant pathogenic fungi, sclerotial bodies of the fungal entomopathogen *Nomuraea rileyi* (Ascomycota: Hypocreales) found in insect cadavers were shown to produce infective conidia *via* sporogenic germination in the following growing season (Sprenkel and Brooks 1977; Speare 1920). Recently, it was shown that the fungal entomopathogen *M. anisopliae* produced high concentrations of microsclerotia (small sclerotia) under specific nutritional conditions during liquid culture fermentation (Fig. 1; Jackson and Jaronski 2009). These microsclerotia were desiccation tolerant with excellent storage stability following air-drying. When air-dried microsclerotial granules of *M. anisopliae* were soil-incorporated, they produced infective conidia *via* sporogenic germination following rehydration (Fig. 1; Jaronski and Jackson 2008). During the production of sclerotia in liquid culture, melanin biosynthesis can be controlled with nutrition or culture age (Jackson and Schisler 1995; Shearer and Jackson 2006; Jackson and Jaronski 2009). Fungal melanins have been shown to have allelopathic and antimicrobial properties, act as anti-desiccants, enhance cell rigidity, and confer fungicide resistance, all properties that would enhance the vigour of sclerotial propagules for use as a mycoinsecticide in the rhizosphere (Butler and Day 1998).

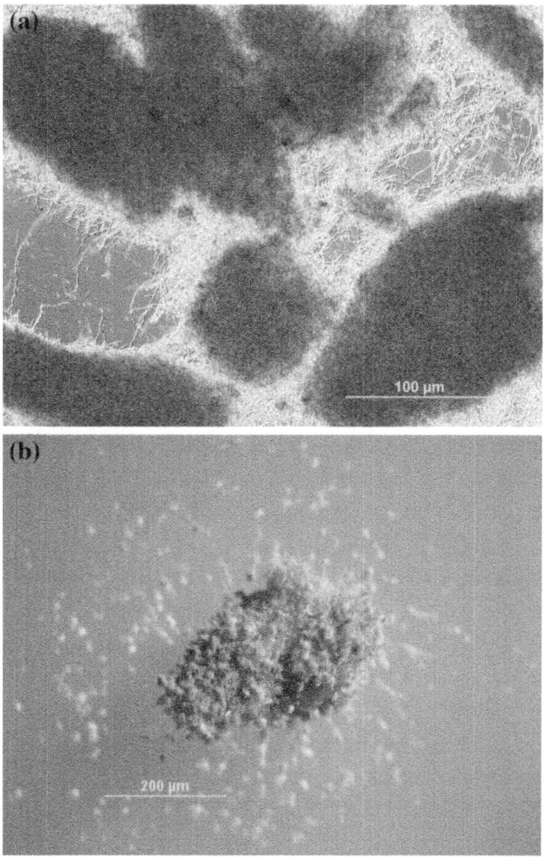

Fig. 1 Photomicrographs of melanized microsclerotia of *Metarhizium anisopliae* produced in liquid culture fermentation (**a**) and conidia production by air-dried microsclerotia-containing granules of *Metarhizium anisopliae* on water agar after incubation for seven days at 28°C (**b**). Microsclerotial granule (**b**) is covered with olive-green conidial masses and hyphal extensions from the granule are producing additional conidial masses. Photomicrographs taken with an Olympus DP70 photosystem, automatic scale calibration, on an Olympus BX60 light microscope with Nomarski optics (**a**) and an Olympus SZH10 stereo microscope (**b**)

The formation of sclerotial propagules by *M. anisopliae* in liquid culture was unexpected but is likely related to its soil-inhabiting nature (Klingen et al. 2002; Zimmermann 2007). Reports pertaining to the environmental association of *M. anisopliae* with various soil types and not to insect host suggested that the persistence of this fungus in these soils was likely unrelated to the presence of an insect host (Bidochka et al. 2001; Quesada-Moraga et al. 2007). Furthermore, the association of *M. anisopliae* with plant roots and root exudates supports the idea that these fungi may be capable of survival in soils without an insect host (Hu and St. Leger 2002; Bruck 2005).

The ability of *M. anisopliae* to form an overwintering propagule, such as a sclerotium, would certainly provide this fungus an ecological advantage. It has been assumed that conidia were the overwintering propagule for *M. anisopliae*. This assumption is confounded by the fact that microsclerotia produce conidia when rehydrated under environmental conditions conducive to growth. This is particularly true given the fact that most studies concerning the presence of *M. anisopliae* in soil have been conducted by baiting with susceptible insects or by serial soil dilution plating onto *Metarhizium*-selective media to identify colony forming units of *M. anisopliae* (Hu and St. Leger 2002; Klingen et al. 2002; Keller et al. 2003; Meyling and Eilenberg 2006; Quesada-Moraga et al. 2007). Microscopic studies of soil are needed to determine the presence or absence of microsclerotial propagules of *M. anisopliae* and, if present, their ability to produce conidia by sporogenic germination in soil, root exudates, or decaying insect cadavers.

The liquid culture production, desiccation tolerance, and sporogenic germination of microsclerotia of *M. anisopliae* supports their use for control of soil-dwelling insects. The ability of *M. anisopliae* to produce sclerotial bodies may also provide insight into the soil-dwelling nature of this fungus. As with other fungi that produce sclerotia under the controlled conditions inherent to liquid culturing, a model is now available for discerning the processes involved in the differentiation of *M. anisopliae* hyphae to produce sclerotia under gnotobiotic conditions. Understanding and developing this biocontrol approach for soil-dwelling insects should provide microsclerotial preparations of *M. anisopliae* with distinct advantages over the use of spore- or mycelium-based insect biocontrol products.

Formulation of fungal propagules—considerations

Shelf life—environment during storage

The formulation of aerial conidia or other fungal propagules is a somewhat different paradigm than formulation of a chemical active ingredient. First of all, the spore, be it a conidium or a blastospore, must be kept alive until used. For commercial use, a mycoinsecticide must have an "acceptable shelf life" generally considered a minimal loss in spore viability for at least one year at room temperature. A typical conidial viability trend for a commercial *B. bassiana* (isolate GHA, Laverlam International, Butte, MT) is depicted in Fig. 2, where longevity is inversely proportional to temperature. Figure 3 shows the relationship of conidial half-life to storage temperature for a typical lot of commercial *B. bassiana* GHA unformulated technical powder. Storage temperatures above 30°C resulted in commercially unacceptable shelf life (<1 year) while temperatures <20°C allowed multi year storage.

A basic premise regarding the storage stability of a fungal mycopesticides is that shortened shelf life is

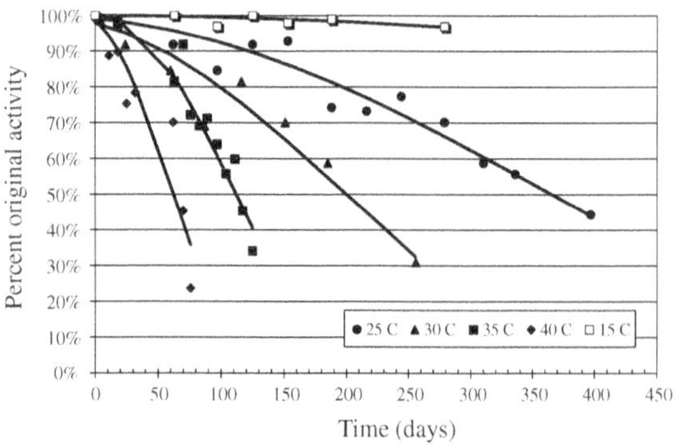

Fig. 2 Viability of *Beauveria bassiana* strain GHA technical grade (unformulated) conidial powder (Mycotech Lot 99-05-2) stored in sealed polypropylene containers at various temperatures. Lines are fitted based on linear regressions of angular transformed data (percent original activity remaining) versus time, backtransformed to percents for use in the graph

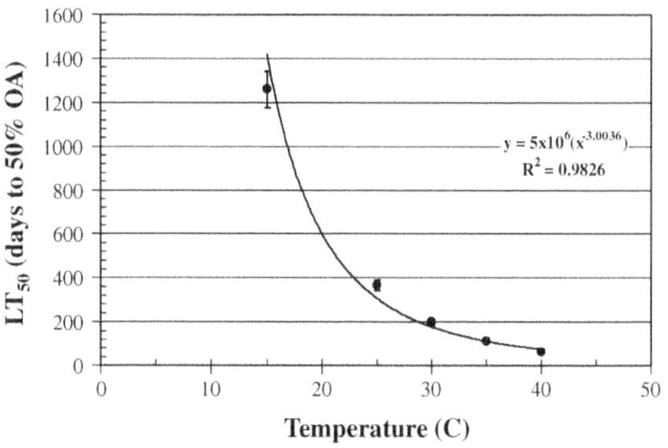

Fig. 3 Effect of storage temperature on the half-life (LT_{50} of original activity remaining, OA) of *Beauveria bassiana* unformulated, conidial powder (Mycotech Lot 99-05-2). Bars equal to one standard deviation

primarily due to spores slowly initiating germination, but dying as the succession of cues and requirements to complete germination are not fulfilled in the storage environment (Jaronski 1997). In addition to being viable, the fungal propagule must also possess the ability to infect and kill the insect host under the environmental conditions where the entomopathogen will be used. How does one keep a fungal spore alive and efficacious, yet dormant, for a satisfactory length of time? Understanding the cues that lead to spore germination and how the storage environment influences these cues is critical to the development of stable mycoinsecticidal products.

An analogy to the requirements for conidial germination is the fire prevention triangle (Anonymous 2009). Three components—fuel, oxygen, and heat, or an ignition source—are necessary for combustion. These three components can be linked conceptually to each other in a triangle. If one corner of the triangle is eliminated, fire can be prevented. Likewise, the three requirements for germination are nutrients, water, and oxygen. Eliminating one component of the germination triangle prevents spore germination. The challenge is to eliminate or inhibit a requirement for germination in the storage environment without killing or reducing the efficacy of the fungal spore.

Nutrients, the first component of the germination triangle, are very difficult to exclude from fungal spore products as they may be endogenous or in the production medium during harvest, thus becoming part of the final product. As little as 6 nM glucose can stimulate and support conidial germination in *B. bassiana* (Smith and Grula 1982). Economics preclude the harvest of mass produced fungal spores free of residual nutrients.

Water is the second component of the germination triangle. Liquid water, at least on the level of a molecular film, is necessary to convey chemical cues to the spore and to initiate germination. Excluding water, or reducing the water activity in the storage environment below a certain level, can prevent germination. This phenomenon has been reported for *B. bassiana*, *L. lecanii*, and *Metarhizium flavoviride* and is the subject of at least one US patent (Jung and Mugnier 1989; Chandler et al. 1994; Hedgecock et al. 1995; Jin et al. 1999). Of course, removing molecular water can damage fungal spores and greatly shorten their longevity (Crowe and Crowe 1986).

The third component of the germination triangle is oxygen. The complete exclusion of oxygen is difficult to achieve and may not be beneficial under some conditions. Living fungal spores continue low-levels of basal metabolic activity even under conditions adverse to growth. Measures that exclude oxygen from the storage environment e.g., vacuum-packing, inclusion of oxygen scavengers, or replacing the air in container head space with nitrogen or carbon dioxide, may be deleterious (Jaronski, unpublished data) or may have a beneficial effect to spore survival (Jin et al. 1999). Moisture levels, temperature, and nutrient availability are conditions present during storage that confound the influence of oxygen on spore survival.

Intrinsic conidial longevity under optimal storage conditions can be unique to a fungal species or even an isolate within a species (Jaronski 1997; Hong et al. 2001). In a study of six *B. bassiana* and three *M. anisopliae var. acridum* isolates from Madagascar, half-life of conidia produced under identical conditions and dried to the same endpoint of 4–6% moisture and stored at 25°C, ranged from <27 d for the *Metarhizium* isolates and >210 d for four of the *Beauveria* isolates (Jaronski 1997). In comparison, 89% of the *B. bassiana* GHA conidia germinated after storage for 266 d. More recently, in a comparison of 27 *B. bassiana* isolates and the commercial *B. bassiana* GHA isolate, the former had LT_{50}s of 19–112 days at 30°C, whereas GHA had an LT_{50} of 215 days (Jaronski, unpublished data).

The shelf life of spores of fungal entomopathogens can be affected by nutritional and environmental conditions present during production and drying. The initial moisture content of the conidial powder and the drying speed of conidia produced on solid substrate culture were shown to influence the shelf life of conidia of *M. flavoviride* and *B. bassiana* (Hedgecock et al. 1995; S. Jaronski, unpublished data). The storage stability and desiccation tolerance of liquid culture produced blastospores of *I. fumosoroseus* were influenced by the speed of drying and the form and quantity of nitrogen provided in the liquid culture medium, respectively (Jackson et al. 1997; Jackson 1999). Work by Hallsworth and Magan (1994, 1995, 1996) suggested that manipulating the polyol content within the conidia of *B. bassiana, M. anisopliae* and *I. farinosus* through nutrition and/or osmotic stress can extend the range of water availability over which fungal propagules can germinate and may provide benefit during desiccation and storage (Hallsworth and Magan 1994, 1995, 1996).

Formulations, adjuvants, adherence, and interactions

Formulation plays an important role in delivering the fungal entomopathogen to the target environment. Formulated fungal entomopathogens are typically prepared as technical concentrates, wettable powders or oil dispersions (Faria and Wraight 2007). Technical concentrates are the fungal propagules combined with production by-products and minimal amendments. In wettable powder formulations, the dried fungal propagules are formulated to be dispersed in water and applied as a suspension. In oil dispersions, the fungal propagules are suspended in a water immiscible liquid which is intended to be diluted in water before use. Oil dispersions are typically limited to use with hydrophobic conidia. Both of these aqueous suspensions would typically be applied with spray applicators. In some cases, however, a wettable powder formulation has advantage over the liquid formulations such as when oil-incompatible materials have been applied to the crop. For example, use of an oil-based formulation on a crop treated with elemental sulphur can cause severe phytotoxicity (Hoy 2008). Recipes of wettable powder formulations must be carefully created to maintain spores and inerts in suspension with minimal agitation during spraying.

There are several important variables to consider when developing formulations which will be applied as aqueous sprays. First, aerial conidia of *Beauveria* spp., *Metarhizium* spp. and *Isaria* spp. are highly hydrophobic due to glycoproteins arranged in overlapping rodlets on the conidial surface (Bidochka et al. 1995). This property makes oil carriers ideal for these conidia. For example, *M. acridum* conidial powder is routinely suspended in groundnut oil, No. 2 diesel or kerosene for ultralow volume application against Orthoptera in Africa and maize oil in Australia. The suspension of aerial conidia in water is very difficult without the use of a wetting agent. A wetting agent must be selected that does not interfere with the infection process, much less kill the fungal propagule. Additionally, consideration should also be given to the fact that some wetting agents and adjuvants have been shown to expand the host range of fungal plant pathogens and may have a similar impact on insect pathogens (Boyette and Abbas 1994; Hoagland et al. 2007).

Secondly, spray formulations are designed to deliver the fungal entomopathogen directly to the insect (contact insecticide), to a location where protection is desired (plant surfaces, post harvest storage areas, etc.) or to areas frequented by the insects. Understanding how the spray droplet and entomopathogenic propagule interact with the target surface can help in guiding formulation decisions. In order for a spray droplet to adhere to a surface, the droplet must first be able to wet the surface. In general terms, for a liquid to wet a solid, the surface tension of the liquid must be lower than the surface energy of the solid. Most of the targets for fungal entomopathogen spray applications are hydrophobic or low surface energy targets (insect cuticles, plant surfaces, etc.). These types of surfaces are commonly referred to as being hydrophobic, since they repel water or the interaction with water is not energetically favourable. Surfaces with low energy are difficult to wet with aqueous solutions, since the surface energy (surface tension for a liquid) of water must be reduced below that of the solid surface for wetting to occur. In order to reduce the surface tension of aqueous solutions low enough to wet these surfaces, surfactants are added, which greatly lower the surface tension of water.

Surfactant selection must meet two criteria for effective use in a fungal entomopathogen formulation: biocompatibility and physical property performance. Biocompatibility is typically tested explicitly with potential surfactants, but there are general guidelines that can narrow one's search. Aerial conidia are generally much more tolerant of surfactants than blastospores, submerged conidia, or hyphal formulations and many successful examples of their use are available in the literature (Daoust et al. 1983; Alves et al. 2002; Akbar et al. 2005; Faria and Wraight 2007; Jin et al. 2008). Blastospores, submerged conidia, and hyphae lack the hydrophobic properties of aerial conidia, which allows surfactants to interact directly with the outer membrane of the cell. These hydrophilic propagules require more discretion when selecting a potential surfactant. The antifungal activity of surfactants is often correlated with lipophilicity of the surfactant (Leal et al. 2009) or more specifically the length of the alkyl chain

(Oros et al. 1999; De Jonghe et al. 2007). The longer the alkyl chain of the surfactant the more lethal they tend to be to yeasts and filamentous fungi. Alternatives with reduced fungicidal activity are available to replace the traditional alkyl chain based surfactants, including surfactants based on branched alkyl chains (Ayala-Zermeno et al. 1999), block co-polymers (Baur et al. 1997) and protein hydrolysates (Dunlap et al. 2007). These surfactants all limit the length of alkyl chains that could enter the membrane, which reduces their toxicity to the fungus. Nevertheless, the effect of wetting agents needs to be empirically determined as compatibility will differ among fungal species (Jaronski 1997). There is potential for differences between isolates in sensitivity to a particular chemical. In some cases, there is a concentration dependent effect of the emulsifier or dispersant on shelf life. Lastly, chemical interactions among formulation ingredients can have an effect on the conidia. In one example, the addition of an inert ingredient into a wettable powder formulation countered the deleterious effect of a dispersant on the shelf life of a commercial *B. bassiana* at 30°C and 35°C, although the dispersant had no effect on these conidia at 5–25°C (Jaronski 1997).

Physical property performance in the selection of a surfactant is often guided by the ability of the surfactant to reduce the surface tension of aqueous solutions. In addition to the ability to lower the equilibrium surface tension of water, an important parameter in surfactant selection is the dynamic surface tension. Dynamic surface tension is important in spray applications because during the spraying/droplet-forming process, new droplet surfaces are constantly being created and the surfactant must diffuse to the surface to reduce the surface tension. The time window between the droplet leaving the sprayer and hitting the target is often very short. If the surfactant has not sufficiently diffused to the surface of the droplet before impact, the surface tension will not be lowered and the surfactant will have provided little to no benefit. A universal spray droplet adhesion model has been proposed for the leaf surfaces (Forster et al. 2005) and its concepts are extendable to other surfaces (i.e., insect cuticle).

Understanding how fungal entomopathogen propagules interact with their insect host or respond to their target environment are important considerations when developing formulations. The target environment may be the insect cuticle or the physical environment in which the propagules are applied, such as, the phyllosphere, rhizosphere, insect nest, etc. There are distinct sequential events required for successful infection: initial attachment through non-specific interactions, adhesion through specific or induced interactions, conidial germination (which has several phases), chemotaxis of the hyphal tip on the cuticle, appressorium formation, and penetration into the cuticle (St. Leger 1991). During the infection process, fungal entomopathogenic propagules interact with their environment (the insect cuticle) through specific and non-specific interactions. Non-specific interactions mediate the initial contact of the propagule with surfaces. Such interactions arise from the physicochemical properties of the propagule surface and include hydrophobic, polar, and electrostatic properties. Specific interactions occur during germination and penetration and are directed responses of the fungus to specific cues on and in the cuticle. Formulation considerations are usually limited to the non-specific interactions. Knowledge of the surface physicochemical properties provides a basis for predicting how these propagules will interact with their insect hosts and their hosts' environment. These physicochemical properties have been reported for three entomopathogens, *I. fumosorosea*, *N. rileyi*, and *B. bassiana* (Pendland et al. 1994; Dunlap et al. 2005; Holder et al. 2007).

The interactions between a microbe and a surface can be described under defined conditions (e.g., pH and ionic strength) using Derjaguin–Landau–Verwey–Overbeek (DLVO) theory (Derjaugin and Landau 1941; Verwey and Overbeek 1948), extended DLVO theory or a thermodynamic approach (van Oss 1995). This information can be the basis for understanding interactions with formulation adjuvants and for choosing formulation conditions which improve adhesion. It will also be useful in predicting the transport properties of propagules, once applied to the host environment, such as transport in soils (Horn et al. 2001), mulches (Sun et al. 2008) or the phyllosphere (Bora et al. 1994). If the surface energy of the target surface (insect cuticle, soil, etc.) is known, propagule-surface interactions can be quantified and possibly optimized through growth conditions (Jana et al. 2000; Shah et al. 2007) or formulation (Webb et al. 1999). Specific interactions typically occur after the initial adhesion of conidia to the insect cuticle. Relatively little is known about these specific interactions. In *M. anisopliae*, the presence of a protein, MAD1, mediates the adhesion of conidia to insect cuticle while the MAD2 protein mediates adhesion to plant cuticle (Wang and St. Leger 2007). The use of microarrays to determine which genes are activated during the infection process should lead to a clearer understanding of these specific interactions (Wang and St. Leger 2007).

Persistence in the insects habitat

Secondary acquisition of infective propagules by insects from sprayed plant surfaces is often equally or more important than direct propagule contact from the spray (Fernandez et al. 2001; van der Valk 2007). The persistence of fungal propagules in the environment can be a very important factor in the overall efficacy of a mycoinsecticide, yet the persistence of propagules of fungal entomopathogens in the environment is generally poor. Estimates of persistence in field situations with UV exposure vary from a few hours to a few days with an exponential decay relationship (Inglis et al. 1993: McCoy et al. 2002; van der Valk 2007). Recently, a number of materials have been found to have value in improving persistence (Leland et al. 2004; Leland and Behle 2005; Reddy et al. 2008; Villamizar et al. 2009). The most promising technology to date is an organo-clay matrix containing one of several food grade dyes (Cohen and Tammar 2009)

Another factor affecting fungal propagule persistence is rain. When applied to leaf surfaces as an aqueous suspension, conidia readily washed off with simulated rain (Inglis et al. 2000). Conidia in an oil formulation, however, greatly reduced conidia loss from the leaf surface; oil-in-water emulsions were intermediate in effect. The use of polymeric stickers and spreaders may increase rainfastness but be counter productive if they prevent the transfer of conidia from the contacted surface to the insect cuticle.

Insect behaviour-based mycoinsecticide delivery

Formulation can be used to improve the delivery of fungal entomopathogens to their host or host environment. Many insect pests reside in difficult-to-reach locations, such as on the undersides of leaves, inside branches, stems or fruit, or in underground nests. While the application of liquid formulations is limited by gravity, foams can expand and deliver fungal propagules to hard-to-reach areas. For example, a foam formulation was developed to deliver blastospores of *I. fumosorosea* to termite nests located in trees or building structures (Fig. 4; Dunlap et al. 2007).

Fig. 4 Application of a biocompatible foam formulation of *Isaria fumosorosea* blastospores (**a**) through a hole drilled in the base of a Formosan termite-infested tree. Blastospores of *I. fumosorosea* are carried upward through termite galleries within the tree, emerging from pre-drilled holes above the injection site (**b**), for improved contact, infection, and control of Formosan subterranean termites

Another potential area to exploit in formulation development is the blocking or interfering with the insect's ability to detect the fungal entomopathogen within its body or to initiate a defence response. Insect defences are based on the innate immune system of the insect and consist of humoral and cellular responses (Lavine and Strand 2002). Insects have evolved receptors that bind to conserved molecules presented by microbial pathogens to identify the specific pathogen attacking them (Fearon 1997). These receptor- based systems identify the specific pathogen through recognition of specific pathogen-associated molecular pattern motifs. The detected molecular motifs are usually cell wall components of the pathogen such as, lipopolysaccharides, peptidoglycans and β (1,3)-D-glucans (Wang and Ligoxygakis 2006; Müller et al. 2008). This strategy was recently used to improve the virulence of *M. anisopliae* against termites. Bulmer et al. (2009) demonstrated that termites exhibit a unique β (1, 3)-glucanase activity in their tissues, cuticular washes, and nest material. The β (1, 3)-glucanase activity was proposed to have two functions related to termite defence. It acts as an environmental sensor by cleaving and releasing pathogen components, which activate the termite defence systems. The second function is to cleave and weaken the pathogen cell wall, making the pathogen more susceptible to the termites' antimicrobial peptides. By combining *M. anisopliae* with a β (1, 3)-glucanase inhibitor, improved biocontrol efficacy was demonstrated (Bulmer et al. 2009). It is easy to envision formulations of fungal entomopathogens combined with small molecules that inhibit insect defence and recognition pathways.

Formulation technology also plays a role in "bait and kill" or "lure and kill" applications as summarized by Vega et al. (2007) and Baverstock et al. (2009). These applications exist in a wide variety of formats. The bait/lure mechanism takes advantage of innate insect behaviour in response to various cues. The use of environmental stimuli such as color (Avery et al. 2008) and preferred habitat, ex. clay water pots to attract mosquitoes (Farenhorst et al. 2008), have been used to attract insects for the dissemination of fungal entomopathogen propagules. Semiochemicals have also been used to lure various insects to a fungal entomopathogen infection site (Vega et al. 2007). Insects targeted using this biocontrol approach include ticks [*Amblyomma vaiegatum* (Maranga et al. 2006; Nchu et al. 2009)], aphids [*Phorodon humuli* (Hartfield et al. 2001)], Japanese beetle [*Popilla japonica* (Klein and Lacey 1999)], grain borers [*Prostephanus truncatus* (Smith et al. 1999)], and the diamond-back moth [*Plutella xylostella* (Furlong et al. 1995)].

Autodissemination devices based on a commercial, pheromone-based Japanese beetle trap are currently being deployed on an operational basis in the Azores (S. Jaronski, unpublished data). Food attractants have been combined with fungal entomopathogens and have had some success controlling a variety of insects, such as termites [*Reticulitermes flavipes* (Wang and Powell 2004)], ants [*Atta cephalotes* (Lopez and Orduz 2003)], locusts [*Schistocerca gregaria* (Caudwell and Gatehouse 1996)] and house flies [*Musca domestica* L. (Renn et al. 1999)]. Vegetable oils rich in oleic, linoleic and linolenic acids, that stimulate necrophagy among grasshoppers (Orthoptera: Acrididae), have been used to attract the insects to a toxicant or to an infective fungal propagule (Bomar and Lockwood 1994a, b; Latchininsky et al. 2007). Linoleic and linolenic acids stimulate necrophagy in many species of grasshoppers and an oil carrier, rich in these compounds, can serve as the basis of a mycoinsecticide attracticide. These attracticides are useful in strip treatments thus reducing the overall rate of mycopesticide application per protected area (Lockwood et al. 2001). Canola olive and flax oils are rich in these compounds. Canola oil has been used to validate the principle in the field with carbaryl (Bomar and Lockwood 1994b). These findings are being extended to mycoacaricide use on US rangeland (Jaronski et al. unpublished).

Formulations may also be used to reduce the repellence of fungal materials or reduce the stimulation of behavioural responses by insects. For example, the conidia of fungal entomopathogens stimulated diverse defensive behaviours in termites that served to eliminate or minimize the impact of the pathogen on the colony (Rosengaus et al. 1998; Fefferman et al. 2007). These nest hygiene behaviours include intense grooming of workers, disposal or isolation of infected workers and sporulating cadavers, etc. Blastospores of *I. fumosorosea* were shown to be much less repellent to the Formosan subterranean termites *Coptotermes formosanus* when compared to solid substrate produced conidia of *I. fumosorosea* (Wright et al. 2003). In addition, vegetative mycelium has been shown to be readily accepted by termites and taken into termite nests where it sporulates (Stamets 2006). Evidently, presporogenic mycelium of *Metarhizium* spp., *Beauveria* spp. or *Cordyceps* spp. contain attractant volatiles that induce social insects such as ants and termites to graze on the mycelium, scattering the mycelium around feeding areas and nesting chambers, after which the fungus sporulates to produce infectious conidia (Stamets 2006). This presporogenic mycelium can be introduced to termites or ants on grain-based solid substrate or freeze-dried mycelium.

These "lure and kill" strategies ultimately rely on the transfer the infective propagule of an entomopathogenic fungus to the insect when it arrives at the dispenser. In this strategy, the dispenser can be any device the insect is lured to which contains the fungal entomopathogen propagules. Little is known about optimizing parameters to improve the transfer of viable fungal propagules to the insect. A basic understanding of the physical and environmental parameters that impact fungal propagule transfer to the insect cuticle would benefit the use of this strategy

Conclusion

The control of insect pests with fungal entomopathogenic fungi requires a significant basic understanding of the interactions between the target insect, fungal entomopathogen, and environment. The production and formulation goals for these fungal entomopathogens must consider economic realities but also be mindful of the ecological constraints and requirements for consistent insect infection and control. An understanding of the critical biotic and abiotic constraints for each particular insect pest–environment will guide the selection and development of these agents. New approaches to the production of efficacious fungal propagules and the development of formulations tailored to overcome specific environmental constraints will lead to the availability of dependable entomopathogen-based bioinsecticides for controlling insects in a wide variety of habitats.

References

Akbar W, Lord JC, Nechols JR, Loughin TM (2005) Efficacy of *Beauveria bassiana* for red flour beetle when applied with plant essential oils or in mineral oil and organosilicone carriers. J Econ Entomol 98:683–688

Alves RT, Bateman RP, Gunn J, Prior C, Leather SR (2002) Effects of different formulations on viability and medium-term storage of *Metarhizium anisopliae* conidia. Neotrop Entomol 31:91–99

Anonymous (2009) Fire triangle. Wikipedia. Downloaded from http://en.wikipedia.org/wiki/Fire_triangle

Askary H, Benhamou N, Brodeur J (1999) Ultrastructural and cytochemical characterization of aphid invasion by the hyphomycete *Verticillium lecanii*. J Invertebr Pathol 74:1–13

Avery PB, Hunter WB, Hall DG, Jackson MA, Rogers ME, Powell CA (2008) Novel delivery of the fungi *Paecilomyces fumosoroseus* (Deuteromycotina: Hyphomycetes) for managing the Asian citrus psyllid (Psyllidae: Hemiptera). In: The 41st annual meeting of the society for invertebrate pathology. Coventry, United Kingdom, p 65

Ayala-Zermeno MA, Navarro-Barranco H, Mier T, Toriello C (1999) Effect of agro-chemicals on in vitro growth of the entomopathogenic fungi *Metarhizium anisopliae* (Metschnikoff) Sorokin and *Paecilomyces fumosoroseus* (Wize) Brown and Smith. Rev Latinoam de Microbiol 41:223–229

Bartlett MC, Jaronski ST (1988) Mass production of entomogenous fungi for biological control of insects. In: Burge MN (ed) Fungi in biological control systems. Manchester University Press, Manchester, pp 61–85

Baur ME, Kaya HK, Gaugler R, Tabashnik B (1997) Effects of adjuvants on entomopathogenic nematode persistence and efficacy against *Plutella xylostella*. Biocontrol Sci Technol 7:513–525

Baverstock J, Roy H, Pell J (2009) Entomopathogenic fungi and insect behaviour: from unsuspecting hosts to targeted vectors. BioControl. doi:10.1007/s10526-009-9238-5 (this SI)

Behle RW, Garcia-Guitierrez C, Tamez-Guerra P, McGuire MR, Jackson MA (2006) Pathogenicity of blastospores and conidia of *Paecilomyces fumosoroseus* against larvae of the Mexican bean beetle, *Epilachna varivestis* Mulsant. Southwestern Entomol 31:289–295

Bidochka MJ, Leger RJ, Joshi L, Roberts DW (1995) The rodlet layer from aerial and submerged conidia of the entomopathogenic fungus *Beauveria bassiana* contains hydrophobin. Mycol Res 99:403–406

Bidochka MJ, Kamp AM, Lavender M, Dekoning J, Amritha de Croos JN (2001) Habit association in two genetic groups of the insect-pathogenic fungus *Metarhizium anisopliae*: uncovering cryptic species. Appl Environ Microbiol 67:1335–1342

Bomar CR, Lockwood JA (1994a) Olfactory basis of cannibalism in grasshoppers (Orthoptera: Acrididae): I. Laboratory assessment of attractants. J Chem Ecol 20:2249–2260

Bomar CR, Lockwood JA (1994b) Olfactory basis of cannibalism in grasshoppers (Orthoptera: Acrididae): II. Field assessment of attractants. J Chem Ecol 20:2263–2271

Bora RS, Murty MG, Shenbagarathai R, Sekar V (1994) Introduction of a lepidopteran-specific insecticidal crystal protein gene of *Bacillus thuringiensis* subsp. kurstaki by conjugal transfer into a *Bacillus megaterium* strain that persists in the cotton phyllosphere. Appl Environ Microbiol 60:214–222

Boucias DG, Pendland JC (1991) Attachment of mycopathogens to cuticle: the initial event of mycosis in arthropod hosts. In: Cole GT, Hoch HC (eds) The fungal spore and disease initiation in plants and animals. Plenum, New York, pp 101–127

Boyette CD, Abbas HK (1994) Host range alteration of the bioherbicidal fungus *Alternaria crassa* with fruit pectin and plant filtrates. Weed Sci 42:487–491

Bruck DJ (2005) Ecology of *Metarhizium anisopliae* in soilless potting media and the rhizosphere implications for pest management. Biol Control 32:155–163

Bulmer MS, Bachelet I, Raman R, Rosengaus RB, Sasisekharan R (2009) Targeting an antimicrobial effector function in insect immunity as a pest control strategy. Proc Nat Acad Sci 106:12652–12657

Butler MJ, Day AW (1998) Fungal melanins: a review. Can J Microbiol 44:1115–1136

Butt TM, Jackson C, Magan N (2001) Introduction—fungal biological control agents: progress, problems and potential. In: Butt TM, Jackson C, Magan N (eds) Fungi as biocontrol agents: progress, problems and potential. CAB International Publishing, Wallingford, pp 1–8

Castrillo LA, Roberts DW, Vandenberg JD (2005) The fungal past, present, and future: germination, ramification, and reproduction. J Invertebr Pathol 89:46–56

Caudwell RW, Gatehouse AG (1996) Formulation of grasshopper and locust entomopathogens in baits using starch extrusion technology. Crop Prot 15:33–37

Chandler D, Heal JB, Gillespie AT (1994) Effect of osmotic potential on the germination of conidia and colony growth of *Verticillium lecanii*. Mycol Res 98:384–388

Charudattan R (2001) Biological control of weeds by means of plant pathogens: significance for integrated weed management in modern agro-ecology. BioControl 46:229–260

Cohen E, Tammar J (2009) Photostabilization of *Beauveria bassiana* conidia using anionic dyes. Appl Clay Sci 42:569–574

Coley-Smith JR, Cooke RC (1971) Survival and germination of fungal sclerotia. In: Horsfall JG, Baker KF, Zentmyer GA (eds) Annu Rev Phytopathol 9:65–92

Costa SD, Grassano S, Li J (2008) Sweet whey based biopesticide composition. World Patent Application #2008/73843

Crowe JH, Crowe LM (1986) Stabilization of membranes in anhydrobiotic organisms. In: Leopold AC (ed) Membranes, metabolism and dry organisms. Comstock Publishing Co., Ithaca, pp 188–209

Daoust RA, Ward MG, Roberts DW (1983) Effect of formulation on the viability of *Metarhizium anisopliae* conidia. J Invertebr Pathol 41:151–160

De Jonghe K, Hermans D, Hafte M (2007) Efficacy of alcohol alkoxylate surfactants differing in the molecular structure of the hydrophilic portion to control *Phytophthora nicotianae* in tomato substrate culture. Crop Prot 26:1524–1531

Derjaugin BW, Landau L (1941) Theory of the stability of strongly charged lyophobic sols and the adhesion of strongly charged particles in solutions of electrolytes. Acta Physiochim (USSR) 14:633–662

Dunlap CA, Biresaw G, Jackson MA (2005) Hydrophobic and electrostatic cell surface properties of blastospores of the entomopathogenic fungus *Paecilomyces fumosoroseus*. Colloids Surf B Interfaces 46:261–266

Dunlap CA, Jackson MA, Wright MS (2007) A foam formulation of *Paecilomyces fumosoroseus*, an entomopathogenic biocontrol agent. Biocontrol Sci Technol 17:513–523

Eilenberg J, Hajek A, Lomer C (2001) Suggestions for unifying the terminology in biological control. BioControl 46:387–400

Evans HC, Samson RA (1982) *Cordyceps* species and their anamorphs pathogenic on ants (Formicidae) in tropical forest ecosystems I. The Cephalotes (Myrmicinae) complex. Trans Br Mycol Soc 79:431–453

Farenhorst M, Farina D, Scholte E-J, Takken W, Hunt RH, Coetzee M, Knols BGJ (2008) African water storage pots for the delivery of the entomopathogenic fungus *Metarhizium anisopliae* to the malaria vectors *Anopheles gambiae s.s.* and *Anopheles funestus*. Am J Trop Med Hyg 78:910–916

Faria M, Wraight SP (2001) Biological control of *Bemisia tabaci* with fungi. Crop Prot 20:767–778

Faria MR, Wraight SP (2007) Mycoinsecticides and mycoacaricides: a comprehensive list with worldwide coverage and international classification of formulation types. Biol Control 43:237–256

Fearon DT (1997) Seeking wisdom in innate immunity. Nature 388:323–324

Fefferman NH, Traniello JFA, Rosengaus RB, Calleri DV (2007) Disease prevention and resistance in social insects: modelling the survival consequences of immunity, hygienic behaviour, and colony organization. Behav Ecol Sociobiol 61:565–577

Fernandez S, Groden E, Vandenberg JD, Furlong MJ (2001) The effect of mode of exposure to *Beauveria bassiana* on conidia acquisition and host mortality of Colorado potato beetle, *Leptinotarsa decemlineata*. J Invertebr Pathol 77:217–226

Forster WA, Kimberley MO, Zabkiewicz JA (2005) A universal spray droplet adhesion model. Trans Am Soc Agric Eng 48:1321–1330

Fravel DR (2005) Commercialization and implementation of biocontrol. Ann Rev Phytopathol 43:337–359

Friesen TJ, Holloway G, Hill GA, Pugsley TS (2006) Effect of conditions and protectants on the survival of *Penicillium bilaiae* during storage. Biocontrol Sci Technol 16:89–98

Furlong MJ, Pell JK, Choo OngPek, Rahman SA (1995) Field and laboratory evaluation of a sex pheremone trap for the autodissemination of the fungal entomopathogen *Zoophthora radicans* (Entomophthorales) by the diamond-back moth, *Plutella xylostella* (Leptidoptera: Yponomeutidae). Bull Entomol Res 85:331–337

Goettel MS, Roberts DW (1992) Mass production, formulation and field application of entomopathogenic fungi. In: Lomer CJ, Prior C (eds) Biological control of locusts and grasshoppers. CAB International, Wallingford, pp 230–238

Hallsworth JE, Magan N (1994) Effect of carbohydrate type and concentration on polyhydroxy alcohol and trehalose content of conidia of three entomopathogenic fungi. Microbiology 140:2705–2713

Hallsworth JE, Magan N (1995) Manipulation of intracellular glycerol and erythritol enhances germination of conidia at low water availability. Microbiology 141:1109–1115

Hallsworth JE, Magan N (1996) Culture age, temperature, and pH affect the polyol and trehalose contents of fungal propagules. Appl Environ Microbiol 62:2435–2442

Hartfield CM, Campbell CAM, Hardie J, Pickett JA, Wadhams LJ (2001) Pheromone traps for the dissemination of an entomopathogen by the damson-hop aphid *Phorodon humuli*. Biocontrol Sci Technol 11:401–410

Hedgecock S, Moore D, Higgins PM, Prior C (1995) Influence of moisture content on temperature tolerance and storage of *Metarhizium flavoviride* conidia in an oil formulation. Biocontrol Sci Technol 5:371–377

Hegedus DD, Bidochka MJ, Khachatourians GG (1990) *Beauveria bassiana* submerged conidia production in a defined medium containing chitin, two hexosamines or glucose. Appl Microbiol Biotechnol 33:641–647

Hoagland RE, Boyette CD, Weaver MA, Abbas HK (2007) Bioherbicides: research and risk. Toxin Rev 26:313–342

Holder DJ, Kirkland BH, Lewis MW, Keyhani NO (2007) Surface characteristics of the entomopathogenic fungus *Beauveria* (*Cordyceps*) *bassiana*. Microbiology 153: 3448–3457

Hong TD, Gunn J, Ellis RH, Jenkins NE, Moore D (2001) The effect of storage environment on the longevity of conidia of *Beauveria bassiana*. Mycol Res 105:597–602

Horn BW, Greene RL, Sorensen RB, Blankenship PD, Dorner JW (2001) Conidial movement of nontoxigenic *Aspergillus flavus* and *A. parasiticus* in peanut fields following application to soil. Mycopathologia 151:81–92

Hoy MA (2008) Acaricides and Miticides In: Capinera JL (ed) Encyclopedia of entomology 2nd ed. Springer Science+Media BV, 3:9–21

Hu G, Leger RJ (2002) Field studies using a recombinant mycoinsecticide (*Metarhizium anisopliae*) reveal that it is rhizosphere competent. Appl Environ Microbiol 68:6383–6387

Inglis GD, Goettel MS, Johnson DL (1993) Persistence of the entomopathogenic fungus, *Beauveria bassiana*, on phylloplanes of crested wheatgrass and alfalfa. Biol Control 3:258–270

Inglis GD, Johnson DL, Goettel MS (1996) Effects of temperature on mycosis by *Beauveria bassiana* in grasshoppers. Biol Control 7:131–139

Inglis GD, Ivie TJ, Duke GM, Goettel MS (2000) Influence of rain and conidial formulation on persistence of *Beauveria bassiana* on potato leaves and Colorado potato beetle larvae. Biol Control 18:55–64

Jackson MA (1997) Optimizing nutritional conditions for the liquid culture production of effective fungal biological control agents. J Ind Microbiol Biotechnol 19:180–187

Jackson MA (1999) Method for producing desiccation tolerant *Paecilomyces fumosoroseus* spores. US Patent number 5,968,808

Jackson MA (2007) The biotechnology of producing and stabilizing living, microbial biological control agents for insect and weed control. In: Hou CT, Shaw FJ (eds) Biocatalysis and biotechnology: functional foods and industrial products. CRC Press, Boca Raton, pp 533–543

Jackson MA, Jaronski ST (2009) Production of microsclerotia of the fungal entomopathogen *Metarhizium anisopliae* and their potential for use as a biocontrol agent for soil-inhabiting insects. Mycol Res 113:842–850

Jackson MA, Schisler DA (1992) The composition and attributes of *Colletotrichum truncatum* spores are altered by the nutritional environment. Appl Environ Microbiol 58:2260–2265

Jackson MA, Schisler DA (1995) Liquid culture production of microsclerotia of *Colletotrichum truncatum* for use as bioherbicidal propagules. Mycol Res 99:879–884

Jackson MA, Schisler DA (2002) Selecting fungal biocontrol agents amenable to production by liquid culture fermentation. Proceedings of 7th biocontrol working group meeting "Influence of a-biotic and biotic factors on biocontrol agents". IOBC/WPRS Bull, pp 387–391

Jackson MA, Schisler DA, Slininger PJ, Boyette CD, Silman RW, Bothast RJ (1996) Fermentation strategies for improving the fitness of a bioherbicide. Weed Technol 10:645–650

Jackson MA, McGuire MR, Lacey LA, Wraight SP (1997) Liquid culture production of desiccation tolerant blastospores of the bioinsecticidal fungus *Paecilomyces fumosoroseus*. Mycol Res 101:35–41

Jackson MA, Cliquet S, Iten LB (2003) Media and fermentation processes for the rapid production of high concentrations of stable blastospores of the bioinsecticidal fungus *Paecilomyces fumosoroseus*. Biocontrol Sci Technol 13:23–33

Jackson MA, Erhan S, Poprawski TJ (2006) Influence of formulation additives on the desiccation tolerance and storage stability of blastospores of the entomopathogenic fungus *Paecilomyces fumosoroseus* (Deuteromycotina: Hyphomycetes). Biocontrol Sci Technol 16:61–75

Jana TK, Srivastava AK, Arora DK, Csery K (2000) Influence of growth and environmental conditions on cell surface hydrophobicity of *Pseudomonas fluorescens* in non-specific adhesion. Can J Microbiol 46:28–37

Jaronski ST (1997) New paradigms in formulating mycoinsecticides. In: Goss GR, Hopkinson MJ, Collins HM (eds) Pesticide formulations and applications systems: 17th Volume, ASTM STP 1328. Am Soc for Testing and Materials, pp 99–112

Jaronski ST (2007) Soil ecology of the entomopathogenic ascomycetes: a critical examination of what we (think) we know. In: Maniana K, Ekesi S (eds) Use of entomopathogenic fungi in biological pest management. Research SignPost, Trivandrum, pp 91–144

Jaronski ST, Jackson MA (2008) Efficacy of *Metarhizium anisopliae* microsclerotial granules. Biocontrol Sci Technol 18:849–863

Jenkins NE, Prior C (1993) Growth and formation of true conidia by *Metarhizium flavoviride* in a simple liquid medium. Mycol Res 97:1489–1494

Jin X, Grigas KE, Johnson CA, Perry P, Miller DW (1999) Method for storing fungal conidia. US Patent number 5,989,898

Jin X, Streett DA, Dunlap CA, Lyn ME (2008) Application of hydrophilic-lipophilic balance (HLB) number to optimize a compatible non-ionic surfactant for dried aerial conidia of *Beauveria bassiana*. Biol Control 46:226–233

Jung G, Mugnier J (1989) Low-water-activity inocula for biological control. US Patent number 4,886,664

Kassa A, Vidal SD, Zimmerman G (2004) Production and processing of *Metarhizium anisopliae var. acridum* submerged conidia for locust and grasshopper control. Mycol Res 108:93–100

Keller S, Kessler P, Schweizer C (2003) Distribution of insect pathogenic soil fungi in Switzerland with special reference to *Beauveria brongniartii* and *Metarhizium anisopliae*. BioControl 48:307–319

Klein MG, Lacey LA (1999) An attractant trap for autodissemination of entomopathogenic fungi into populations of the Japanese beetle *Popillia japonica* (Coleoptera: Scarabaeidae). Biocontrol Sci Technol 9:151–158

Klingen I, Eilenberg J, Meadow R (2002) Effects of farming system, field margin, and bait insect on the occurrence of insect pathogenic fungi in soils. Agric Ecosyst Environ 91:191–198

Lacey LA, Shapiro-Ilan DI (2008) Microbial control of insect pests in temperate orchard systems: potential for incorporation into IPM. Ann Rev Entomol 53:121–144

Latchininsky AV, Schell SP, Lockwood JA (2007) Laboratory bioassays of vegetable oils as kairomonal phagostimulants for grasshoppers (Orthoptera: Acrididae). J Chem Ecol 33:1856–1866

Lavine MD, Strand MR (2002) Insect hemocytes and their role in immunity. Insect Biochem Mol Biol 32:1295–1309

Leal PC, Mascarello A, Derita M, Zuljan F, Nunes RJ, Zacchino S, Yunes RA (2009) Relation between lipophilicity of alkyl gallates and antifungal activity against yeasts and filamentous fungi. Bioorganic Medicinal Chem Lett 19:1793–1796

Leger RJ (1991) Integument as a barrier to microbial infections. In: Buinnington K, Retnakaran A (eds) Physiology of the insect epidermis. CSIRO Publications, Melbourne, pp 284–306

Leland JE, Behle RW (2005) Coating *Beauveria bassiana* with lignin for protection from solar radiation and effects on pathogenicity to *Lygus lineolaris* (Heteroptera: Miridae). Biocontrol Sci Technol 15:309–320

Leland JE, Mullins DE, Vaughn LJ (2004) Water soluble uv-protective coatings for biological pesticides and process for making same. US Patent Application number 20040038825

Leland JE, Mullins DE, Vaughan LJ, Warren HL (2005) Effects of media composition on submerged culture spores of the entomopathogenic fungus, *Metarhizium anisopliae var. acridum*, part 1: comparison of cell wall characteristics and drying stability among three spore types. Biocontrol Sci Technol 15:379–392

Li J, Feng MG (2009) Intraspecific tolerance of *Metarhizium anisopliae* conidia to the upper thermal limits of summer with a description of a quantitative assay system. Mycol Res 113:93–99

Lockwood JA, Narisu JA, Schell SP, Lockwood DR (2001) Canola oil as a kairomonal attractant of rangeland grasshoppers: an economical liquid bait for insecticide formulation. Intl J Pest Manag 47:185–194

Lopez E, Orduz S (2003) *Metarhizium anisopliae* and *Trichoderma viride* for control of nests of the fungus-growing ant, *Atta cephalotes*. Biol Control 27:194–200

Magan N (2001) Physiological approaches to improving the ecological fitness of fungal biocontrol agents. In: Butt TM, Jackson C, Magan N (eds) Fungi as biocontrol agents: progress, problems and potential. CAB International Publishing, Wallingford, pp 239–252

Maranga RO, Hassanali A, Kaaya GP, Mueke JM (2006) Performance of a prototype baited-trap in attracting and infecting the tick *Amblyomma variegatum* (Acari: Ixodidae) in field experiments. Exp Appl Acarol 38:211–218

McCoy CM, Quintela ED, de Faria M (2002) Environmental persistence of entomopathogenic fungi. Factors affecting survival of entomopathogens. South Coop Ser Bull 400:20–30

Meyling NV, Eilenberg J (2006) Occurrence and distribution of soil borne entomopathogenic fungi within a single organic agroecosystem. Agric Ecosyst Environ 113:336–341

Müller U, Vogel P, Alber G, Schaub G (2008) The innate immune system of mammals and insects. In: Egesten A, Schmidt A, Herwald H (eds) Trends in innate immunity, contributions to microbiology. Karger AG, Basel, pp 21–44

Nchu F, Maniania NK, Touré A, Hassanali A, Eloff JN (2009) The use of a semiochemical bait to enhance exposure of *Amblyomma variegatum* (Acari: Ixodidae) to *Metarhizium anisopliae* (Ascomycota: Hypocreales). Vet Parasitol 160:279–284

Oros G, Cserhàti T, Vrbanovà A (1999) Relationship between the physicochemical parameters and biological activity of sulfosuccinic acid ester surfactants. Biochem Mol Biol Int 47:79–88

Pendland JC, Lopez-Lastra C, Boucias DG (1994) Laminin-binding sites on cell walls of the entomopathogen *Nomuraea rileyi* associated with growth and adherence to host tissues. Mycologia 86:327–335

Poprawski TJ, Jackson MA (1999) Laboratory activity of blastospores of *Paecilomyces fumosoroseus* on *Bemisia argentifolii* nymphs, 1997. Arthropod Manag Tests 24:399–400

Quesada-Moraga E, Navas-Cortez JA, Maranhao AA, Ortiz-Urquiza A, Santiago-Alvarez C (2007) Factors affecting the occurrence and distribution of entomopathogenic fungi in natural and cultivated soils. Mycol Res 111:947–966

Reddy NP, Khan PAA, Devi KU, Victor JS, Sharma HC (2008) Assessment of the suitability of Tinopal as an enhancing adjuvant in formulations of the insect pathogenic fungus *Beauveria bassiana* (Bals.) Vuillemin. Pest Manag Sci 64:909–915

Renn N, Bywater AF, Barson G (1999) A bait formulated with *Metarhizium anisopliae* for the control of *Musca domestica L.* (Dipt., Muscidae) assessed in large-scale laboratory enclosures. J Appl Entomol 123:309–314

Rosengaus RB, Maxmen AB, Coates LE, Traniello JFA (1998) Disease resistance: a benefit of sociality in the dampwood termite *Zootermopsis angusticollis* (Isoptera: Termopsidae). Behav Ecol Sociobiol 44:125–134

Sanchez-Murillo RI, de la Torre-Martinez M, Aguirre-Linares J, Herrera-Estrella A (2004) Light-regulated asexual reproduction in *Paecilomyces fumosoroseus*. Microbiology 150:311–319

Shah PA, Pell JK (2003) Entomopathogenic fungi as biological control agents. Appl Microbiol Biotechnol 61:413–423

Shah FA, Allen N, Wright CJ, Butt TM (2007) Repeated in vitro subculturing alters spore surface properties and virulence of *Metarhizium anisopliae*. FEMS Microbiol Lett 276:60–66

Shapiro-Ilan DI, Cottrell TE, Jackson MA, Wood BW (2008) Virulence of Hypocreales fungi to pecan aphids (Hemiptera: Aphididae) in the laboratory. J Invertebr Pathol 99:312–317

Shearer JF, Jackson MA (2006) Liquid culturing of microsclerotia of *Mycoleptodiscus terrestris*, a potential biological control agent for the management of hydrilla. Biol Control 38:298–306

Shimuzu S, Tsuchitani Y, Matsumoto T (1993) Production of an extracellular protease by *Beauveria bassiana* in the haemolymph of the silkworm, *Bombyx mori*. Lett Appl Microbiol 16:291–294

Sieglaff DH, Pereira RM, Capinera JL (1997) Pathogenicity of *Beauveria bassiana* and *Metarhizium flavoviride* (Deuteromycotina) to *Schistocerca americana* (Orthoptera: Acrididae). J Econ Entomol 90:1539–1545

Smith RJ, Grula EA (1982) Toxic components of the larval surface of the corn earworm (*Heliothis zea*). J Invertebr Pathol 39:15–22

Smith SM, Moore D, Karanja LW, Chandi EA (1999) Formulation of vegetable fat pellets with pheromone and *Beauveria bassiana* to control the larger grain borer, *Prostephanus truncatus* (Horn). Pestic Sci 55:711–718

Speare AT (1920) On certain entomogenous fungi. Mycologia 12:62–76

Sprenkel RK, Brooks WM (1977) Winter survival of the entomogenous fungus *Nomuraea rileyi* in North Carolina. J Invertebr Pathol 29:262–266

Stamets P (2006) Mycoattractants and mycopesticides. US Patent number 7,122,176

Sun J-Z, Fuxa JR, Richter A, Ring D (2008) Interactions of *Metarhizium anisopliae* and tree-based mulches in repellence and mycoses against *Coptotermes formosanus* (Isoptera: Rhinotermitidae). Environ Entomol 37:755–763

Thomas MB (1999) Ecological approaches and the development of "truly integrated" pest management. Proc Nat Acad Sci USA 96:5944–5951

Thomas KC, Khachatourians GG, Ingledew WM (1986) Production and properties of *Beauveria bassiana* conidia cultivated in submerged culture. Can J Microbiol 33:12–20

van der Valk H (2007) Desert locust technical series. Review of the efficacy of *Metarhizium anisopliae* var. acridum against the desert locust. Plant production and protection division, Locusts and other migratory pest group No. AGP/DL/TS/34, p 81

van Oss CJ (1995) Hydrophobicity of biosurfaces—origin, quantitative determination and interaction energies. Colloids Surf B Biointerfaces 5:91–110

Vega FE, Jackson MA, McGuire MR (1999) Germination of conidia and blastospores of *Paecilomyces fumosoroseus* on the cuticle of the silverleaf whitefly, *Bemisia argentifolii*. Mycopathologia 147:33–35

Vega FE, Dowd PF, Lacey LA, Pell JK, Jackson DM, Klein M (2007) Dissemination of beneficial microbial agents by insects. In: Lacey LA, Kaya H (eds) Field manual of techniques in invertebrate pathology. Second edition. Application and evaluation of pathogens for control of insects and other invertebrate pests. Springer, The Netherlands, pp 127–146

Vega FE, Goettel MS, Blackwell M, Chandler D, Jackson MA, Keller S, Koike M, Maniania NK, Monzón A, Ownley BH, Pell JK, Rangel DEH, Roy HE (2009) Fungal entomopathogens: new insights on their ecology. Fungal Ecol 2:149–159

Verwey EJW, Overbeek JG (1948) Theory of the stability of lyophobic colloids. Elsevier, Amsterdam

Vestergaard S, Butt TM, Bresciani J, Gillespie AT, Eilenberg J (1999) Light and electron microscopy studies of the infection of the western flower thrips *Frankliniella occidentalis* (Thysanoptera: Thripidae) by the entomopathogenic fungus *Metarhizium anisopliae*. J Invertebr Pathol 73:25–33

Villamizar L, Grijalba E, Zuluaga V, Gomez M, Cotes AM (2009) Evaluation of some parameters influencing the activity of a fungal biocontrol agent used for *Bemisia tabaci* control. IOBC/WPRS Bull 45:327–330

Wang C, Leger RJ (2007) The MAD1 adhesin of *Metarhizium anisopliae* links adhesion with blastospore production and virulence to insects, and the MAD2 adhesin enables attachment to plants. Eukaryot Cell 6:808–816

Wang L, Ligoxygakis P (2006) Pathogen recognition and signaling in the *Drosophila* innate immune response. Immunobiol 211:251–261

Wang C, Powell JE (2004) Cellulose bait improves the effectiveness of *Metarhizium anisopliae* as a microbial control of termites (Isoptera: Rhinotermitidae). Biol Control 30:523–529

Webb JS, Robson GD, Handley PS, Nixon M, Eastwood IM, Greenhalgh M (1999) Plasticizers increase adhesion of the deteriogenic fungus *Aureobasidium pullulans* to polyvinyl chloride. Appl Environ Microbiol 65:3575–3581

Wraight SP, Jackson MA, De Kock SL (2001) Production, stabilization and formulation of fungal biocontrol agents. In: Butt TM, Jackson C, Magan N (eds) Fungi as biocontrol agents: progress, problems and potential. CABI Publishing, Wallingford, pp 253–288

Wright MS, Connick WJ Jr, Jackson MA (2003) Use of *Paecilomyces* spp. as pathogenic agents against subterranean termites. US Patent number 6,660,291

Yanagawa A, Yokohari F, Shimizu S (2008) Defence mechanism of the termite, *Coptotermes formosanus* Shiraki, to entomopathogenic fungi. J Invertebr Pathol 97:165–170

Yeo H, Pell JK, Alderson PG, Clark SJ, Pye BJ (2003) Laboratory evaluation of temperature effects on the germination and growth of entomopathogenic fungi and on their pathogenicity to two aphid species. Pest Manag Sci 59:156–165

Ying SH, Feng MG (2006) Medium components and culture conditions affect the thermotolerance of aerial conidia of the fungal biocontrol agent *Beauveria bassiana*. Lett Appl Microbiol 43:331–335

Zimmermann G (2007) Review on safety of the entomopathogenic fungus *Metarhizium anisopliae*. Biocontrol Sci Technol 17:879–920

Zimmermann G (2008) The entomopathogenic fungi *Isaria farinosa* (formerly *Paecilomyces farinosus*) and the *Isaria fumosorosea* species complex (formerly *Paecilomyces fumosoroseus*): biology, ecology and use in biological control. Biocontrol Sci Technol 18:865–901

Author Biographies

Mark A. Jackson is a Research Microbiologist with the USDA, Agricultural Research Service at the National Center for Agricultural Utilization Research in Peoria, IL, USA. He received his PhD in Microbiology from the University of Arkansas in 1987. For the past 20 years, Dr. Jackson's research has focused on optimizing production and stabilization processes for fungal biological control agents. Liquid culture production systems were used to examine the impact of nutrition on fungal propagule formation, stability and biocontrol efficacy. This work has resulted in the development of novel methods for producing stable blastospore- and sclerotia-based insect biocontrol preparations.

Chris A. Dunlap received his Ph.D. in Chemistry from the Ohio State University, USA. For the past five years, he has provided formulation expertise to several USDA, Agricultural Research Service biological control projects as a Research Chemist in the Crop Bioprotection Research Unit located in Peoria, IL, USA. This work has spanned projects covering the control of insects, weeds and fungal plant pathogens. Research has been focused on: improving the drying of microbial organisms, improving adhesion of the biological control agents, and increasing field efficacy.

Stefan T. Jaronski is a Research Entomologist with the Agricultural Research Service, U.S. Department of Agriculture in Sidney Montana USA. His current research concerns development of microbial control of grasshoppers, and soil dwelling insect pests of sugar beets, part of bio-based integrated pest and plant pathogen management system. Before joining USDA in 2000, Dr. Jaronski spent 17 years in industry commercializing bacterial and fungal pest control agents, including *Beauveria bassiana* GHA. His research experience includes basic microbiology, mass production, formulation, bioassay systems, field trials, and regulatory aspects of entomopathogenic Hypocreales. He has a Ph.D. in Insect Pathology from Cornell University (1978).

BioControl (2010) 55:147–158
DOI 10.1007/s10526-009-9253-6

Fungal pathogens as classical biological control agents against arthropods

Ann E. Hajek · Italo Delalibera Jr.

Received: 28 July 2009 / Accepted: 26 October 2009 / Published online: 12 November 2009

Abstract Fungal entomopathogens have been used more frequently than other types of pathogens for classical biological control. Among 136 programs using different groups of arthropod pathogens, 49.3% have introduced fungal pathogens (including both the traditional fungi and microsporidia). The most commonly introduced species was *Metarhizium anisopliae* (Metschnikoff) Sorokin, with 13 introductions, followed by *Entomophaga maimaiga* Humber, Shimazu & Soper, which was released seven times. The majority of introduction programs have focused on controlling invasive species of insects or mites (70.7%) rather than on native hosts (29.4%). Almost half of the introductions of traditional fungi targeted species of Hemiptera and 75% of the microsporidia introduced have been introduced against lepidopteran species. The United States was the country where most introductions of fungi took place ($n = 24$). From 1993 to 2007, no arthropod pathogens were released in the US due to the rigorous regulatory structure, but in 2008 two species of microsporidia were introduced against the gypsy moth, *Lymantria dispar* (L.). Establishment of entomopathogenic fungi in programs introducing traditional fungi was 32.1% and establishment was 50.0% for programs introducing microsporidia. In some programs, releases have resulted in permanent successful establishment with no non-target effects. In summary, classical biological control using fungal entomopathogens can provide a successful and environmentally friendly avenue for controlling arthropod pests, including the increasing numbers of invasive non-native species.

Keywords Biological control · Microbial control · Entomopathogens · Fungi · Microsporidia

Handling Editor: Helen Roy.

A. E. Hajek (✉)
Department of Entomology, Cornell University, Ithaca, NY 14853-2601, USA
e-mail: aeh4@cornell.edu

I. Delalibera Jr.
Department of Entomology and Acarology, ESALQ-University of São Paulo, Piracicaba, SP 13418-900, Brazil

Introduction

Classical biological control has been defined as "the intentional introduction of an exotic biological control agent for permanent establishment and long-term pest control" (Eilenberg et al. 2001). Parasitoids, predators, pathogens and nematodes have all been used for control of arthropods. However, pathogens and nematodes have been introduced much less frequently than insect parasitoids and predators, although some introduced pathogens and nematodes have been successful in providing long-term pest control (Hajek et al. 2007a). Among the different groups of pathogens that have been used, a relatively large percentage of programs have introduced fungal entomopathogens against a diversity of insect and mite pests (Hajek et al. 2005, 2007a).

Classical biological control has frequently been used when invasive non-native species (subsequently we will call these 'invasive species') establish in new areas and reach high densities. It is generally assumed that pest populations reach high densities because no natural enemies are present (i.e., the enemy release hypothesis) (Keane and Crawley 2002). The general objective of classical biological control introductions has been to establish in the area of concern the natural enemies that regulate the populations of the pest in its area of origin.

Many fungal entomopathogens that have been used for classical biological control have characteristics that make them well suited for this use. For example, some fungal species are well known for their potential to cause epizootics and such species can be very appropriate for classical biological control introductions where only a relatively small amount of inoculum can be released and is expected to increase naturally by causing mortality to the host population. For example, releases of relatively small amounts of inocula of *Entomophaga maimaiga* Humber, Shimazu & Soper against the gypsy moth, *Lymantria dispar* (L.) (Hajek et al. 1996), and *Neozygites tanajoae* Delalibera, Hajek & Humber against cassava green mite, *Mononychellus tanajoa* (Bondar) (Hountondji et al. 2002), resulted in epizootics. Many types of insect pathogens must be eaten to infect, such as the microsporidia, but most fungal entomopathogens infect by direct penetration through arthropod cuticle and therefore are especially well-suited agents for controlling a wide range of pests.

A catalogue referencing and briefly describing classical biological control introductions of pathogens and nematodes against insects and mites was published in 2005 (Hajek et al. 2005) and summarized in 2007 (Hajek et al. 2007a). Several classical biological control programs introducing arthropod-pathogenic fungi that resulted in control of target pests have been described in detail, focusing mostly on the environmental impacts of these introductions (Hajek et al. 2003). In this paper, we will specifically describe and discuss the information on all classical biological control introductions of fungal pathogens of insects and mites.

Methods

We will use the same criteria for categorizing introductions and organisms as described in Hajek et al. (2005, 2007a) and will discuss important criteria as well as exceptions below. We have not evaluated programs by the numbers of pest species being targeted because in some instances, a pest complex was targeted. Thus, we refer to the numbers of programs instead of the numbers of target species, although in the majority of programs, only one pest species was targeted. For a significant percentage of introduction programs, reports on whether the natural enemy became established or not could not be found. Reports regarding whether control was successful are even scarcer so we will only discuss whether the fungus or microsporidia became established. By 'established', we mean that a pathogen was "recovered over a time period after release that would have been adequate for reproduction and reinfection to have occurred in the host population" (Hajek et al. 2007b). For analyses of establishment, releases after 1999 are not considered to have been conducted long enough ago to determine establishment and therefore are excluded.

As recent phylogenetic data suggests that the microsporidia are closely related to fungi (Fischer and Palmer 2005; James et al. 2006; Hibbett et al. 2007), this group will be included in our summary, but will be discussed separately due to the distinct differences in the general biology of microsporidia compared with groups that have traditionally been regarded as fungal entomopathogens. Microsporidia are obligate, intracellular eukaryotic pathogens that infect animal cells while the other groups of fungi included are extracellular when acting as entomopathogens. All fungi outside of microsporidia will henceforth be referred to as traditional fungi in this paper.

Alternatively, the mosquito pathogen *Lagenidium giganteum* Couch (Class Oomycota), has historically been included with fungal pathogens in the Opisthokonta but now is not considered a fungus and belongs to the Heterokontophyta (or Stramenopiles) (Adl et al. 2005). *L. giganteum* was introduced only once for classical biological control, from North Carolina, USA to California, USA in 1972 (Hajek et al. 2005) but this introduction will not be included in the summaries presented below.

Additions to the catalogue

Since the catalogue of classical biological control introductions of pathogens and nematodes against arthropods was summarized and updated in 2007 (Hajek et al. 2007a), we have found two reports of introductions of traditional fungal entomopathogens and two reports of introductions of microsporidia that were not included previously. In 1893, it was reported that *Botrytis* (probably *Beauveria brongniartii* (Saccardo) Petch) from white grubs had been introduced from France to New Zealand to control 'grubs' and codling moth (Anonymous 1893), although it is uncertain whether these releases resulted in establishment (Glare and Inwood 1998). In spring 1992, *Zoophthora radicans* (Brefeld) Batko isolated from the leafhopper *Empoasca vitis* Gothe was introduced from Yugoslavia into wheat fields in Idaho (United States) against the Russian wheat aphid, *Diuraphis noxia* (Mordvilko) (Poprawski and Wraight 1998; Nielsen and Wraight 2009). Samples collected after release documented little infection with no fungal spread and surveys to document establishment were not conducted in subsequent years. The microsporidia *Vairimorpha disparis* (Timofejeva) (= *V. lymantriae*) and *Nosema lymantriae* Weiser from Rupite and Levishte, Bulgaria, respectively, were released against *L. dispar* in northern Illinois, USA in May, 2008 (L. Solter, personal communication; see below). In addition, we learned that information on one program included in the catalogue was incorrect. *Metarhizium anisopliae* (Metschnikoff) Sorokin (strain KVL-00-37) was introduced from Iceland to the Faroe Islands in 2004 instead of the release reported to Iceland in 2003 (E. Oddsdottir, personal communication).

Summaries of introduction programs

Numbers of programs and levels of establishment

In 2007, a total of 131 classical biological control introductions of pathogens and nematodes attacking insects and mites were summarized (Hajek et al. 2007a). Since then, we have added the four introductions described above plus one introduction of the nematode *Steinernema scapterisci* Nguyen & Smart from Florida to Puerto Rico in 2001 to control *Scapteriscus* spp. mole crickets (Frank 2009). Among these 136 introductions, 59 were introductions of traditional fungal entomopathogens, with an additional eight introductions of microsporidia. Thus, an overall total of 49.3% of the programs introducing pathogens and nematodes introduced species of fungi (both traditional fungi and microsporidia).

Among programs for which establishment was reported, 57.6% of traditional fungal agents and 75.0% of microsporidia became established. However, among the total programs there is information regarding establishment of the introduced agent for only 33 programs introducing traditional fungi and four programs introducing microsporidia. When all programs before 2000 (with establishment reported or not) are considered, 32.1% of introductions of traditional fungi resulted in establishment and 37.5% of introductions of microsporidia resulted in establishment.

Releases across time

The first example of a classical biological control release of an insect pathogenic organism was the introduction of a fungus to New Zealand reported in 1893 (see above; Glare and Inwood 1998). Almost half (49.1%) of the total introductions of traditional fungi were conducted before 1949. The numbers of introductions of traditional fungi per decade averaged 4.9 ± 0.7 (mean ± SE), ranging from a minimum of one release in 1940–1949 to a maximum of nine releases between 1980 and 1989 (Fig. 1).

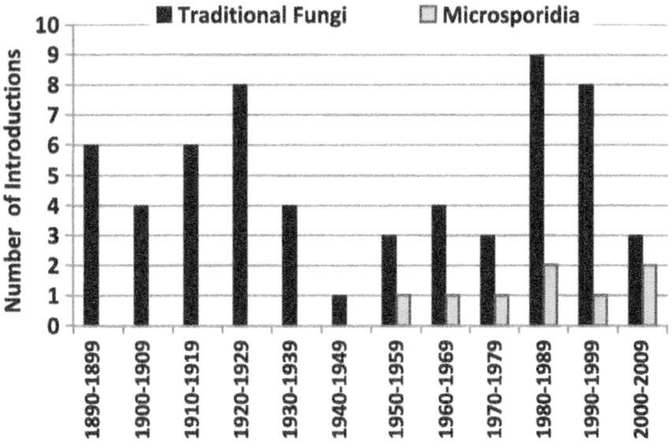

Fig. 1 Numbers of classical biological control programs introducing traditional fungi and microsporidia by decade. Programs spanning more than one decade are counted in the decade when the program was initiated. One program reported in 1893 as 'some time ago' is included in the 1890–1899 category (Anonymous 1893)

Microsporidia have been released less frequently than traditional fungi and releases only began between 1952 and 1960 (Hajek et al. 2005). However, every decade since then, at least one or two microsporidian species have been introduced for classical biological control worldwide.

Merging microsporidia with traditional fungi, percent successful establishment did not vary through time ($\chi^2 = 1.0094$, $df = 1$, $P = 0.3150$), with 66.7% establishment for programs before 1950 ($n = 15$ programs with results reported) and 50.0% establishment between 1950 and 1999 ($n = 22$ programs with results reported).

Types of fungi introduced

An estimated total of 20 species of fungal entomopathogens have been introduced in classical biological control efforts (Table 1). Species in the Order Hypocreales (Ascomycota) have been used more than any other fungal group, with 33 introductions of nine species (plus two introductions of unidentified *Aschersonia* spp. and including introductions of two different varieties of *Hirsutella thompsonii* Fisher). With recent significant changes in mycological taxonomy (e.g., Bischoff et al. 2009), it is possible that this number of species is actually larger. Overall, the most commonly introduced species was *M. anisopliae*, with 13 introductions from 1914 to 2003, but only three programs were undertaken during the last 30 years. The experience gained with this fungal entomopathogen suggests that this species is less likely to keep insect populations below the economic damage level when used for classical biological control. Two of the introductions, in Tonga and Kiribati, resulted in establishment and we could not find results regarding establishment of the remaining introductions. The species *M. anisopliae* has recently been redefined to be composed of nine separate species (Bischoff et al. 2009), leaving the actual identities of the fungal isolates that were introduced in question. Thus, it is possible that many different species in the genus *Metarhizium* were introduced. The next most commonly introduced hypocrealean species was the diaspidid scale pathogen *Fusarium coccophilum* (Desmazieres) Wollenweber, with six introduction programs between 1897 and 1926.

The other large fungal group introduced has been the Order Entomophthorales, with 18 introductions of seven species (Table 1). The species introduced the most was the gypsy moth pathogen *E. maimaiga*, which was released seven times, with one program conducted in 1910–1911 and the remaining programs conducted from 1985 to 2002 (Hajek et al. 2005). Entomophthoralean fungi became established in only four of 18 introductions and failed to establish in 53.0% of the introductions for which results are known. Considering that many entomophthoralean fungi possess good attributes as classical biological control agents, such as strict host specificity, ability to cause epizootics, and specialized long-lived resting spores (Pell et al. 2001), one might expect that the success rate of this group should be higher. However, many of these programs were undertaken without a good understanding of the fungal life cycles and the effects of the environmental conditions on resting spore formation, persistence and germination. Most entomophthoraleans are difficult to grow in vitro (Papierok 2007) and cannot be mass produced (Pell et al. 2001), so in programs releasing species of Entomophthorales, low densities of inoculum (usually in the form of infected hosts) were usually released. If the pathogen is released in low densities the chances of contacting a susceptible host in the area of release is low. If some hosts are infected in the release area but the environmental conditions are not appropriate for resting spore production and hosts are not present year-round, the fungi may not survive until susceptible hosts are present again. Survival strategies for prolonged periods in the absence of hosts are very important for long-term establishment of classical biological control agents.

In addition, two introductions of Ascomycetes not in the Hypocreales (*Podonectria coccicola* Petch and *Triblidium caespitosum* Cooke & Massee; Table 1) and one introduction of the chytrid *Coelomomyces stegomyiae* Keilin were made, as well as three other introductions of fungal entomopathogens that were not identified. Among the Microsporidia, seven species have been introduced in eight introduction programs.

Types of hosts

Traditional fungi were most commonly introduced against species of Hemiptera (47.3% of total introductions) (Fig. 2). Among the 27 introductions

Table 1 Species of fungi that have been introduced for classical biological control of arthropods

Phylum Blastocladiomycota
Class Blastocladiomycetes
Order Blastocladiales
Family Coelomomycetaceae
Coelomomyces stegomyiae Keilin
Phylum (at present undetermined)
Subphylum Entomophthoromycotina
Order Entomophthorales
Family Entomophthoraceae
Entomophaga grylli (Fresenius) Batko, pathotype I
Entomophaga grylli (Fresenius) Batko, pathotype III
Entomophaga maimaiga Humber, Shimazu & Soper
Pandora neoaphidis (Remaudière & Hennebert) Humber
Zoophthora radicans (Brefeld) Batko
Family Neozygitaceae
Neozygites fresenii (Nowakowski) Batko
Neozygites parvispora (MacLeod & Carl) Remaudière & Keller
Neozygites tanajoae Delalibera, Hajek & Humber
Phylum Ascomycota
Class Dothideomycetes
Order Pleosporales
Family Tubeufiaceae
Podonectria coccicola Petch
Subphylum Pezizomycotina
Order Triblidiales (at present no class has been designated)
Family Triblidiaceae
Triblidium caespitosum Cooke & Massee
Class Sordariomycetes
Order Hypocreales
Family Nectriaceae
Fusarium coccophilum (Desmazieres) Wollenweber & Reinking
Fusarium juruanum P. Hennings
Family Clavicipitaceae
Aschersonia aleyrodis Webber
Aschersonia goldiana Saccardo & Ellis
Aschersonia spp.
Metarhizium anisopliae (Metschnikoff) Sorokin
Family Cordycipitaceae
Beauveria bassiana (Balsamo) Vuillemin
Beauveria brongniartii (Saccardo) Petch
Lecanicillium lecanii (Zimmermann) Zare & W. Gams
Family Ophiocordycipitaceae
Hirsutella thompsonii Fisher var. *synnematosa* Samson, McCoy & O'Donnell
Hirsutella thompsonii Fisher var. *vinacea* Samson, McCoy & O'Donnell
Phylum Microsporidia
Class Microsporea
Order Nosematidida
Family Nosematidae
Nosema lymantriae Weiser
Nosema portugal Maddox & Vávra
Nosema pyrausta (Paillot)
Paranosema locustae (Canning)
Family Burenellidae
Vairimorpha disparis (Timofejeva)
Order Microsporida
Family Pleistophoridae
Endoreticulatus sp.
Pleistophora culicis Weiser

Especially for earlier introductions, names for fungal species cited in the literature have frequently been changed. The text of this paper and Hajek et al. (2005) provide synonymies for the affected species

Taxonomy of traditional fungi is in agreement with Hibbett et al. (2007) and taxonomy of microsporidia is according to J. Becnel (personal communication)

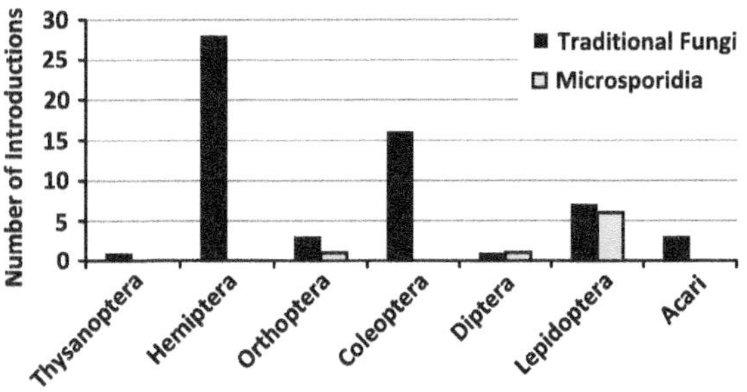

Fig. 2 Numbers of programs from different orders of arthropod hosts targeted by classical biological control introductions of traditional fungi and microsporidia

against hemipterans, 19 were against species with immobile life stages (four introductions against coccids, nine against diaspidids and six against aleyrodids). The next most common group of hosts targeted for introductions was the Coleoptera (26.3%

of total introductions). For the 15 introductions against Coleoptera, 13 targeted scarab beetles. Microsporidia have predominantly been introduced against species of Lepidoptera (75.0% of the eight introductions). Interestingly, products for inundative control of arthropod pests based on traditional entomopathogenic fungi also target these same host groups in the same order of prevalence: Hemiptera (59.6%), Coleoptera (40.9%) and then Lepidoptera (17.5%) (de Faria and Wraight 2007). These latter percentages add to >100% because products often target more than one group of pests.

For classical biological control using traditional fungi, the single most common target species was *L. dispar* (seven introductions), followed by four introductions each against the rhinoceros beetle *Oryctes rhinoceros* (L.) and the diaspidid scale *Cornuaspis beckii* (Newman). *Oryctes rhinoceros* is especially well known as the target for successful classical biological control introductions of the *O. rhinoceros* virus. However, the less frequent introduction programs of *M. anisopliae* against *O. rhinoceros* were never reported as providing successful control (Hajek et al. 2007a).

For both traditional fungi and microsporidia, the majority of introduction programs focused on invasive species of insects or mites (70.7%) rather than native hosts (29.4%). Among those programs with sufficient information (only 50.7%), 61.5% of introductions against invasive arthropods resulted in establishment while 50.0% of introductions against natives resulted in establishment.

Locations

Programs introducing traditional fungi and microsporidia were conducted on all continents and the Pacific Islands. Continents and major areas with the most introduction programs were North America (25 introductions) and the Pacific Islands (17 introductions). South America used this control approach with fungal pathogens only four times. Programs introducing microsporidia were predominantly conducted in North America. The United States (including five introductions to Hawaii) was the country where most introductions took place (24 introductions) and it was also the place of origin for 23 traditional fungal entomopathogens and microsporidia introduced in other countries of the world.

Comparison of classical biological control introductions of fungi and microsporidia with introductions of other pathogens and nematodes attacking arthropods

As demonstrated previously, arthropod pathogens and entomopathogenic nematodes have not been used as much as insect parasitoids and predators (Hajek et al. 2007a). Among the pathogens, traditional fungal entomopathogens were the only types of pathogens introduced worldwide until about the 1950s. Although the number of programs introducing traditional fungi is much greater than programs releasing viruses, bacteria or nematodes, for a high percentage of programs introducing traditional fungi we have no reports on whether establishment of the fungus was successful (for general trend see Fig. 2b in Hajek et al. 2007a). Microsporidia have been used for classical biological control infrequently, compared with the other groups of pathogens (i.e., a total of eight introduction programs out of 136).

There is great variation in successful establishment among different types of pathogens. While 91.0% of programs releasing viruses resulted in establishment (Hajek et al. 2007a) only 32.1% of the programs with traditional fungi and 50.0% of microsporidia introductions resulted in establishment. On the other hand, only one of the four (25.0%) bacterial release programs resulted in documented establishment (*Paenibacillus popilliae* (Dutky) introduced from Papua New Guinea and the Solomon Islands to Kiribati; Theunis and Teuriari 1998).

Some classical biological control programs with fungal entomopathogens have been considered successful, causing drastic reductions in pest populations and the pathogen was subsequently released elsewhere. As an example, after the occurrence of epizootics caused by *E. maimaiga* in northeastern North America, this pathogen was subsequently distributed to other countries (i.e., Bulgaria, Russia). However, to date, the impact of these later introductions has not been as remarkable as in the initial location although *E. maimaiga* became established in Bulgaria (Pilarska et al. 2000). In contrast, extensive establishment and control have been recorded in numerous release locations for other groups of pathogens, e.g., programs to control the coconut pest, *O. rhinoceros* using the *O. rhinoceros* virus (Jackson 2009) and to control of the woodwasp *Sirex noctilio*

F. with the nematode *Deladenus* (= *Beddingia*) *siricidicola* Bedding (2009).

On-going classical biological control projects

A few projects on classical biological control using fungal entomopathogens are currently being implemented or are under post-release monitoring. Three programs are discussed here: the introduction of two fungal species from Brazil that are being investigated for control of two species of mites invasive to Africa and two microsporidia species introduced in the United States against *L. dispar*.

Neozygites tanajoae Delalibera, Hajek & Humber against the cassava green mite (CGM), *Mononychellus tanajoa* (Bondar) in Africa

A long term example of a classical biological control project using fungi is the program targeting CGM. This project was conceived in 1979, eight years after this mite was first discovered attacking cassava in Uganda (Yaninek and Herren 1988). Initially, research was focused on introduction of phytoseiid mite predators of CGM from South America to Africa. It was only in 1988, that exploration for potential natural enemies in Brazil revealed that the entomophthoralean *N. tanajoae* (= *Neozygites* sp. in early publications) was one of the most important natural enemies of CGM in northeastern Brazil (Delalibera et al. 1992). During the last 20 years, a series of studies was undertaken to make the release of this pathogen in Africa possible. The history of these studies has been summarized recently by Delalibera (2009).

It was not until 1999 that two Brazilian isolates of *N. tanajoae* were released in Benin (Hountondji et al. 2002). The releases were conducted in farmers' fields in five locations. In each location four cassava fields separated by at least 0.5 km were selected with a different treatment applied at each field. Treatments consisted of releases of three isolates of *N. tanajoae* (two Brazilian and one from Cotonou, Benin) and an untreated control. Forty-eight weeks after the releases, higher infection levels were recorded in fields where Brazilian isolates were released compared to locations where the Beninese isolate had been inoculated. The highest infection level for the Beninese isolate was 4.5%, while the highest infection levels caused by the two Brazilian isolates were 36.5% and 34.0%. In control fields, infection by *N. tanajoae* was not observed during this evaluation. Another observation indicating establishment of the introduced Brazilian isolate occurred in one of the fields where up to 36.5% infection by *N. tanajoae* was detected three months after the crop had been replanted. The authors of this study reported that the principal constraint of their findings was their inability to distinguish between isolates, but they presented evidence (i.e., the slow rate of spread of infection and erratic epizootic development in the isolated release fields; Hountondji et al. 2002) indicating that the higher infections levels observed were due to the Brazilian isolates and not the Beninese isolate. Later, molecular techniques were developed for differentiation between *N. tanajoae* isolates from Brazil and Africa (Delalibera 2009; Agboton et al. 2009). For development of these molecular probes, samples of *N. tanajoae* strains from Benin that had been collected before the releases as well as isolates from Brazil were used. Random amplification of polymorphic DNA (RAPD) markers were converted into sequence characterized amplified regions (SCARs) and specific oligonucleotide primers were designed for the detection and differentiation of indigenous and exotic isolates. These probes were validated using a collection of isolates from several locations in Brazil and indigenous strains from Benin, Ghana and Tanzania, collected before the introduction of Brazilian strains of *N. tanajoae* to West Africa (Agboton et al. 2009). The two oligonucleotide primer pairs are presently being used to follow the establishment and spread of Brazilian isolates already introduced into Benin and Tanzania (R. Hanna et al., unpublished data) and will be used in the future for following the establishment and spread of Brazilian *N. tanajoae* isolates that will be introduced into other countries in sub-Saharan Africa.

Neozygites floridana Weiser and Muma against *Tetranychus evansi* Baker & Pritchard in Africa

Another spider mite species, the tomato red spider mite, *Tetranychus evansi* Baker & Pritchard, became an important pest of commercial crops in Africa soon after it was first detected in Zimbabwe in 1979 (Blair 1983). This pest reached many southern and eastern African countries (Knapp et al. 2003; Smith Meyer 1996; Bonato 1999; El-Jaouani 1988). The place of origin of *T. evansi* is unknown but it has been hypothesized for a long time that *T. evansi* could have originated in South America (Gutierrez and Etienne 1986) and current studies based on molecular data corroborate this hypothesis (M. Navajas, personal communication). *T. evansi* specializes on solanaceous crops (Moraes et al. 1987; Moraes and Flechtmann 1981; Jeppson et al. 1975) and reaches very high population densities on tomato and nightshade, causing tomato yield losses of up to 90% in some African countries (Sarr et al. 2002).

A collaborative project to study *T. evansi* was undertaken between the African Insect Science for Food and Health Institute (ICIPE, Kenya) and two Brazilian universities: Universidade de São Paulo (ESALQ/USP) and Universidade Federal Rural de Pernambuco (UFRPE). The project was initiated in the early 2000s and investigations are being carried out in Brazil to determine the potential of natural enemies for controlling this pest in Africa through classical biological control. This project adopted similar steps and methods to those developed for the CGM project. In Brazil, *T. evansi* is not considered an important pest and populations rarely reach high densities, suggesting that probably *T. evansi* is kept under control by natural enemies. Surveys for natural enemies were conducted mostly in Brazil but also in Argentina, Paraguay and Peru, with locations selected based on similarity of climatic conditions to locations where *T. evansi* is a pest in Africa (Fiaboe et al. 2006). These surveys revealed the presence of a predatory mite, *Phytoseiulus longipes* Evans, and the fungal pathogen *N. floridana* infecting *T. evansi*. Both of these natural enemies showed promise for classical biological control use. However, the predator had a limited distribution as it was only found in the southern part of the State of Rio Grande do Sul, Brazil. In contrast, infected *T. evansi* were found in many fields from the north of Argentina to northeastern Brazil, suggesting that epizootics can occur over a broad range of climates. Actually, epizootics caused by *N. floridana* had previously been reported on *T. evansi* in northeastern Brazil over 28 years ago (Humber et al. 1981).

The impact of this fungus on *T. evansi* populations was demonstrated in the field and under screenhouses during four crop cycles of tomato and nightshade by Duarte et al. (2009) in Piracicaba, SP, Brazil. *N. floridana* was the only natural enemy found associated with *T. evansi* in all crop cycles under screenhouses but only in the last crop cycle in the field. In the treatments where the fungus appeared, reduction of mite populations was drastic. *N. floridana* appeared in tomato plants even when the population density of *T. evansi* was relatively low (less than 10 mites/3.14 cm^2 of leaf area) and even at this low population density, the fungus maintained infection rates greater than 50%. The application of pesticides directly affected the fungus by delaying epizootic initiation and contributing to lower infection levels than in unsprayed treatments (Duarte et al. 2009).

Studies aiming to determine the risk of the pathogen to non-target species focused on the predatory mite *P. longipes* (Furtado et al. 2007) because this natural enemy had already been shipped from Brazil to Kenya for experimental releases. Several tests were conducted to determine the compatibility and possible impact of *N. floridana* on performance of this predatory mite. *N. floridana* is not pathogenic to *P. longipes* and no effect of the pathogen was detected on the life cycle parameters of *P. longipes*. The only effect of *N. floridana* on *P. longipes* was reduced predation of *T. evansi* eggs and increased time spent grooming on leaf discs with capilliconidia. *P. longipes* did not avoid areas with capilliconidia, and it was efficient in removing most capilliconidia attached to its body through self-grooming behavior. However, the increased grooming time may have accounted for the lower egg predation rates.

To be incorporated in the tomato production system, *N. floridana* has to be compatible with the pesticides used for control of other pests and diseases. Several pesticides used in tomato production have been tested for their effect on *N. floridana* to determine their selectivity and efficacy for use in an integrated pest management program (Wekesa et al. 2008).

In addition, the influence of host plant species on conidial contamination, infection, host mortality and mummification of *T. evansi* with *N. floridana* was evaluated (Wekesa 2008). This study showed that efficiency of *N. floridana* in the control of *T. evansi* may vary with the host plant and demonstrates the need to select suitable host plants for laboratory production and field release. High mummification and sporulation of cadavers were observed in tomato and eggplant and rapid development of epizootics is expected on these plants. The high mummification accompanied by poor sporulation in pepper would lead to decreased rates of transmission just as in nightshade and cherry tomato host plants, which displayed poor mummification and sporulation. The effect of temperature on the *N. floridana* life cycle was compared among two isolates from Brazil and one from Argentina (Wekesa 2008). The main purpose of these investigations was to determine the best *N. floridana* isolate for introduction in distinct climatic regions in Africa and the appropriate host plants for release, to increase chances of establishment of the fungus in Africa. Plans are under way to import *N. floridana* to Kenya as a possible agent for the classical biological control of *T. evansi*.

Releases of two species of microsporidia against *Lymantria dispar* in the United States

The program releasing two species of microsporidia against *L. dispar* in 2008 is the first introduction program in the United States since 1993. As described in 2007 (Hajek et al. 2007a), the regulatory structure has made classical biological control introductions of arthropod pathogens very difficult in the United States. To conduct the recent introductions of microsporidia against *L. dispar*, a battery of non-target tests were conducted. These results demonstrated that *N. lymantriae* did not infect any of the non-target species tested and *V. disparis* killed a small number of non-targets directly but did not persist in the non-target populations the following years (Solter and Maddox 1998, 1999; Solter et al. 1997, 2000). To introduce *V. disparis* and *N. lymantriae* in 2008, permission was sought and granted from four different agencies: the United States Department of Agriculture-Animal and Plant Health Inspection Service-Plant Protection and Quarantine, the Environmental Protection Agency, the North American Plant Protection Organization and the Illinois Department of Agriculture. Thus, a large effort was necessary to satisfy requirements for these introductions and obtain permission for release.

V. disparis and *N. lymantriae* were introduced in two different plots (≤10 acres) in northern Illinois in May 2008 by releasing 10,000 infected third instar *L. dispar* larvae at each plot (L. Solter, personal communication). After releases in 2008, many of the *L. dispar* at release sites died due to an *E. maimaiga* epizootic but, since interactions between *E. maimaiga* and these microsporidian species are unknown, the impact of the epizootic on microsporidian establishment has yet to be determined. Release sites were monitored in 2009 for microsporidian persistence and non-target effects.

The future of classical biological control using arthropod pathogenic fungi

Global trade has resulted in increased numbers of invasive non-native species being introduced to new areas. Controlling these invasive species presents an unparalleled challenge worldwide. Once an invasive species has become established, classical biological control might be the only available method for providing long-term control without regular control-based manipulations by humans. Thus, classical biological control programs can be extremely beneficial for combating invasive non-native species. One of the reasons classical biological control has not been used more, especially in recent years, has been the concern that classical biological control must be environmentally safe. As reported in Hajek et al. (2007a), no documented case has been found in the literature where a fungal pathogen introduced for classical biological control of an insect pest caused substantial mortality to a non-target species or caused negative effects to the environment. In the United States, until 2008, regulatory procedures were not in place for deciding whether a potential introduction was safe enough. However, in 2008 two species of microsporidia were approved for release in the United States, indicating that a regulatory procedure, although complex, is now in place.

The number of releases of entomopathogens is very low when compared with releases of parasitoids and predators for classical biological control of arthropods (Hajek et al. 2007a). The poor reporting in the past regarding impacts of classical biological control introductions of entomopathogenic fungi could have hindered utilization of this approach for control (see Hajek et al. 2007b for a description of how the impact of a pathogen should be monitored). However, classical biological control successes with entomopathogens, including fungal entomopathogens, indicate that pathogens should definitely be more commonly considered for classical biological control programs. Some traditional fungal species are known to cause epizootics resulting in drastic reductions of arthropod pest populations, while other species, such as the microsporidia, often cause sublethal effects throughout the life cycles of their hosts. Both groups of microbes can contribute to regulation of host populations and thus can be well-suited for classical biological control introductions that will potentially result in permanent establishment and long-term pest control.

Acknowledgments We thank R. Humber, J. Becnel, L. Solter and W. Fry for their taxonomic assistance. We also thank L. Solter, E. Oddsdottir, C. Nielsen and J. Eilenberg for sharing unpublished data on introductions.

References

Adl SM, Simpson AGB, Farmer MA, Andersen RA, Anderson OR, Barta JR, Bowser SS, Brugerolle GUY, Fensome RA, Fredericq S, James TY, Karpov S, Kugrens P, Krug J, Lane CE, Lewis LA, Lodge J, Lynn DH, Mann DG, McCourt RM, Mendoza L, Moestrup O, Mozley-Standridge SE, Nerad TA, Shearer CA, Smirnov AV, Spiegel FW, Taylor MFJR (2005) The new higher level classification of eukaryotes with emphasis on the taxonomy of protists. J Euk Microbiol 52:399–451

Agboton BV, Delalibera I Jr, Hanna R, Tiedemann AV (2009) Molecular detection and differentiation of Brazilian and African isolates of the entomopathogen *Neozygites tanajoae* (Entomophthorales: Neozygitaceae) with PCR using specific primers. Biocontrol Sci Technol 19:67–79

Anonymous (1893) White grub culture. New Zealand Farmer, Bee and Poultry Journal February, p 45

Bedding RA (2009) Controlling the pine-killing woodwasp, *Sirex noctilio*, with nematodes. In: Hajek AE, Glare TR, O'Callaghan M (eds) Use of microbes for control and eradication of invasive arthropods. Springer, Dordrecht, pp 213–235

Bischoff JF, Rehner SA, Humber RA (2009) A multilocus phylogeny of the *Metarhizium anisopliae* lineage. Mycologia 101:508–528

Blair BW (1983) *Tetranychus evansi* Baker & Pritchard (Acari: Tetranychidae): a new pest of tobacco in Zimbabwe. Bergerac, France: CORESTA Phytopathol and Agron Study Group pp 1–6

Bonato O (1999) The effect of temperature on the life history parameters of *Tetranychus evansi* (Acari: Tetranychidae). Exp Appl Acarol 23:11–19

de Faria M, Wraight SP (2007) Mycoinsecticides and mycoacaricides: a comprehensive list with worldwide coverage and international classification of formulation types. Biol Control 43:237–256

Delalibera I Jr (2009) Biological control of the cassava green mite in Africa with Brazilian isolates of the fungal pathogen *Neozygites tanajoae*. In: Hajek AE, Glare TR, O' Callaghan M (eds) Use of microbes for control and eradication of invasive arthropods. Springer, Dordrecht, pp 259–269

Delalibera I Jr, Gomez DRS, de Moraes GJ, de Alencar JA, Araújo WF (1992) Infection of *Mononychellus tanajoa* (Acari: Tetranychidae) by the fungus *Neozygites* sp. (Zygomycetes: Entomophthorales) in northeastern Brazil. Fla Entomol 75:145–147

Duarte V, Silva RA, Wekesa VW, Rizzato FB, Dias CTS, Delalibera I Jr (2009) Impact of natural epizootics of the fungal pathogen *Neozygites floridana* (Zygomycetes: Entomophthorales) on population dynamics of *Tetranychus evansi* (Acari: Tetranychidae) in tomato and nightshade. Biol Control 51:81–90

Eilenberg J, Hajek A, Lomer C (2001) Suggestions for unifying the terminology of biological control. BioControl 46:387–400

El-Jaouani N (1988) Contribution à la connaisance des acariens phytophages au Maroc et étude bio-écologique de *Tetranychus evansi* Baker et Pritchard (Acarina: Tetranychidae). 230 p. Thesis—Institut Agronomique et Veterinaire Hassan II, Rabat, Morocco

Fiaboe KKM, Fonseca RL, Moraes GJ, Ogol CKPO, Knapp M (2006) Identification of priority areas in South America for exploration of natural enemies for classical biological control of *Tetranychus evansi* (Acari: Tetranychidae) in Africa. Biol Control 38:373–379

Fischer WM, Palmer JD (2005) Evidence from small-subunit ribosomal RNA sequences for a fungal origin of Microsporidia. Mol Phylogenet Evol 36:606–622

Frank JH (2009) *Steinernema scapterisci* as a biological control agent of *Scapteriscus* mole crickets. In: Hajek AE, Glare TR, O'Callaghan M (eds) Use of microbes for control and eradication of invasive arthropods. Springer, Dordrecht, pp 115–131

Furtado IP, Moraes GJ, Kreiter S, Tixier MS, Knapp M (2007) Potential of a Brazilian population of the predatory mite *Phytoseiulus longipes* as a biological control agent of *Tetranychus evansi* (Acari: Phytoseiidae, Tetranychidae). Biol Control 42:139–147

Glare TR, Inwood AJ (1998) Morphological and genetic characterisation of *Beauveria* spp. from New Zealand. Mycol Res 102:250–256

Gutierrez J, Etienne J (1986) Les Tetranychidae de l'île de la Réunion et quelques uns de leurs prédateurs. Agron Tropical 41:84–91

Hajek AE, Elkinton JS, Witcosky JJ (1996) Introduction and spread of the fungal pathogen *Entomophaga maimaiga* along the leading edge of gypsy moth spread. Environ Entomol 25:1235–1247

Hajek AE, Delalibera I Jr, Butler L (2003) Entomopathogenic fungi as classical biological control agents. In: Hokkanen HMT, Hajek AE (eds) Environmental impacts of microbial insecticides. Kluwer, Dordrecht, pp 15–34

Hajek AE, McManus ML, Delalibera Jr I (2005) Catalogue of introductions of pathogens and nematodes for classical biological control of insects and mites. USDA, Forest Service FHTET-2005-05

Hajek AE, McManus ML, Delalibera I Jr (2007a) A review of introductions of pathogens and nematodes for classical biological control of insects and mites. Biol Control 41:1–13

Hajek AE, Delalibera I Jr, McManus ML (2007b) Introduction of exotic pathogens and documentation of their establishment and impact. In: Lacey LA, Kaya HK (eds) Field manual of techniques in invertebrate pathology: application and evaluation of pathogens for control of insects and other invertebrate pests, 2nd edn. Kluwer, Dordrecht, pp 299–325

Hibbett DS, Binder M, Bischoff JF, Blackwell M, Canon PF, Eriksson OE, Huhndorf S, James T, Kirk PM, Lücking R, Lumbsch HT, Lutzoni F, Matheny PB, McLaughlin DJ,

Powell MJ, Redhead S, Schoch CL, Spatafora JW, Stalpers JA, Vilgalys R, Aime MC, Aptroot A, Bauer R, Begerow D, Benny GL, Castlebury LA, Crous PW, Dai YC, Gams W, Geiser DM, Griffith GW, Gueidan C, Hawksworth DL, Hestmark G, Hosaka K, Humber RA, Hyde KD, Ironside JE, Kõljalg U, Kurtzman CP, Larsson KH, Lichtwardt R, Longcore J, Miadlikowska J, Miller A, Moncalvo JM, Mozley-Standridge S, Oberwinkler F, Parmasto E, Reeb V, Rogers JD, Roux C, Ryvarden L, Sampaio JP, Schuessler A, Sugiyama J, Thorn RG, Tibell L, Untereiner WA, Walker C, Wang Z, Weir A, Weiss M, White MM, Winka K, Yao YJ, Zhang N (2007) A higher-level phylogenetic classification of the fungi. Mycol Res 111:509–547

Hountondji FCC, Lomer CJ, Hanna R, Cherry AJ, Dara SK (2002) Field evaluation of Brazilian isolates of *Neozygites floridana* (Entomophthorales: Neozygitaceae) for the microbial control of cassava green mite in Benin, West Africa. Biocontrol Sci Technol 12:361–370

Humber RA, de Moraes GJ, dos Santos JM (1981) Natural infection of *Tetranychus evansi* (Acarina: Tetranychidae) by a *Triplosporium* sp. (Zygomycetes: Entomophthorales) in Northeastern Brazil. Entomophaga 26:421–425

Jackson TA (2009) The use of *Oryctes* virus for control of rhinoceros beetle in the Pacific Islands. In: Hajek AE, Glare TR, O'Callaghan M (eds) Use of microbes for control and eradication of invasive arthropods. Springer, Dordrecht, pp 133–140

James TY, Kauff F, Schoch CL, Matheny PB, Hofstetter V, Cox CJ, Celio G, Gueidan C, Fraker E, Miadlikowska J, Lumbsch HT, Rauhut A, Reeb V, Arnold AE, Amtoft A, Stajich JE, Hosaka K, Sung G-H, Johnson D, O'Rourke B, Crockett M, Binder M, Curtis JM, Slot JC, Wang Z, Wilson AW, Schüler A, Longcore JE, O'Donnell K, Mozley-Standridge S, Porter D, Letcher PM, Powell MJ, Taylor JW, White MM, Griffith GW, Davies DR, Humber RA, Morton JB, Sugiyama J, Rossman AY, Rogers JD, Pfister DH, Hewitt D, Hansen K, Hambleton S, Shoemaker RA, Kohlmeyer J, Volkmann-Kohlmeyer B, Spotts RA, Serdani M, Crous PW, Hughes KW, Matsuura K, Langer E, Langer G, Untereiner WA, Lücking R, Büdel B, Geiser DM, Aptroot A, Diederich P, Schmitt I, Schultz M, Yahr R, Hibbett DS, Lutzoni F, McLaughlin DJ, Spatafora JW, Vilgalys R (2006) Reconstructing the early evolution of fungi using a six-gene phylogeny. Nature 443:818–822

Jeppson LR, Keifer HH, Baker EW (1975) Mites injurious to economic plants. University of California Press, Berkeley, 614 p

Keane RM, Crawley MJ (2002) Exotic plant invasions and the enemy release hypothesis. Trends Ecol Evol 17:164–170

Knapp M, Wagener B, Navajas M (2003) Molecular discrimination between the spider mite *Tetranychus evansi* Baker and Pritchard, an important pest of tomatoes in Southern Africa, and the closely related species *T. urticae* Koch. African Entomol 11:300–304

Moraes GJ, Flechtmann CHW (1981) Acaros fitófagos do nordeste do Brasil. Pesq Agropec Brasil 16:177–186

Moraes GJ, McMurtry JA, Baker EW (1987) Redistribution and distributions of the spider mites *Tetranychus evansi* and *T. marianae*. Acarologia 28:333–343

Nielsen C, Wraight SP (2009) Exotic aphid control with pathogens. In: Hajek AE, Glare TR, O'Callaghan M (eds) Use of microbes for control and eradication of invasive arthropods. Springer, Dordrecht, pp 93–113

Papierok B (2007) Isolating, growing and storing arthropod-pathogenic Entomophthorales. In: Keller S (Ed) Arthropod-pathogenic Entomophthorales: biology, ecology; identification. COST Action 842. Luxembourg, pp 66–81

Pell JK, Eilenberg J, Hajek AE, Steinkraus DC (2001) Biology, ecology and pest management potential of Entomophthorales. In: Butt TM, Jackson CW, Magan N (eds) Fungi as biocontrol agents: progress, problems and potential. CABI International, Oxon, pp 71–153

Pilarska D, McManus M, Hajek AE, Hérard F, Vega FE, Pilarska P, Markova G (2000) Introduction of the entomopathogenic fungus *Entomophaga maimaiga* Hum., Shim. & Sop. (Zygomycetes: Entomophthorales) to a *Lymantria dispar* (L.) (Lepidoptera: Lymantriidae) population in Bulgaria. Anz Schadlingsk 73:125–126

Poprawski TJ, Wraight SP (1998) Fungal pathogens of Russian wheat aphid (Homoptera: Aphididae). In: Quisenberry SS, Peairs FB (eds) Response model for an introduced pest—the Russian wheat aphid. Entomological Society of America, Lanham, pp 71–167

Sarr I, Knapp M, Ogol CKP, Baumgärtner J (2002) Impact of predators on *Tetranychus evansi* Baker and Pritchard populations and damage on tomatoes (*Lycopersicon esculentum* Mill.) in Kenya. In: International Congress of Acarology, 11 Merida, Mexico. Abstract book, p 271

Smith Meyer MKP (1996) Mite pests and their predators on cultivated plants in Southern Africa. Vegetables and berries. Protection Research Handbook, 6. Pretoria: Plant ARC-Plant Protection Research Institute, Biosystematics Division, pp 47–69

Solter LF, Maddox JV (1998) Physiological host specificity of microsporidia as an indicator of ecological host specificity. J Invertebr Pathol 71:207–216

Solter LF, Maddox JV (1999) Strategies for evaluating the host specificity of lepidopteran microsporidia. Rev Soc Entomol Argent 58:146–150

Solter LF, Maddox JV, McManus ML (1997) Host specificity of microsporidia (Protista: Microspora) from European populations of *Lymantria dispar* (Lepidoptera: Lymantriidae) to indigenous North American Lepidoptera. J Invertebr Pathol 69:135–150

Solter LF, Pilarska DK, Vossbrinck CF (2000) Host specificity of microsporidia pathogenic to forest Lepidoptera. Biol Control 19:48–56

Theunis W, Teuriari N (1998) Biological control of *Papuana huebneri* (Coleoptera: Scarabaeidae) in Kiribati: field trials with *Metarhizium anisopliae* and *Bacillus popilliae*. J So Pac Agric 15:46–51

Wekesa VW (2008) Avaliação do potencial de *Neozygites floridana* (Entomophthorales: Neozygitaceae) para o controle biológico clássico de *Tetranychus evansi* (Acari: Tetranychidae) na África. PhD dissertation, Escola Superior de Agricultura Luiz de Queiros, USP, Piracicaba, 114 p

Wekesa VW, Knapp M, Delalibera I Jr (2008) Side-effects of pesticides on the life cycle of the mite pathogenic fungus *Neozygites floridana*. Exp Appl Acarol 46:287–297

Yaninek JS, Herren HR (1988) Introduction and spread of the cassava green mite *Mononychellus tanajoa* (Bondar) (Acari: Tetranychidae) an exotic pest in Africa and the search for appropriate control methods: a review. Bull Entomol Res 78:1–13

Author Biographies

Ann E. Hajek is a professor studying insect pathology in the Department of Entomology at Cornell University in Ithaca, New York. She teaches undergraduate courses in biological control and invasive species, a graduate course in invertebrate pathology and graduate seminars in ecology and evolution of infectious disease and biological control. Her research is focused on the epizootiology of insect diseases, ecology and evolution of entomopathogens and use of entomopathogens for control of insects, with particular emphasis on fungal and viral pathogens and parasitic nematodes attacking invasive insect species.

Italo Delalibera Jr. is a professor in the Department of Entomology and Acarology at the University of São Paulo in Piracicaba-SP, Brazil. He teaches graduate courses in Arthropod Pathology and Microbial Control of Pests. His main research is microbe—arthropod interactions, with emphasis on microbial control of agricultural pests and insect-gut microbial symbioses. He has been an associate editor of Neotropical Entomology and Bioassay since 2005 and 2006, respectively.

BioControl (2010) 55:159–185
DOI 10.1007/s10526-009-9248-3

Ecological factors in the inundative use of fungal entomopathogens

Stefan T. Jaronski

Received: 1 September 2009 / Accepted: 19 October 2009 / Published online: 24 November 2009

Abstract Fungal entomopathogens have been developed in numerous countries as biocontrol agents with more than 100 mycoinsecticide products commercially available in 2006. The chief, perhaps sole, use of these mycoinsecticides has been as inundative agents, within a chemical paradigm. Large numbers of propagules are applied in an attempt to overwhelm by brute force many of the factors that keep a pathogen in nonepizootic equilibrium with its host. This review attempts to summarize what we know about the abiotic and biotic factors that affect the efficacy of these mycoinsecticides in both foliar and soil applications. Sunlight, humidity, temperature, and phylloplane-associated factors can affect both immediate efficacy and persistence on plants. Likewise, soil texture-moisture interactions, temperature, and a host of biotic factors can affect mycoinsecticides in the soil. Despite much research, our understanding of these ecological aspects is imperfect, especially in a holistic, dynamic sense.

Keywords *Metarhizium* · *Beauveria* · Persistence · Efficacy · UV · Humidity · Temperature · Phylloplane · Soil

Handling Editor: Helen Roy.

Mention of trade names or commercial products in this publication is solely for the purpose of providing specific information and does not imply recommendation or endorsement by the US Department of Agriculture.

S. T. Jaronski (✉)
US Department of Agriculture, Agricultural Research Service, Northern Plains Agricultural Research Laboratory, 1500 N. Central Ave., Sidney, MT 59270, USA
e-mail: stefan.jaronski@ars.usda.gov

Introduction

The advent of chemical insecticides in the mid twentieth century created the concept that insect pests could be all but eliminated from threatened crops. A succession of compounds has appeared since then. Initially, many were quite toxic and environmentally damaging. In recent years, however, new materials have appeared, which address human and environmental safety concerns caused by the earlier materials. In parallel, we have realized the inadvisability of using chemicals as stand-alone, catastrophic mortality factors, and integrated pest management schemes have evolved to employ a variety of cultural, chemical, and biological tools to manage (not eradicate) pest invasion to a point below an economic threshold. Biological tools, including microbial agents, have received increasing attention as alternatives to chemicals within this context. Nevertheless, the chemical paradigm, in which a material is used to efficiently, simply, and quickly eradicate a pest problem, still persists. Microbials are too often merely substituted for chemical pesticides.

Among entomopathogens, fungi have attracted a lot of attention as biologically based pesticides. Per Faria and Wraight (2007), 129 mycoinsecticide products (fungus-based formulations targeting insects) were commercially available worldwide in 2006. Fungi used in these products are primarily ascomycetes including: *Beauveria bassiana* (Bals.) Vuill.; *B. brongniartti* (Sacc.) Petch; *Metarhizium anisopliae* (Metsch.) Sorokin., *sensuo latum, M. acridum* (formerly *M. anisopliae* var. *acridum*) (Driver and Milner) J.F. Bischof., Rehner and Humber stat. nov.; *Isaria fumosorosea* Wize (formerly *Paecilomyces fumosoroseus*); *Lecanicillium longisporum* and *muscarium* (Petch) R. Zare and W. Gams (formerly *Verticillium lecanii*); and *Hirsutella thompsoni* F.E. Fisher. In addition, the Oomycete *Lagenidium giganteum* Couch has been commercialized in the US. Inundative use of *Nomuraea rileyi* (Farl.) Samson and *Aschersonia aleyrodis* Webber has been studied and the latter fungus was briefly commercialized in Europe. Despite the number of products, mycoinsecticides have not captured a significant market share of the biopesticide market, especially in the US and EU. A principle reason is that, compared to chemical pesticides, these mycoinsecticides lack consistent, speedy efficacy in combating insect pest problems, and are more complicated to use, despite their obvious safety. The chemical paradigm too often pervades the use of biopesticides.

Fungal entomopathogens, as well as other microbial agents, can be used in several ways (Fuxa 1987; Eilenberg et al. 2001). Classical biological control involves introducing a novel fungus for permanent establishment and long term pest control. This subject is discussed by Hajek and Delalibera (2009). Inoculative biological control has the expectation that the agent will multiply, spread, and provide extended control of an insect pest, but only for a finite period. These approaches require several key characteristics, notably the ability for reproduction and horizontal transmission to create epizootic spread. Alternatively, the crop environment can be manipulated to enhance resident microbial agent populations (conservation biocontrol), as discussed by Pell et al. (2009).

The fourth approach is to inundate a crop with a microbial agent in much the same manner as a chemical pesticide. Insect control is achieved only by the organisms that were applied; there is little or no epizootic spread. Several reviews on the use of fungal entomopathogens in field crops are those of Feng et al. (1994), Wraight and Carruthers (1999), Inglis et al. (2001), Shah and Pell (2003), and Goettel et al. (2005). My review will focus on the ecological aspects affecting the efficacy of these fungi when they are used in an inundative approach in foliar and soil arenas. Understanding these factors, which affect both efficacy and persistence, will allow intelligent manipulation of insect, pathogen, crop, and especially their environment, to achieve satisfactory management of an insect pest population within the context of inundative use.

Why inundate a crop with an entomopathogenic fungus? There are regulatory and economic, as well as technical, reasons. Pesticide regulations, which include microbial agents, require generation of human and environmental safety, and in some countries, replicated verification of efficacy, a process that can require approximately US$1–1.5 M, in addition to normal development costs. If money is to be invested in commercializing a microbial pest control agent, there must be return on investment, which in turn means repeat sales. Inoculation of a crop with a self-replicating organism (classical biocontrol) defeats this purpose.

There are technical reasons for employing inundative use. Unlike the Entomophthorales, the Hypocreales, particularly *Beauveria* spp. and *Metarhizium* spp., do not commonly cause natural, large-scale epizootics among insects in annual crops, nor have many classical or inoculative biological control introductions been successful, with the possible exception of *Lecanicillium* spp. (Hajek et al. 2005). Most cropping systems and their insect pests are transient in nature, being present for only one growing season, sometimes for only a few weeks. In addition, widespread adoption of crop rotation on large areas of monoculture creates a very temporally unstable environment for any microbial agent. Annual disruption in habitat not only removes the insect hosts, but in many cases directly destroys the microbial agent. Inundation with a microbial agent, sometimes repeatedly, is therefore necessary. Inundation attempts to overcome many of the factors that keep a pathogen in nonepizootic equilibrium with its host, by overwhelming the habitat with sheer numbers of infectious propagules. Inundative use also fits, for better or worse, into the familiar chemical paradigm—farmers simply apply the fungus as they

would a chemical pesticide with the expectation of rapid, extreme efficacy.

The entomopathogenic Hypocreales have, in particular, received considerable commercial attention over the past 30–35 years because they lend themselves to in vitro mass production of sufficient quantities of infective propagules (aerial conidia, submerged conidia, or blastospores) for use in an inundative approach. In most or all cases, the propagules need to be formulated with additives to provide shelf life, inert diluents, spreaders, stickers, and emulsifiers (Jaronski 1997; Burges 1998). The propagule types and ecological considerations in formulating entomopathogenic fungi are discussed by Jackson et al. (2009). For the sake of simplicity, I will use "spore" as a generic description of the different propagules in this review, unless a specific propagule type is indicated in a cited example.

Inundative use against foliar pests

A mycopesticide can be employed inundatively using a variety of delivery methods: ground or aerial ultralow volume (ULV) sprays, medium to high-volume broadcast or directed sprays, as dusts, as granules, or distributed via autodissemination devices. Application technology has been thoroughly discussed by a number of authors (Bateman et al. 2007; Chapple et al. 2007; Mierzejewski et al. 2007; Vega et al. 2007).

Winning the numbers game

Inundative use of fungal entomopathogens, as well as other pathogens, is a "numbers game," in which one applies sufficient numbers of spores to overwhelm an insect population. Unlike other insect pathogens, these fungi are percutaneously infectious agents. They act by contact. An insect can acquire spores directly from impingement of a spray or indirectly from contact with a fungus-contaminated surface. Behavior of the insect and the nature of the plant canopy determines which of these two routes is more important. Immature whiteflies, being sessile, need the fungus "to come to them." Large insects positioned prominently in a habitat, such as gregarious locusts or migrating Mormon crickets (*Anabrus simplex* Haldeman), also present a direct target for spray applications. For insects that actively move about their habitat, e.g., thrips and Heteroptera, acquisition of spores from the environment can be more important. Some insects, e.g., nongregarious locusts and grasshoppers, present a mixed picture, in which both direct and indirect acquisition of propagules are important (Lobo Lima et al. 1992; Johnson et al. 1992). In other cases, the situation is more complex. An example is larval *Trichoplusia ni* (Hübner) on cabbage versus beans. With cabbage, applications of a *B. bassiana* resulted in nearly equal mortalities among insects exposed to direct spray contact or exposed to spray residue, whereas on beans, direct spray contact provided significant insect mortality, but mortality due to residual contact was ineffective (Behle 2006). Nevertheless, efficacy is based on the number of propagules that end up contacting the host cuticle.

Typically, inundative use of fungal entomopathogens in a field or glasshouse crop involves application of at least 10^{13}–10^{14} propagules ha^{-1} (Wraight and Carruthers 1999), although standard rates of the commercial *M. acridum* against African locusts and grasshoppers are 2–2.5 × 10^{12} conidia ha^{-1} (van der Valk 2007), with successful control being achieved under certain circumstances with 1–1.25 × 10^{12} ha^{-1} in Australia (D. Hunter personal communication). Broadcast application of 1 × 10^{13} conidia ha^{-1} translates to 1 × 10^5 conidia cm^{-2} on a planar surface, or, theoretically, 2.5 × 10^4 conidia cm^{-2} on a crop with a typical leaf area index of 4 (Scurlock et al. 2001).

Why so many spores? Target insects have to acquire a sufficient number of conidia for infection to occur. Tens to thousands of conidia are needed per insect for a median effective dose (LC_{50} or LD_{50}). For example, in laboratory bioassays using larval *Plutella xylostella* (L.), LC_{50}s of 11–6,500 conidia cm^{-2} of sprayed surface were observed among 41 isolates of *B. bassiana* (Wraight et al. 2009).

Additionally, the dose-mortality response with fungal entomopathogens typically has a low regression slope value. The implication of this phenomenon is that very large increments in the number of spores are needed for commensurate increase in efficacy. We can gain some insights into this phenomenon from published bioassay and field efficacy data. Data from field trials of *M. acridum* IMI330189 (Green Muscle^{TM}) and FI985 (Green Guard^{TM}) against various acridids indicate that a consistently efficacious (>80% insect

mortality) field rate is 2.5 × 10^{12} conidia ha^{-1} (Hunter et al. 2001; van der Valk 2007), which translates to 2.5 × 10^4 conidia cm^{-2} planar surface. A locust with a cross-sectional target area of a approximately 6 cm^2 could be expected to acquire approximately 1.5 × 10^5 conidia from a direct spray at the above rate; a small grasshopper, approximately 5 × 10^4 conidia. In direct, topical bioassays of *M. acridum* FI985 and the wingless grasshopper, *Phaulacridium vittatum* (Sjöstedt), the log dose-response regression slope was 2.08 (Milner 1997). An eightfold increase in dose was needed to go from the LD_{50} (1,212 conidia per insect) to the estimated LD_{95} (9,240 conidia per insect). For the Migratory Locust, *Locusta migratoria* L., the slope was 1.30, requiring a 19-fold increase from the LD_{50} (4363 conidia) to the estimated LD_{95} (7.94 × 10^4 conidia). Using a spray application onto various larval Lepidoptera and substrate, Wraight et al. (2009) reported one *B. bassiana* isolate had a regression slope of 0.97. The result of this low slope was that a 57-fold increase was needed to go from the LC_{50} to the LC_{95} (2.5 × 10^6 conidia cm^{-2}). This last concentration corresponds to 2.5 × 10^{14} conidia ha^{-1} on a planar surface, higher when leaf area index is included.

There have been very few field studies where propagules per unit area of leaf surface have actually been measured as a basis for understanding efficacious rates. Notable among these are Poprawski et al. (1997), Wraight et al. (2000), and Wraight and Ramos (2002). Wraight et al. (2000) applied *B. bassiana* GHA against *Bemisia tabaci* (Gennadius) in various crops, monitoring spray coverage with plastic coverslips on which conidial deposition rates could be determined. Their application of 5 × 10^{13} conidia ha^{-1} with an air-assisted electrostatic sprayer achieved 1.7–2.8 × 10^5 conidia cm^{-2} on the lower surfaces of cucumber (*Cucumis sativus* L.) or melon (*Cucumis melo* L.) leaves. A parallel application of 1 × 10^{14} conidia ha^{-1} yielded 3.9–4.8 × 10^5 conidia cm^{-2}. They subsequently observed a 69% reduction in large nymphs after one application of 1 × 10^{14} conidia ha^{-1}, and 90% reduction after two sprays of the higher rate and four sprays at the lower rate, at 4-day intervals. To place these data in context, Wraight et al. (1998) observed an LC_{50} of approximately 2.5 × 10^4 cm^{-2} for this *B. bassiana* isolate in laboratory bioassays. Based on their mean log-dose regression slope for this fungus (1.09), a theoretical LC_{95} would be on the order of 8 × 10^5 conidia cm^{-2} of sprayed surface. Wraight and Ramos (2002) also monitored spray coverage using plastic coverslips when applying commercial *B. bassiana* formulations at 1.25, 2.5 or 5 × 10^{13} conidia ha^{-1} for the control of Colorado potato beetle (*Leptinotarsa decemlineata* Say). When the fungus was applied using upward pointing spray nozzles placed below the canopy, they observed mean conidial deposition rates on upper and lower leaf surfaces of 7.31–11.4 × 10^4 and 2.6–6.5 × 10^4 cm^{-2}, respectively. These rates of fungus yielded 10–65% beetle reduction depending on frequency of application. An LC_{95} of 2.3 × 10^4 conidia cm^{-2} *B. bassiana* GHA for second instar larvae was based on an LC_{50} of 1,460 conidia cm^{-2} and a regression slope of 1.37 (Furlong and Groden 2003). In summary, much of the published data indicates that considerable numbers of spores have to be applied for good efficacy, and that large additional increments of fungus are needed to achieve increasing levels of efficacy, when control relies upon infections from only the applications. There is an economic context to these inundative rates. An internet survey in 2009 of prices of the commercial *B. bassiana* product in the US (Jaronski unpublished data) yielded an average sales price of US$25 per 1 × 10^{13} conidia (plus shipping). Efficacious rates, such as discussed earlier, imply a user cost of US$25–250 ha^{-1} per spray using this product, clearly restricting use to very high margin crops. The Green Muscle *M. acridum,* with its greater infectivity and virulence for Acrididae, cost US$9–18 ha^{-1} in 2007 at the 2.5–5.0 × 10^{12} conidia ha^{-1} rate (Jaronski unpublished data).

How can we change the mathematics of application rates? One way is to concentrate the conidia into a narrower, targeted zone. An example is application of *B. bassiana* against larval whitefly species in cucurbits. The target insect resides on the underside of the plant's leaves, which are generally a layer of umbrella-like structures. Figure 1 illustrates a directed spray application using conventional spray equipment that directs most of the spray into the cucurbit canopy. Similarly, by using a backpack sprayer with hydraulic drop nozzles pointing upwards, Wraight and Ramos (2002) were able to increase the conidial deposition on leaf undersides 6- to 30-fold. In cotton, fungus sprays can be applied to the undersides of the leaves by use of a horizontal bar preceding the hydraulic spray boom. This bar bends the cotton

Fig. 1 Directed spray application of mycoinsecticide using conventional spray equipment that directs most of the spray into the cucurbit canopy. (Top) Spray designed to treat two rows of cucurbits. (Bottom) Close-up of drop tube and nozzle arrangement. Note height of nozzles in relation to plant canopy and rearward direction of spray

plants and exposes the leaf undersides to the spray (Jaronski, unpublished data).

A second method is to concentrate a broadcast application of spores into a directed band over the row of crop ("band over row"). For instance, sugarbeet root maggot (*Tetanops myopaeformis* Röder) adults oviposit into the upper soil surface within 3 cm of the emerging seedling. One approach in controlling this insect has been to apply conidial suspensions of *M. anisopliae* in a 12.5-cm band-over-row application just before oviposition begins (Jaronski et al. 2007). If applied in a broadcast spray, 5×10^{13} conidia ha^{-1} would result in a level of 4.9×10^5 conidia cm^{-2} of soil surface. With the banded application soil, levels become 2.4×10^6 cm^{-2}, a fivefold increase at the same rate per hectare, and confined to the actual oviposition site. Similarly, spores can be placed in the path of insects, such as on fiber bands wrapped around tree trunks to control the Asian longhorned beetle (*Anoplophora glabripennis* Britton and Sun) (DuBois et al. 2004; Shanley et al. 2009), affording considerable economies.

A different approach is to bring the insect to the fungus, using a bait or attractant formulation. For example, linoleic and linolenic acid-rich vegetable oils can be used to draw grasshoppers to fungus-treated strips spaced every 20–30 m rather than applying the fungus broadcast). This approach has been validated on rangeland using chemical insecticide (Lockwood et al. 2001) and is being pursued with fungi. Also, pheromones can be used to draw target insects into trap stations where the insects become dusted with conidia. This approach, thoroughly reviewed by Vega et al. (2007), is being used operationally against Japanese beetle (*Popillia japonica* L.) in the Azores. Alternatively, another insect can be used to carry fungal conidia specifically to the target insect's habitat. An example is use of bumblebees (*Bombus* spp.) and honey bees (*Apis mellifera* L.) to vector *B. bassiana* conidia to greenhouse crops and canola (*Brassica rapae* L.) (Al-Mazra'Awi et al. 2006a, b) to control thrips and *Lygus lineolaris* Palisot de Beauvois. Baverstock et al. (2009) discuss using insect behavior to enhance fungal entomopathogen efficacy. Lastly, efficacy could be increased by combination with other microbials such as *Bacillus thuringiensis* Berliner (Wraight and Ramos 2005) or chemical pesticides.

Environmental factors affecting the fungi in foliar use

Once spores are applied to foliage, their levels decline, the rate of decline affected by a number of factors: sunlight, rain, temperature, humidity, leaf surface chemistry, and phylloplane microbiota. The fungal spores, once on the insect cuticle, usually invade the body of their hosts within 24 h. During the initial infection process—spore activation, germination, initial cuticular penetration—the fungi continue to be susceptible to many of the same environmental factors. Once inside the host's body, the fungi continue to be affected by temperature, and, indirectly, humidity, via its effects on overall insect health, and become exposed to new, humoral factors, which themselves can be affected by food plant suitability, partitioning of insect resources among reproduction, movement and immunity, temperature, etc. During the past few years new discoveries (e.g., Lemaitre and Hoffmann 2007; Müller et al. 2008) indicate insect humoral immunity may be more important than previously thought, even with fungal entomopathogens. These diverse factors can combine to limit the efficacy of mycoinsecticides applied at economically acceptable rates. At the same time, certain of these aspects can be manipulated, at least theoretically, to enhance efficacy.

Sunlight

It is generally recognized that sunlight, particularly the UV-A and UV-B components, is a major mortality factor of fungal propagules on the phylloplane and is largely responsible for short persistence of mycoinsecticides in the epigeal habitat. There have been an number of laboratory studies using artificial UV sources (e.g., Ignoffo et al. 1977b; Hunt et al. 1994; Inglis et al. 1993, 1995a; Alves et al. 1998; Lee et al. 2002). Outdoor studies have been far fewer (Gardner et al. 1977; Inglis et al. 1995a, 1997a, b; Smits et al. 1996; Braga et al. 2001b; and Behle 2006). In general, the half-life of fungal conidia under natural, outdoor sunlight, in terms of percent viability or viable numbers per unit area, is 3–4 h (Roberts and Campbell 1977; Braga et al. 2001a), although Inglis et al. (1997a) observed a half-life of approximately one day in a North American short-grass prairie, and Sabbahi et al. (2008) observed viable conidia on sprayed strawberry foliage for up to six days. As measured by insect efficacy during field trials, however, persistence may be as long as 8–14 days, at least in the case of *M. acridum* under African subtropical, semi-arid conditions (summarized by van der Valk 2007). The UV-A component (320–400 nm) represents about 95% of total solar UV and is associated with conidial death and delayed germination (Braga et al. 2001a), but the UV-B component (280–320 nm) is considered more damaging (e.g., Moore et al. 1993) and has been the general focus of most studies. Both components have to be considered because they have different modes of action, while more realistic, outdoor studies need to be conducted.

In addition to outright conidial mortality, several authors have noted delayed germination as well (Braga et al. 2001a). This latter aspect has bearing on overall efficacy because it gives advantage to rapidly molting insects such as aphids, and earlier instars of many Lepidoptera and Coleoptera. Some sort of recovery from UV damage may be possible (Braga et al. 2001a), but this aspect is not well understood and needs further inquiry.

The effect of sunlight on persistence can be affected by location of the spores (abaxial vs. adaxial leaf surfaces) (Jaronski, unpublished data), formulation (Alves et al. 1998; Edgington et al. 2000; Cohen and Joseph 2009; Thompson et al. 2006), and fungal species or strain (Ignoffo and Garcia 1992; Fargues et al. 1996; Fernandes et al. 2007). With regards to the last aspect, conidia of *I. fumosorosea* were the most susceptible, while *M. acridum* were the most resistant to UV irradiation followed by *B. bassiana* and *M. anisopliae* (Fargues et al. 1996). Significant differences also existed among isolates within each species (Fargues et al. 1996; Fernandes et al. 2007). There may also be an interaction between temperature and sensitivity to UV radiation (Smits et al. 1996). A caveat about some of the published studies on this subject is that photodegradation of conidia on glass is faster than on leaf or agar surface (Inglis et al. 1997a) so that some data have to be interpreted with caution.

Survival on the lower surface of leaves, especially when there is considerable lateral shading by adjacent canopy, can be considerable (Fig. 2). During persistence studies of *B. bassiana* GHA (as Mycotrol™ 22WP) in southern California, viabilities of conidia applied to the lower and upper surfaces of melon (*Cucumis melo* L.) leaves were followed on a daily basis using germination tests of spores washed off leaf surfaces (Jaronski unpublished data). Conidial viability on leaf undersides decreased approximately 9–11% day^{-1}. On upper leaf surfaces viabilities dropped by 47% day^{-1}. As the melon canopy grew and expanded, the rate of conidial death on adaxial surfaces decreased to 1.2–1.6% day^{-1} although neither the amount of daily solar radiation nor air temperatures varied. Adjacent canopy increasingly protected conidia on lower surfaces. The host plant itself may also have a role in photoinactivation of conidia. In parallel tests, conidial viability on adaxial

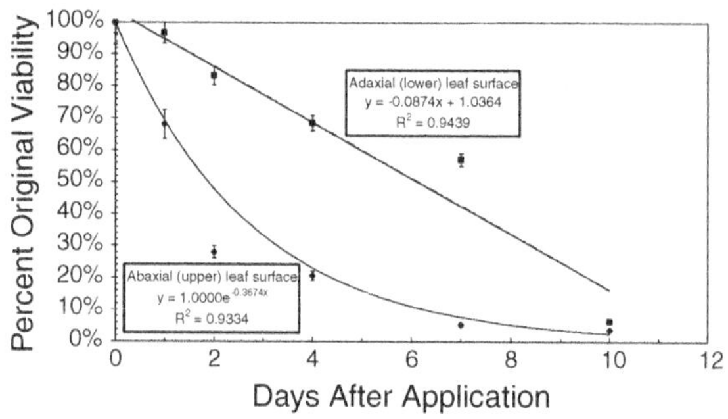

Fig. 2 Conidial residual life of *Beauveria bassiana* GHA as Mycotrol™ 22WP on abaxial and adaxial leaf surfaces of *Cucumis melo* at Brawley CA, USA late May 1995. Each point is the mean of three replicate leaves. Error bars represent SD

cotton leaf surfaces decreased 24.3% vs. 4.2% day^{-1} for melon leaves (Jaronski unpublished data).

A number of UV protectants have been evaluated and a few with practical potential identified (Jackson et al. 2009). Inglis et al. (1995a) identified a number of water soluble and oil-soluble UV protectants in laboratory tests. When the candidates were tested outdoors, however, the degree of protection was greatly reduced for all protectants and was inconsistent between two replicate trials. Nevertheless, there was a quantitative indication of the potential of photoprotection under realistic outdoor conditions—25–37%. More recently, Reddy et al. (2008) identified 1–10 g Tinopal UNPA-GX^{TM} L^{-1} of carrier as providing significant UV protection of a *B. bassiana*. The LT_{50}, in terms of hours of exposure to natural sunlight, was increased by 26%. While these results seem encouraging, one has to retain a sense of practicality. Use of 1 g Tinopal L^{-1}, as was tested by Reddy et al. (2008), in an aqueous spray applied at 187 L ha^{-1}, which is a low but common application rate for insect control on US vegetables, would cost US$99 when the optical brightener is obtained as 85% technical grade material from the manufacturer (2009 prices). Such an additional cost is rarely practical. The above rates of photoprotectant in an ultra low volume (ULV) oil formulation, which is typically applied at 1–2 L ha^{-1} could be more feasible. But such low rates, although common in some instances, for example locust control in Africa and Australasia, are rare in the United States and the EU.

A critical question remains: how much must persistence be improved for a significant increase in field efficacy? Twofold? Fourfold? More? In many intended uses of a commercial mycoinsecticide, farmers apply fungus to their crops repeatedly over the crop cycle. This is certainly the case with *B. tabaci* in the southwestern US, where season-long (6–8 weeks) weekly applications of a mycoinsecticide, for example, *B. bassiana*, can be necessary. One implication of the interaction of repeated mycoinsecticide sprays with a constant loss of viable conidia is a fluctuating, "sawtooth" variation in the overall levels of fungus in the crop (Fig. 3). (The simple models presented here ignore other sources of conidial losses, such as physical loss from canopy due to rain or wind, or dilution of conidial concentrations on leaf surfaces due to canopy growth.) A more-or-less constant (9%) daily loss occurs in the levels of viable conidia as affected by application intervals of five vs. ten days for six weeks during the melon-growing season (Fig. 4). More frequent replenishment of fungal conidial levels may serve to overcome loss in conidial viability over time. With a 10-day schedule, the "average" conidial titers plateau at a low level, which may not be sufficient for control, while more frequent applications rapidly increase overall titers. In 1995 Wraight et al. (unpublished) observed considerably reduced efficacy from a 10-day application schedule of Mycotrol versus a 5-day schedule, although continuing oviposition by adult whiteflies probably was also a factor.

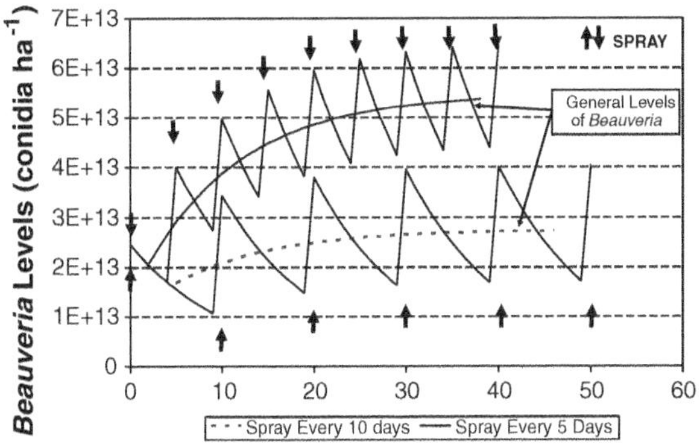

Fig. 3 Effect of a 9% day^{-1} loss of *Beauveria bassiana* GHA due to sunlight at two spray schedules of 1 Kg Mycotrol^{TM} 22WP (2.5 × 10^{13} conidia) ha^{-1}. Applications of *B. bassiana* (arrows) are either every five or ten days. At each spray the fungus titers increase but then decrease due to solar radiation. Degradation rate is based on observations of Mycotrol 22WP persistence on *Cucumis melo* adaxial leaf surfaces at Brawley CA, USA late May–June 1995 and 1996. Trends assume no expansion of plant canopy

The persistence studies of *B. bassiana* GHA in cucurbits, presented earlier, may provide some insight into the degree of photoprotection necessary. The data were used to model fluctuations in conidial densities on plant leaf surfaces. Trends in conidial levels resulting from a change in conidial loss, from 24 to 12% day^{-1} ("50% protection" from UV), with reapplication of fungus every seven days (as recommended by the company for whiteflies in cucurbits) are depicted in Fig. 4. With 24% daily loss in conidial viability, weekly applications of fungus are needed to maintain conidial levels in the crop (ignoring new growth and canopy expansion). The initial level of 2.5 × 10^{13} conidia ha^{-1} quickly drops by more than one-half and then fluctuates around an

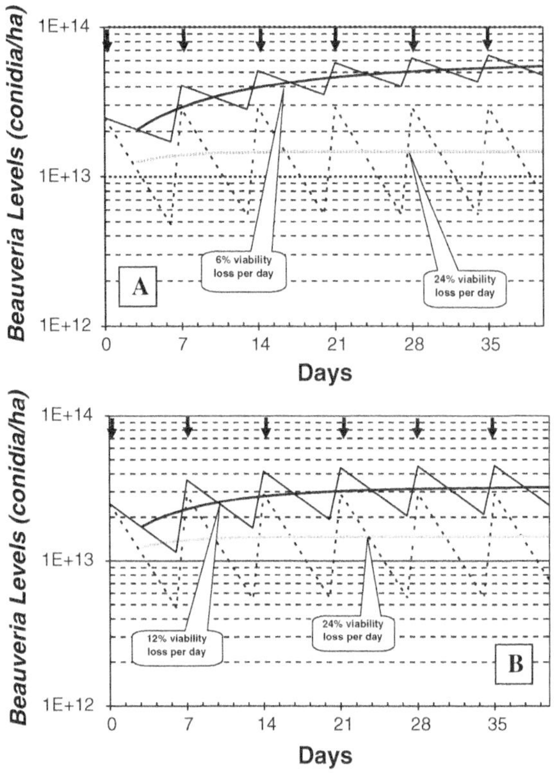

Fig. 4 Effect of reducing *Beauveria* GHA photodegradation rate from 24% day^{-1} to 6% (**A**) or 12% day^{-1} (**B**) on conidial levels in a crop (e.g., melons for whitefly control) subject to weekly sprays of 2.5 × 10^{13} conidia ha^{-1}. Applications of *Beauveria* (arrows) are weekly. At each spray the fungus titers increase but then subsequently decrease due to solar radiation at a daily rate of 24, 12 or 6%, accordingly. Trends assume no expansion of plant canopy. Legend: spaced dash 24% viability loss day^{-1}, dash 6% or 12% viability loss day^{-1}

equilibrium level of 1.4–1.5 × 10^{13} conidia ha^{-1} with the repeated weekly sprays. If a UV protectant decreases the daily loss in viability by one-half, to 12%, the weekly applications may result in greatly decreased fluctuations and a gradual accumulation of fungus to levels slightly greater than the original application (Fig. 4B). If, instead, 75% protection was achieved, a reduction in the rate of viability loss to 6% day^{-1} (Fig. 4A), the trends in conidial numbers might increase through the growing season to about twice the original level. This model is, of course, a gross simplification and ignores a number of factors, including conidial landing on non-target plant surfaces, physical loss of conidia from leaf surfaces, canopy expansion, insect movement to or oviposition on new foliage flush, and critical concentrations of conidia and rate response of efficacy (slope of the rate-efficacy regression, discussed in a previous section). But the model is a starting place. Obviously, carefully designed field experiments with effective photoprotectants are a critical need to resolve this question.

Rainfall

Rain events following application of fungal propagules can be catastrophic for efficacy. Relatively few controlled studies have been conducted regarding this aspect, notably those of Inglis et al. (1995b, 2000) and Inyang et al. (1998). In their earlier work, Inglis and his coworkers observed that *B. bassiana* conidia suffered rates of removal of 25–47% from alfalfa (*Medicago sativa* L.) and 51–56% from wheat (*Triticum* spp.) leaflets with as little as 30 min of simulated rain (either 27 or 113 mm h^{-1}). The conidia were applied as aqueous suspensions without wetting agents. Later on, they examined the rainfastness of a series of commercial and experimental formulations (Inglis et al. 2000). Only conidia in a nonemulsifiable oil carrier resisted simulated rainfall of 77 mm h^{-1}. Several emulsifiable suspensions (ES) and a wettable powder formulation washed off leaves as readily as conidia applied in water only (Fig. 5). What is instructive is that the total volume of spray had an effect on rainfastness for the commercial Mycotrol ES formulation—the 0.8% spray had greater rainfastness than 0.125% spray. The inference is that the volume of oil confers a degree of rainfastness. Inyang et al. (2000) observed that 39–76% of *M. anisopliae* conidia applied in three different formulations were lost from oilseed rape leaves after 1 h of simulated rainfall (rate of simulated rain not given). The least removal of conidia occurred with a safflower oil-Shellsol T^{TM} carrier, once again indicating that oil-based formulations may be more rainfast. Similarly, Wraight and Ramos (2002) observed better efficacy of an ES formulation of *B. bassiana* than a wettable powder formulation against Colorado potato beetle in potatoes when their field trial was beset by frequent rainfall during the fungus application phase. Nevertheless, significant rain or overhead irrigation, after a spray application of a mycopesticide, may be a major detriment to efficacy. At the same time in certain situations, rainfall can act in the dispersal of conidia especially

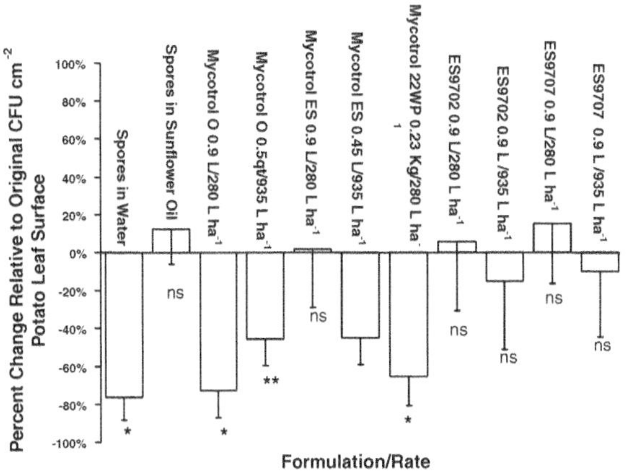

Fig. 5 Rainfastness of commercial and experimental formulations of *Beauveria bassiana* on potato leaves after a simulated rainfall of 77 mm h^{-1} for 30 min. The rate, 280 L ha^{-1}, represents typical application rate onto vegetables in US; 935 L ha^{-1} was the manufacturer's recommended spray volume for whiteflies in cucurbits and cotton. Symbols: * significantly different from 0% change, at $P = .05$; ns not significant. Error bars represent SE. (Data adapted from Inglis et al. 2000)

to the soil beneath the plant canopy (Bruck and Lewis 2002) and thus enhance efficacy. There also seem to be differences among plants in terms of persistence of spores on foliage, differences that may be mediated by leaf cuticle chemistry. There were significant differences in retention of conidia on lettuce (*Lactuca sativa* L.) and celery (*Apium graveolens* L.) following significant natural rainfall during a field trial (Kouassi et al. 2003). Rain reduced the numbers of CFU on celery by 92% but only 10% on lettuce. This latter aspect is an area that needs further research.

Temperature

Ambient temperatures can affect fungal entomopathogen field efficacy. For example, efficacy of *B. bassiana* GHA against *Lygus hesperus* Knight was greatly reduced in small plot field tests in July but not June of the same year, even though the insect is quite susceptible to this fungus (Noma and Strickler 1999). While optimal germination and growth rates of fungal entomopathogens range between 23°C and 28°C, growth, in general, rapidly slows above 30°C, and ceases for most isolates at 34–37°C. Similarly, conidial germination is adversely affected by temperatures above 30°C. In the Noma and Strickler (1999) study, temperatures within the plant canopy were observed to be in excess of 30°C during the July trials; the upper thermal limit for this strain is 32–34°C. In subsequent laboratory experiments, Noma and Strickler (2000) demonstrated that infections were greatly reduced at 35°C, compared to 25°C. Temperatures below 16°C increasingly slow germination and growth rates for most of the fungal entomopathogens, and thus affect efficacy in terms of a longer survival of the target population (Inglis et al. 1999; Ihara et al. 2008). This can have important bearing on mycopesticide field efficacy in northern climes and also on temperate rangeland where night time insect body temperatures can be <10°C for more than 6 h day^{-1}. Night time ground temperatures even reached 5–6°C, in South and North Dakota during the Summer of 2003 during a grasshopper field trial (Jaronski unpublished data). The temperature data shown in Fig. 6, using thermal surrogates (Lactin and Johnson 1998), represent the maximum temperatures that could be achieved during a 24 h cycle by grasshoppers on the ground and in the plant canopy. Not only are mid-day body temperatures in excess of the upper thermal limit for *B. bassiana* GHA, due to normal basking as well as 'behavioral fever' thermoregulation (see below), but night time temperatures are cold enough to greatly slow fungal growth within the insects. The result is that there can be only a few hours each day during which temperatures are permissive for fungal growth. For example, the temperature observations as represented in the top graph in Fig. 6, and made during successive days during the same field trial. They were used to construct a heat budget for *B. bassiana* GHA (Fig. 7). As can be seen in Fig. 7, only 6–7 h each day were permissive for growth (based on upper and lower cutoff temperatures for 50% of fastest fungal growth) on sunny days and 11–16 h on partly cloudy days. The end result is that considerable time can elapse before infected insects succumb to infection, time for the insects to damage a crop. An example is *B. bassiana* GHA used against grasshoppers. Johnson and Goettel (1993) and Inglis et al. (1997b) observed that even though a considerable proportion of the targeted grasshopper population was infected, few died in the field within the observation period.

There are considerable differences in temperature tolerances among the fungal entomopathogens, even among isolates of the same species (e.g., Fargues et al. 1996; Bugeme et al. 2009), so that a candidate

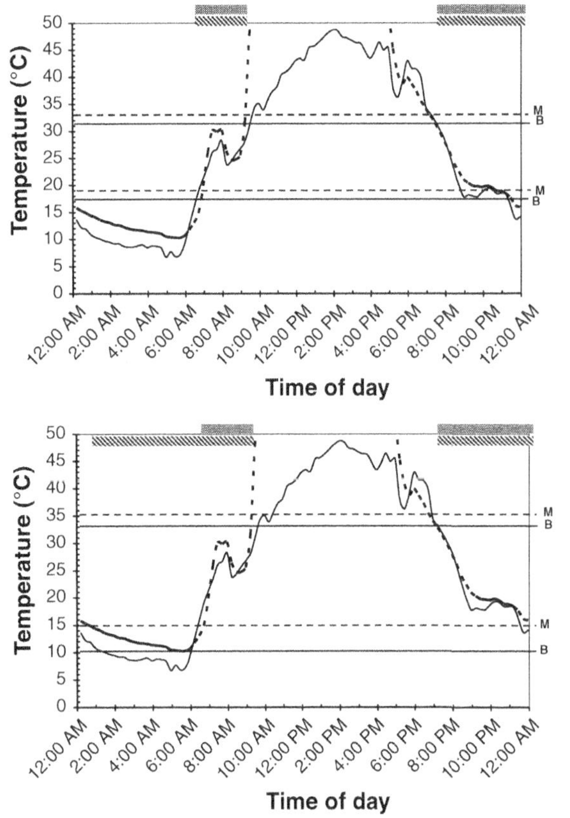

Fig. 6 Maximum potential grasshopper body temperatures based on heat absorbance of thermal surrogates July 10, 2003, on mixed grass prairie, Edgemont, South Dakota USA, and parallel effects on entomopathogenic fungus growth. Legends: dash temperatures recorded in surrogate in plant canopy 10 cm above ground; spaced dash temperatures of a thermal surrogate placed on ground simulating a basking grasshopper; horizontal lines associated with "B" and "M" are upper and lower temperature thresholds for 50% (top graph) or 20% (bottom graph) of maximum growth rate of *Beauveria bassiana* GHA and *Metarhizium acridum* IMI33189 respectively; patterned horizontal bars represent duration of permissive temperatures for fungus growth, ▒ *Metarhizium acridum* IMI330189; ▧ *Beauveria bassiana* GHA

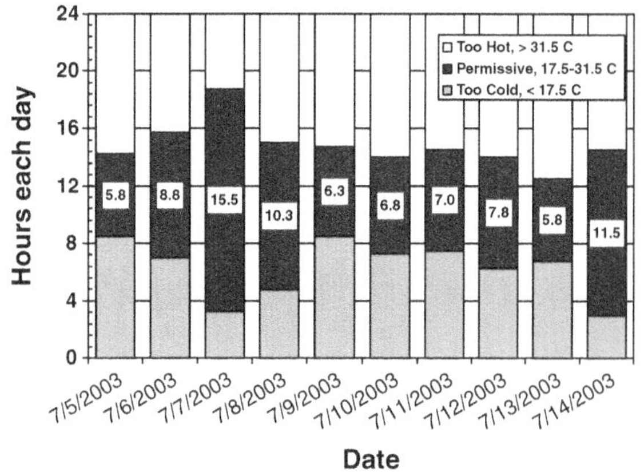

Fig. 7 Heat budget for *Beauveria bassiana* GHA based on temperature observations of thermal surrogate placed in grass canopy, in July 2003 at Edgemont, South Dakota, USA, and on the upper and lower temperature limits for 50% fastest growth rate for this strain

fungus may be identified for better heat tolerance to suit intended use, either by itself or to complement a second fungus with the opposite temperature tolerance. This latter approach was tried by Inglis et al. (1999). There are isolates with some degree of cold tolerance (Rath et al. 1995; Li and Feng 2009) including ones that grow at 8°C (De Croos and Bidochka 1999) and this attribute may make them superior to others for the control of insects in colder situations. Temperature tolerance should be one of the criteria for candidate selection if proposed uses so justify the concern. The downside is that ability of a microorganism to grow at 36–37°C raises concerns about pathogenicity for homeothermic vertebrates. Whether these differences are reflected in differences in field efficacy remains to be determined.

There is an additional aspect. Almost all thermal tolerance work has been done using constant temperatures, typically examining radial growth of colonies on agar media at a range of temperatures, e.g., Fargues et al. (1997). Only a few researchers have examined fluctuating temperatures (Inglis et al. 1999; Fargues and Luz 2000; Devi et al. 2005). In nature, temperatures fluctuate to a considerable extent in some habitats. Furthermore, certain insects (grasshoppers, houseflies, cockroaches) demonstrate active thermoregulation whereby they maintain their body temperatures several degrees above ambient by absorbing heat directly from the sun as well as from warm substrate (Carruthers et al. 1992). This thermoregulatory behavior can be pronounced upon infection with a pathogen, a phenomenon termed "behavioral fever" (Watson et al. 1993; Inglis et al. 1996; Kalsbeek et al. 2001). Thus insects, for example grasshoppers, can be infected following inundative application of a fungal pathogen, but do not die unless they are prevented from thermoregulating (Inglis et al. 1997a, b; Ouedraogo et al. 2004). An assumption in this phenomenon is that fungal growth resumes when temperatures become permissive. This is not always the case. Many isolates of

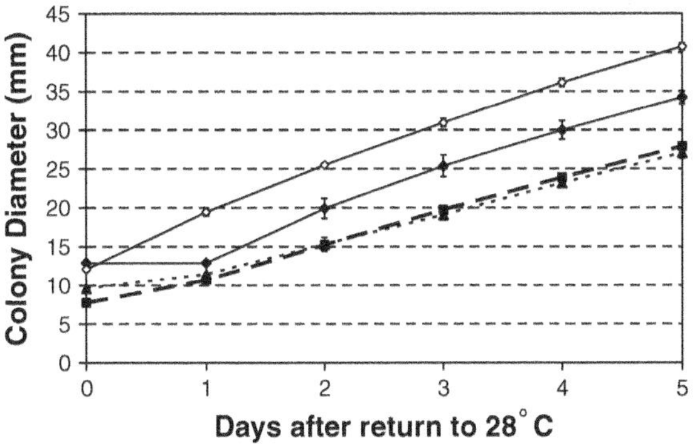

Fig. 8 Effect of a transient, 6 h exposure to 41°C on the subsequent radial growth at 28°C of *Beauveria bassiana* Strain GHA, *Metarhizium anisopliae* Strain F52, and *M. acridum* IMI330189. The mean colony diameter of *B. bassiana* in the 41°C treatment was significantly smaller one day after return of cultures to 27°C than for counterpart cultures grown at constant 27°C (*T* test statistic 23.66, 4 *df*, $P < .001$). Subsequent rate of growth (slope) was not significantly different from the 27°C treatment. Neither *M. anisopliae* F52 nor *M. acridum* IMI330189 displayed a significant lag and their rates of growth were not significantly different in the two treatments (data for the two *Metarhizium* at 27°C not shown). [diamond] *B. bassiana* 27°C; [small filled diamond] *B. bassiana* transient 41°C; [large filled triangle] *M. anisopliae* F52 transient 41°C; [filled square] *M. acridum* IMI330189 transient 41°C. Error bars represent 95% Confidence Limits

B. bassiana and *M. anisopliae* demonstrate a delayed resumption of normal growth after exposure to short periods of temperatures above their normal threshold (Jaronski, Keyser and Roberts, unpublished data). Figure 8 represents the effect of a 6-h exposure to 41°C on subsequent in vitro radial growth of a *B. bassiana* and two *M. anisopliae* isolates. This exposure time and temperature would be encountered by fungi infecting the Mormon cricket (Turnbow 1998). The commercial *B. bassiana* GHA displays a 1-day delay before normal growth rate is resumed. This delay becomes more pronounced with higher temperature and increased exposure time. Neither *M. anisopliae* F52 nor *M. acridum* isolates IMI330189 (Green Muscle^{TM}) and FI985 (Green Guard^{TM}) demonstrate a growth delay after 6 h at 41°C. F52 shows delays in resuming growth only after 9 h exposure to 41°C or 3 h at 44°C. In contrast, IMI330189 and FI985 require more than 6 h at 44°C or 18 h at 41°C before they show delayed resumption of growth. Thus, fluctuating temperatures can have more than a simple subtractive effect on efficacy and at least partially explain underestimates in time to onset of Mormon cricket or grasshopper field mortality predicted by simple heat budgets (Jaronski and Foster unpublished data).

Another aspect is indirect effect of temperature, especially high temperatures, on efficacy, as mediated through the insect's defense system. Larval *Galleria mellonella* L. exposed to 38°C for 30 min, then injected with *B. bassiana* blastospores may have a longer survival time than non-heat shocked larvae (Wojda et al. 2009). Heat-shocked larvae had elevated humoral anti-yeast and lysozyme activity and galiomycin expression in response to subsequent infection. While the purpose of inundative application of a mycopesticide is to overcome such defenses, the latter may still be manifested through slowed efficacy.

Humidity

While there is a requirement for high humidity for spore germination in vitro (e.g., Lazzarini et al. 2006), insects can become infected at much lower humidity. It is generally thought that infection is independent of ambient relative humidity (Ferron 1977; Marcandier and Khachatourians 1987; James et al. 1998; Lord 2005). But this is not true in all cases, viz., Luz and Fargues (1999), who observed a humidity threshold of >96% for efficacy of *B. bassiana* against *Rhodnius prolixus* Stål. Similarly, Yasuda et al. (1997) observed reduced efficacy of against *Cylas formicarius* Fabricius at <43% relative humidity. There are other examples in the literature, e.g., Altre and Vandenberg (2001), Lazzarini et al. (2006). The fungi *H. thompsonii* and *Lecanicillium* spp. may represent an extreme example of high humidity requirement for efficacy. Key to efficacy of *H. thompsonii* is very high humidity for at least 24 h (McCoy 1981). The current recommendations for commercial *Lecanicillium* spp. are application with subsequent relative humidity of at least 80–95% at the leaf surface, for 10–12 h per day for several days (Koppert 2009a, b). Thus the dependence of infection on humidity depends upon the insect, and its ecology, especially in relation to the phylloplane and its microclimate. Oil based formulations seem to overcome this problem (Ibrahim et al. 1999).

Phylloplane microhabitat vs. macrohabitat as it affects environmental variables

In considering environmental effects on a mycoinsecticide, one must differentiate the ambient environment, within canopy habitat, and, especially, leaf-surface microhabitat, especially for small target insects such as whitefly nymphs, aphids, thrips, and mites. Ambient temperature and humidity measurements, taken above the crop canopy can have little relationship to conditions within the canopy. For example, Shipp et al. (2003) observed that ambient humidity had little effect on *B. bassiana* activity against aphids, thrips and whiteflies on cucumber leaves under greenhouse conditions.

Each leaf on a plant and even different parts of a leaf have their own equilibrium temperature with the environment, based on sensible and latent heat losses vs. net heat gain from irradiation, and thus have a unique microclimate. During the day, upper leaf surfaces can be 10°C greater than ambient, while lower surfaces can be 1–2°C below ambient (Burrage 1971). Plant geometry affects leaf temperatures. Sunlight penetrates plant canopy in a reduced intensity and changed spectrum as determined by leaf angle and leaf area distribution. With crops having vertical leaves, the angle of the sun is most important with greatest penetration being at mid-day. The sunlight in turn affects leaf temperature. For example, leaves of 24-cm rye grass (*Lolium* spp.) can vary by as much as 6–7°C from air temperature at mid-day. On plants with horizontally held leaves, e.g., beans (*Phaseolus vulgaris* L.), the upper leaf surfaces were 2.5°C higher than the air while the lower surfaces were 3°C lower than ambient (Willmer 1986). In addition, temperatures can vary by as much as 2–3°C across a leaf surface (Burrage 1971). Above 33°C, leaf evapotranspiration can keep the leaf cooler than the surrounding air, but this is affected by leaf canopy and the leaf's position therein. Ferro et al. (1979) recorded abaxial apple (*Malus domestica* Borkh.) leaf temperatures 12°C lower than air temperature when the air was 38°C. Similarly, Chu et al. (1994) observed that leaves of cotton could be 5–7°C cooler than ambient under hot desert conditions, which may explain the efficacy of *B. bassiana* GHA against whitefly nymphs in Arizona cotton when air temperatures were in excess of 48°C (Jaronski et al. 1997).

The critical factor in humidity microclimate is the leaf boundary layer (LBL), which can be defined as the transition zone above the leaf surface in which wind speed increases with distance from the surface. The LBL can be 1–10 mm thick (Bonan 2008; Willmer 1986) although other sources cite a thickness of 2–3 cm in greenhouse tomato (*Solanum lycopersicum* L.) leaves (Boulard et al. 2002). Vesala (1998) examined the complexity of the factors affecting the thickness of the boundary layer. He divided the leaf boundary layer into two regions, an upper "adhering air layer" and a "lower superstomatal air layer." The former is affected by the size and shape of the leaf, presence of other leaves and wind velocity. The latter is affected by number of stomata per cuticle area, pore radius, leaf radius, wind velocity, and stomatal resistance. The boundary layer of air above a leaf surface is affected by leaf topology, radiation temperature, and air movement. More simply put, the humidity at the leaf surface is affected by evapotranspiration rate and wind velocity which combine to control the rate at which water vapor is transferred through the boundary layer. For more details about the physics of LBL, see Schuepp (1993).

The LBL can cause relative humidity immediately adjacent to the leaf surface to be higher than the ambient humidity. In cabbage (*Brassica oleracea* Linne) leaves, the ambient relative humidity (RH) of 70% increased to 90% 1 cm above both upper and lower leaf surfaces, and increased from 56% RH to 70% within 5 mm of waterlily (*Nymphaea* spp.) leaf (Willmer 1986). Within the immediate proximity of leaf stomata, RH could be 95–99% at 1 mm above leaf surface. Ramsay et al. (1938) observed an RH of 40% at 1 mm above the leaf surface of dock (*Rumex* spp.) vs. 10% ambient RH, and 95% vs. 50% ambient with a tulip (*Tulipa* sp.) leaf. A study by Boulard et al. (2002) is the most detailed and potentially relevant to the use of entomopathogenic fungi. They observed a 20–30% increase above ambient RH at 5 mm above tomato leaf surface in the morning, 7–10% at the end of the day. As the wind speed in the immediate vicinity of the leaf surface exceeds 0.36 km h^{-1}, however, turbulence disrupts the LBL and relative humidity approaches ambient (Gates 1968). This aspect creates a very complex and dynamic situation on the leaf surface especially in outdoor crops, but even with glasshouse plants. Large insects such as adult beetles, grasshoppers and late instar Lepidoptera, are probably less influenced by the LBL because their size places much of their bodies above it. Nevertheless infection via the tarsi and ventral surfaces of large insects may still be under the influence of the LBL.

There seems to be no comparable information about boundary layers above the insect cuticle. One can infer that there is a boundary layer and sufficiently high humidity to allow conidial germination and penetration into the cuticle from bioassays where the ambient RH during incubation was less than required for in vitro conidial germination, e.g., Ferron (1977), Marcandier and Khachatourians (1987), and Ramoska (1984). Charnley (1989) mentions that infection is often through the cuticle of the mouthparts, intersegmental folds, and spiracles, regions where the humidity may be higher than on other parts of the cuticle. However independence from humidity is not universal. A number of authors report a direct relationship between ambient humidity and infection rate.

Influence of phylloplane chemistry

Plant cuticle comprises a mesh of insoluble polymers, cutin and cutan, infused with a mixture of lipids, mostly long-chain (C20–C40) fatty acids and derivatives. Above this matrix is a layer of epicuticular waxes either crystalline or smooth in appearance. See Beattie (2002) and Andrews and Buck (2002) for more information. Plant cuticular compounds have the potential of affecting spore persistence on the phylloplane, and the susceptibility of insects to infection. The plant can either affect spore acquisition by insects or spore persistence. Persistence can be affected either by simple physical removal from the leaf surface (without rain) or toxicity from chemicals lethal to the spore. Inyang et al. (1998) observed that twice as many mustard beetles became infected when exposed to treated Chinese cabbage leaves than oil seed rape, with turnip leaves being intermediate. According to Poprawski et al. (2000) whitefly nymphs reared on tomatoes were significantly less susceptible to infection by *B. bassiana* and *I. fumosorosea* than whiteflies reared on cucumber. Lygus bug mortalities from *B. bassiana* were significantly different between celery and lettuce (Kouassi et al. 2003). Ugine et al. (2007) described a strong difference—sevenfold—in thrips infection rates from *B. bassiana* on beans and impatiens due to a much greater acquisition of conidia from bean leaves. There are, however, other insect-fungal entomopathogen systems where the plant had no effect, e.g., Colorado potato beetle-*Beauveria* (Costa and Gaugler 1989), *Spodoptera-Nomuraea* (Fargues and Maniania 1992). Such plant-associated inconsistency has important implications for adoption of these fungi for microbial pest control. Insect control on some crops may be far more amenable than others. There is very little information regarding effects of plant hybrid or variety, which effects may be important due to the many varieties of any crop plant in common use.

Leaf topography seems to affect the numbers of spores acquired by insects from treated surfaces (Inyang et al. 1998). Host plant and leaf age had significant influence on conidial attachment to beetle abdomens, less so to thoraces. These effects were paralleled by insect mortality from fungus infection. Notably, insect mortality from mycosis decreased drastically with increasing delay between leaf treatment and addition of insects (73–77% at three days to 0–10% at nine days), with little difference between Chinese cabbage (*Brassica rapa* L.) and oilseed rape (*B. napus* L.). The authors postulated that this effect was due to leaf expansion and thinning of the conidial levels on the leaf surface, but a leaf-surface-associated mortality of conidia could also have been at work. Similarly, Ugine et al. (2007) reported that the LD_{50} of *B. bassiana* GHA in western flower thrips (*Frankliniella occidentalis* L.) was almost sevenfold greater on impatiens (*Impatiens walleriana* Hook.f.) than on beans. This differential effect was paralleled by a different extent to which thrips acquired conidia from leaf surfaces.

A thorough study of the effect of plant cuticular compounds on fungal entomopathogens was reported by Inyang et al. (1999). Leachates of turnip (*Brassica rapa* var. *rapa* L.), Chinese cabbage, and oilseed rape had both stimulatory and inhibitory effects on conidial germination. There were differences among leachates from the three plants and among different solvents used. In vivo germination on leaf cuticle was stimulated by dewaxing the leaf surfaces. There was also significantly higher conidial germination on young (77%) versus old (40%) turnip leaves, but this was not the case with the other two plants. Water treatment, such as might occur during rain or heavy dew periods, resulted in higher germination on

Chinese cabbage and old turnip leaves, but not rape. Fungal virulence for larval mustard beetle (*Phaedon cochleariae* Fabricius) was enhanced by both leaf extracts and cuticle leachates. The authors hypothesized that this enhancement was a result of accelerated germination and higher effective dose. The in vitro fungistasis from cuticular leachates did not occur on the insect cuticle, highlighting the potential pitfalls of in vitro studies. From a practical aspect, any effect of plant cuticular compounds becomes significant if spore germination on insect cuticle and penetration are affected. Ostensibly, spores on the plant cuticle surface remain dormant until picked up by an insect. Fungistasis on the leaf cuticle is beneficial. If fungistasis is absent or lost, as happened when leaves were dewaxed, spores can germinate on the plant cuticle (Inyang et al. 1999), and are potentially lost from the effective dose of fungus presented to the insects. Implications for consistent control by fungal entomopathogens on a range of plants are considerable. This type of research needs to be expanded to other crop plants. For further exploration of this topic see Muller and Riederer (2005), Cory and Hoover (2006), and Cory and Ericsson (2009).

Tritrophic effects on efficacy may also be exerted via plant effects on the host insect. Nutrition has bearing on overall health of an insect, and suboptimal nutrition may mediate effects of fungal entomopathogens. For example, Thungrabeab et al. (2006) reported that two species of thrips reared on cotton or *Saintpaulia* spp. (Gesneriaceae) were much less susceptible to *B. bassiana* than those reared on bean, leek (*Allium ampeloprasum* var. *porrum* L.), cucumber, or daisy (*Bellis perennis* L.). Similarly, *B. tabaci* reared on cucumber, tomato, melon, green pepper (*Capsicum annuum* L.), potato, eggplant (*Solanum melongena* L.), marrow (*Cucurbita* spp.), cabbage, bean, or cotton displayed different susceptibilities to *B. bassiana* with significant differences in median survival times (Santiago-Alvarez et al. 2006). Nutritional differences were suggested as a causal factor. Hare and Andreadis (1983) observed that host plant affected susceptibility of Colorado potato beetle larvae to *B. bassiana*. The plant most suitable for insect growth produced larvae with the least susceptibility to the fungus. Furthermore, potato plants grown in the glasshouse were less suitable for the insect than field-grown plants, resulting in greater susceptibility to *B. bassiana*. This subject is treated in more detail by Cory and Ericcson (2009).

There is evidence that phytopathogens may have an effect on susceptibility of an insect to a fungus (Rostas and Hilker 2003). Treatment of mustard beetle larvae with *M. anisopliae* resulted in 100% mortality when insects were on leaves infected with *Alternaria brassicae* Berk. but only 50% mortality when they fed on uninfected leaves. The beetles were feeding on symptom-free plant parts and displayed slowed development, indicating either suboptimal nutrition or the effects of chemical changes in the plant accompanying Systemic Acquired Resistance (SAR). The influence of multitrophic interactions on fungus efficacy is a largely unexplored area but one with considerable potential bearing on field efficacy, especially enhancing entomopathogen efficacy via SAR.

Influence of pesticide residues on the phylloplane

There is a considerable body of literature on the effects of pesticides on fungal entomopathogens, most recently summarized by Klingen and Haukeland (2006). Most of this data is based on in vitro laboratory assays in which germination and vegetative growth is observed on agar media incorporating an insecticide or fungicide, or tests where spores are incubated with operational concentrations of pesticides for varying periods of time, then plated on germination media (e.g., Clark et al. 1982; Mietkiewski and Gorski 1995; Todorova et al. 1998). While these approaches readily identify innocuous agrochemicals, they can yield "false positives"—chemicals that have an adverse effect in vitro but not in vivo. The critical arena for chemical-spore interaction is on the leaf surface where the spore is dormant under most conditions until it contacts the insect cuticle. Some agrochemicals are rapidly absorbed after application by the leaf. For example, the strobilurin fungicides are toxic in vitro to fungal entomopathogens as well as to a wide range of fungi (da Silva and Neves 2005). The strobilurin derivative fluoxastrobin, however, is absorbed into the plant leaf within 15 min of application (Arysta LifeScience 2009) rendering contact between already present, or concurrently applied fungal spores to a very short exposure. Spores applied after the fungicide should have no contact with the residues. A more realistic strategy is to test the effects of spore deposits on plant surfaces that have been treated before, after, or concurrently with the chemical under study, an approach adopted by Mycotech for their pesticide compatibility recommendations (Laverlam International 2005). The spores are incubated for a set period of time, then washed off, rinsed by centrifugation and plated on agar media for germination.

The insect cuticle as it affects efficacy

While epicuticular substances clearly stimulate spore germination (or otherwise cuticular route of infection would not be possible), some insects possess fungistatic compounds. For example, Smith and Grula (1982) showed that cuticular extracts from larval *Helicoverpa zea* Boddie inhibited *B. bassiana* conidial germination. A more striking example may be *Nezaria viridula* L. SosaGomez et al. (1997) observed conidial germination of *M. anisopliae* on *N. viridula* cuticle was much lower than on other insect cuticle substrates, parallel to reduced infectivity for that insect. Only 5–20% of the conidia on *N. viridula* cuticle produced germ tubes, attributed to presence of the aldehyde, (E)-2-decenal. In addition, the cuticular topography affected conidial binding. Similarly, a pentane extract of *Melolontha melolontha* L. or *Ostrinia nubilalis* Hubner cuticle inhibited conidial germination and hyphal growth of a *B. bassiana* non-pathogenic to each insect (Lecuona et al. 1997), while in each case, a pathogenic isolate was not inhibited. In exploring the basis of differential susceptibility of *G. mellonella*, *Dendrolimus pini* L., and *Calliphora vicina* Robineau-Desvoidy to an entomophthoralean fungus, *Conidiobolus coronatus* (Costantin) Batko, Golebiowski et al. (2008) observed that reduced susceptibility to infection was associated with presence of C14, C16, and C20 fatty acids in *C. vicina*, but direct causation was not proven. In such studies caution should be taken in extrapolating in vitro germination tests with cuticular extracts to in vivo situations. The in vitro situation may not parallel in vivo conditions. The immediate environment for conidial attachment and germination, and the molecular concentration of stimulants or inhibitors cannot be easily duplicated. For example, free fatty acid toxicity to *B. bassiana* conidia was dependent on nutritional conditions (Smith and Grula 1982). Nevertheless, current evidence does indicate that the insect cuticle surface can mediate successful infection and thus the efficacy of inundatively used fungi against some insects. The existence of highly pathogenic isolates for most insects, however, implies that fungi can be found for which these barriers are unimportant.

Interaction with biotic components of the foliar environment

Phylloplane microflora

The phylloplane is replete with a great variety of microorganisms. Biofilms are almost ubiquitous on the phylloplane and are often 20 μm in depth and up to 1 mm in length (Morris et al. 1997). Aerial conidia of species such as *B. bassiana* and *M. anisopliae*, being dormant until they contact insect cuticle, are probably unaffected, at least as inferred from various on-leaf persistence studies where UV effects are absent. The ability of *H. thompsoni* to germinate and grow vegetatively on the leaf surface, then conidiate (McCoy 1981; McCoy and Couch 1982), infers that, at least in the citrus phylloplane microhabitat, the microflora is innocuous. Relatively few studies regarding any interaction between fungal entomopathogen conidia and microorganisms have been reported and all were done in vitro, creating another area ripe for investigation.

Non-target invertebrates and vertebrates

One of the characteristics of most of the fungal entomopathogens is their specificity to the Arthropoda. Published studies (see reviews by Zimmermann 2007a, b, 2008), as well as publicly released registration data, have demonstrated general safety for healthy vertebrates. There is relatively little data about the effect of vertebrates on the inundative release of fungal entomopathogens, but birds do attack locusts infected with *M. acridum* and can have a significant impact on treated populations (Mullie 2009). Nontarget invertebrates, for their part, have the potential of vectoring a fungal entomopathogen. An example are the Collembola, which seem refractive to infection by fungal entomopathogens and able to vector several fungal entomopathogen species to larval *Tenebrio mollitor* larvae, at least in a laboratory setting (Dromph 2003). Many predatory and parasitic insects seem to be ecologically protected from serious impact by inundatively applied fungal entomopathogens (Jaronski et al. 1998) and they have the potential to vector the fungal spores (Roy et al. 2001; Baverstock et al. 2009), as well as to complement the fungi in reducing an insect population. For instance, simultaneous use of predators, parasitoids, and mycopesticides can provide additive effects under greenhouse conditions (Labbe et al. 2009). There was a considerably faster elimination of aphids on leaf disks over which *Lecanicillium longisporum*-treated *Orius laevigatus* Fieber had walked (Down et al. 2009). However this potential dispersal of conidia was not universal. In the same study *F. occidentalis* were only slightly more affected than controls under the same conditions, and *B. tabaci* not at all. Conidia of *I. fumosorosea* were vectored by *Hippodamia convergens* Guérin-Méneville to healthy aphids and caused a variable proportion of the aphids to become infected (Pell and Vandenberg 2002), with greatest vector efficiency after the beetles fed among sporulated aphid cadavers. The authors pointed out this phenomenon might facilitate the spread of mycopesticide application within and between fields and therefore improve the efficacy. Predators might also act as vectors moving fungal inoculum into cryptic feeding sites. However, this potential dispersal of fungal conidia is not universal. See the recent review by Furlong and Pell (2005) for more details about fungal entomopathogen-natural enemy interactions.

Other insect pathogens

There are a few laboratory studies examining interaction of a fungal entomopathogen with another insect pathogen, or two fungal pathogens within the same host. The presence of another pathogen may make the target insects more susceptible to a fungal entomopathogen. For example, Brinkman and Gardner (2000) observed that fire ants (*Solenopsis invicta* L.) from microsporidian-infected colonies were 4.5 times more susceptible to *B. bassiana* than ants from healthy colonies, based on the LD_{50} ratio. Similarly, at low doses, joint infections of *Paranosema* (*Nosema*) *locustae* Canning and *B. bassiana* had a faster onset of mortality than nymphs with single infections and at high doses were synergistic (Tounou et al. 2008). Infections of *M. acridum* and a *B. bassiana* complemented each other under oscillating high and low temperatures (Inglis et al. 1999). But this mutually beneficial situation may not always be the case. A notable exception is a study by Thomas et al. (2003), in which the in vivo interactions of virulent and avirulent fungal entomopathogens in locusts were variable and were affected by the order of infection and by environmental conditions, particularly temperature. The significance of inter-pathogen interactions would depend upon the prevalence of the endemic pathogen in the target population, and would probably be manifested by degree of efficacy from the applied fungus.

Inundative use against soil-dwelling pests

Extensive discussions of abiotic and biotic factors affecting persistence and efficacy of fungal entomopathogens in the soil were published recently (Klingen and Haukeland 2006; Jaronski 2007). Therefore, the present review will only mention highlights.

Use of mycoinsecticides in the soil presents a different situation than foliar applications. All dose transfer to the insect is indirect—the insect pest must come into contact with the fungus spores. The key with all tactics is to create an infectious "minefield" of fungal spores to intercept the insects as they migrate through the soil and around the plant roots (Jaronski et al. 2005). In moving through the minefield the insects must physically contact and acquire sufficient numbers of spores for infection.

One approach is to apply conidia in aqueous suspension or as a dust into the soil or into the plant crown. Soil drenches do not carry conidia very far into all but the coarsest textured soils or potting mixes, limiting this approach (Ignoffo et al. 1977a; Storey and Gardner 1988; Storey et al. 1989). But soil drenches are still useful if spores can be applied to intercept insects dropping into the soil for pupation (e.g., *Curculio caryae* Horn, the pecan weevil; and *Rhagoletis indifferens* Curran, the cherry fruit fly), or neonates hatching from eggs laid on or just in the soil surface (e.g., *T. myopaeformis,* or *Delia* spp.).

The "numbers game" in the soil arena

Use of fungal entomopathogens in the soil arena is also a numbers game. There are several estimates of the LC_{50} and LC_{95} of fungi in terms of colony forming units or spores, per cm^{-3} or g of soil (e.g., Ferron 1981; McDowell et al. 1990; Bruck 2005; Ekesi et al. 2002; Bruck et al. 2005). The exact value, however, depends on the specific soil characteristics (sterile vs. nonsterile, organic content, texture, etc.); size and behavior of the insect; and specific environmental conditions (moisture, temperature). Data from numerous lab assays and field trials indicate an efficacious level is approximately 10^5–10^6 colony forming units (CFU) cm^{-3} or g^{-1} soil with better isolates. In broadcast application of spores with incorporation to a depth of 10 cm, the volume of the arena is 1×10^9 cm^3 ha^{-1}, requiring 10^{14}–10^{15} spores ha^{-1} at the previously mentioned levels. Where the target insect tends to be restricted to a specific location in the soil, e.g., neonate sugarbeet root maggot (Jaronski et al. 2005), requirements can be reduced by concentrating spore application to that specific arena. For example, a 10-cm-wide, banded application of spores in water, centered on the bases of seedling sugar beets, with a target soil penetration of 1 cm (the oviposition zone for the sugarbeet root maggot adult fly) reduces the arena volume to 1.64×10^7 cm^3 ha^{-1} for, potentially, a 100-fold reduction in spores needed per hectare for a given spore concentration in that zone. The distribution of spores in soil, however, is extremely heterogeneous, even with a thorough soil drench. The soil consists of a complex network of soil pores ranging from 5 to 500 μm wide, depending upon the soil texture and upon compaction. On a larger scale, cracks in dry soil will conduct spore suspensions into the larger spaces, leaving large zones devoid of spores. Also, spores do not readily move subsequent to deposition by the aqueous carrier except in the sandiest soils (Ignoffo et al. 1977a; Storey and Gardner 1988).

The fungal entomopathogen often has to persist until the target insects arrive in the specific arena. This persistence may have to range from a few days to a number of weeks or months. For example, efficacious titers of fungi applied at planting for control of corn rootworm must persist for about a month before the eggs of the insect hatch. Sugarbeet root maggots hatch 4–8 weeks after typical planting so that a fungus applied at planting must persist at efficacious levels for at least that long. Fungal persistence in the soil is very variable, from a few weeks (Storey et al. 1989) to more than 40 weeks (Kabaluk et al. 2007).

Application of fungi on granules change the numbers game. Presuming the granule is covered with conidia, a soil insect has merely to brush against it to acquire a large dose of spores. Furthermore, nutritive granules allow for the fungus to germinate, grow and resporulate increasing the titer of spores. A critical concentration of granules is still needed for acceptable efficacy. In replicated bioassays, using third-instar sugarbeet root maggot larvae in a clay soil and at optimal moisture and temperature for the fungus, four or more granules (corn grit granules, 0.5–1 mm diameter, coated with *M. anisopliae* conidia and having a titer of 1,400 granules g^{-1}) per cm^3 of soil were needed for >90% efficacy in laboratory bioassays (Jaronski et al. 2005). If such granules are applied broadcast and incorporated into the top 10 cm of soil, one would need 2,858 kg ha^{-1} to achieve four granules per cm^3 soil. Application amounts decrease to 351 kg granules ha^{-1} if granules are applied in a 15-cm band over the row (with 61 cm row spacing) and incorporated to a depth of 5 cm. The critical concentration of granules could be achieved at 1.9 kg ha^{-1} if the granules are applied in-furrow (essentially a band 1 cm wide, 1 cm thick). In furrow application may not properly intercept the target insects, however. The nature of the granule can also change the numbers game. For example, granular formulations of the newly discovered *M. anisopliae* microsclerotia (Jackson and Jaronski 2009; Jaronski and Jackson 2008, 2009) have a laboratory LC_{90} of 0.5 grains cm^{-3} soil against sugarbeet root maggot, possibly because of attraction (Jaronski and Jackson 2009). Thus, the above amounts needed per hectare would theoretically decrease by a factor of eight.

Effect of soil abiotic and biotic factors

Abiotic factors

Primary abiotic factors affecting the efficacy of fungal entomopathogens are soil texture (pore size distribution), temperature, and moisture. Other physical factors in soil—pH, cation exchange capacity, and inorganic salts—do not seem to have any important impact on fungal entomopathogen infectivity or persistence.

Soil texture and moisture interact in a complex relationship that also involves the size and movement behavior of the insect. Soil texture, particularly the size distribution of pore spaces, affects the infectivity of spores by evidently mediating physical contact. For example, the LC_{50} of *B. bassiana* IL116 for second-instar *Diabrotica undecimpunctata howardii* Barber (southern corn rootworm) ranged from 9.0×10^4 to 2.25×10^6 CFU g^{-1} soil in ten different soils all held at 25% field capacity (Jaronski 2007). There was no correlation with soil type and the differences were not due to differences in conidial viabilities. Similarly, Kabaluk et al. (2007) observed that efficacy of *M. anisopliae* for wireworms (Elateridae) differed significantly among sand, clay, and organic soils at the same moisture level. The moisture content of a soil interacts with soil texture to further complicate effects on efficacy. Infection and mortality of third instar sugarbeet root maggot larvae by *M. anisopliae* F52 were significantly affected by soil type and moisture in five soils and three moisture levels (Table 1) even though levels of fungus (CFU g^{-1} soil) were not different (Jaronski et al. 2005). To complicate matters, different fungal isolates may respond differently to different soil types and moisture levels (Jaronski 2007).

Just as in the foliar arena, temperature in the soil influences efficacy. Soil temperatures, at levels more than 5 cm below the surface, tend to be cool, especially in the mesic, frigid, and cryic regions, which encompass much of North America, the northern half of Europe and Asia, and the southern portion of South America. For example, in Tennessee (Lat. 34–36°N), soil temperatures at 10 cm depth generally are above 15°C only between Julian Day 110 and 270, while in Oregon and southern Michigan (Lat. 41–44°N) that period is Julian Day 150–280 (data drawn from Zheng et al. 1993). In tropical and subtropical regions, soil temperatures are higher and usually within the optimal range for most fungal entomopathogens. Low soil temperatures often prolong the duration before mortality from mycosis is achieved. For example, in evaluating a *M. anisopliae* for the control of the pasture scarab beetle, *Adoryphorous couloni* Burmeister, in Tasmania, Rath et al. (1995), observed that LT_{50} values increased from 36 to 189 days when the treated insects were incubated at 5 vs. 15°C. In a Canadian study, wireworms exposed to *M. anisopliae* and incubated at 12°C escaped infection for over 60 days whereas those incubated at 18°C suffered considerable mortality from mycosis (Kabaluk and Ericsson 2007). The wireworms also needed at least an initial 48 h exposure at 18°C for fatal infection to occur, presumably to allow conidia to germinate and the fungi to penetrate the wireworm cuticle. Soil temperatures in cooler latitudes can thus be a major factor in timely efficacy and extensive selection of appropriate isolates are warranted. Most of the fungal entomopathogens studied, e.g., Fargues et al. (1997), display slowed germination and growth at temperatures below 15°C although there are isolates with better tolerance to cooler temperatures (Bidochka et al. 2001). Temperature tolerances must be contrasted with not only regional and seasonal soil

Table 1 Mortality of third instar sugarbeet root maggot (SBRM) after two weeks of exposure to *Metarhizium anisopliae* F52 (2.5×10^6 conidia g^{-1} dry soil), in six soils at three moisture levels (10, 15, 30% field saturation for each soil type)

Soil	Snd:Cl:Slt	FS (%)	SBRM mortality	
			Mean (%)	SD (%)
Clay	11:56:33	10	11	4c
		15	44	10b
		30	84	12a
Clay	16:51:33	10	22	4c
		15	9	10c
		30	98	4a
Loam	35:19:46	10	56	21b
		15	100	0a
		30	100	0a
Clay loam	39:31:30	10	5	4c
		15	100	0a
		30	98	4b
Sandy clay loam	56:21:23	10	47	18b
		15	80	9a
		30	58	10b
Sandy loam	75:13:12	10	100	0a
		15	100	0a
		30	98	4a

Snd:Slt:Cl is the sand:silt:clay ratio for each soil. Data are means and SD of three replicate assays each with three replicates per treatments. In all cases the numbers of colony forming units g^{-1} soil were not significantly different from each other at start and end of each assay. Mean mortalities followed by different letters are significantly different (Tukey's HSD test, $P = 0.05$)

temperatures but also temperatures in the specific soil arena and time of use.

Agricultural inputs (fertilizer, pH modifiers, pesticides) and practices can have major impacts on soil microbial and macrobial populations (Stewart 1991). However, there have been very few in situ studies with fungal entomopathogens. Most studies have focused on correlation between fungal entomopathogen titers and agricultural practices. There have been very few manipulative in situ studies where agricultural inputs are controlled variables. See Jaronski (2007) and Klingen and Haukeland (2006) for more information on this topic.

Agrichemicals, especially fungicides, can have direct bearing on fungal persistence. Most studies have been concerned with in vitro fungal-pesticide interactions. Using an agar incorporation approach is useful in identifying harmless pesticides but may imply false adverse effects from a particular chemical. One must remember that for the most part, conidia in soil remain ungerminated; germination on an insect's cuticle may be isolated from the effects of a soil pesticide; and, once inside an insect, a fungus may well be insulated from adverse effects of a pesticide in the soil. There is a lack of realistic, in situ studies that examine potential interference of agrochemicals in fungal entomopathogen efficacy. Often the best, most realistic approach is within the context of a field trial or at least outdoor, in-field microcosm. For instance, it was demonstrated that of 13 fungicides toxic in vitro, none had adverse impact on *M. anisopliae* in commercial potting media under realistic conditions, even when applied twice during the observation period (Bruck 2009a).

Biotic factors

Biotic factors, primarily soil microbiota, are important particularly with regards to persistence of fungi. A typical soil can contain 10^8–10^9 bacteria, "several metres of fungi", 10^5 soil protozoa, 10–20 nematodes, and 0–100 arthropods per gram (Tugel et al. 2000). In general, most natural soils exhibit a fungistasis for fungal entomopathogen conidia (e.g., Pereira et al. 1993) as well as other fungi (Stotzky 1972). This fungistasis is removed by soil sterilization, after which fungal titers can increase by several orders of magnitude. In one of the few manipulative experiments, the fungistatic effect of a natural and a synthetic soil was destroyed by autoclaving and restored by inoculation with bacteria, actinomycetes or fungi (Ho and Ko 1986). Interestingly, soil fungistasis does not operate on fungal entomopathogens when they are on nutritive granules or *M. anisopliae* microsclerotia (e.g., Jaronski and Jackson 2008, 2009). This ostensible anomaly indicates that fungistasis may be due more to nutritional factors than antibiosis. The typical subsequent trend in fungal titers is a decline in CFU, the rapidity of which depends on a number of as yet incompletely understood factors. What is known about the effect of microflora underscores the complexity of relationships in time and space (Jaronski 2007). Jaronski et al. (2007) examined the in vitro interactions between 30 sugarbeet rhizoplane bacteria and each of three isolates of *B. bassiana* and *M. anisopliae*. There were qualitative differences among the fungal species and isolates in their response to the various bacteria. A general trend appeared to be greater inhibition of conidial germination by Gram negative (G−) than Gram positive (G+) species. Hyphal growth of the fungi was generally not inhibited by any of the bacteria. More G+ bacteria were inhibited by *M. ansiopliae* than by *B. bassiana*, and fewer G− bacteria were inhibited by either fungus. It should be added that their in vitro observations may not be necessarily reflected in vivo. Certainly, the availability of nutrients in vivo can be much lower and more tightly restricted to minute foci within the soil. At the same time, soil fungistasis does not seem to be completely effective because soil insects do become infected in nature.

When fungi are applied as a seed coat, the key requirement is that the fungi colonize the growing root system and subsequently sporulate. Using green fluorescent protein (gfp)-labeled *M. anisopliae,* Hu and St Leger (2002), observed rhizosphere colonization by *M. anisopliae* in field plots. Bruck (2005) indirectly observed a higher titer of *M anisopliae* in *Picea abies* (L.) H. Karst rhizosphere than in the bulk potting medium, although the observations are complicated by use of fungus on rice grain spent substrate and peat- or bark-based media rather than soil. Using gfp transformants, Jaronski et al. (2007) observed that several isolates of *M. anisopliae* and *B. bassiana* could only colonize the rhizoplane of young sugarbeet seedlings in vitro in an agar-based system. In gnotobiotic media—sterile clay soil, sterile potting mix, vermiculite + 10% Hoagland's Solution—rhizoplane colonization was not observed, regardless of whether conidia were applied to the seed coat or added to the medium itself and seeds thereafter added. Root colonization was also not observed with chard (*Beta vulgaris* var. *cicla*), bean, or maize (*Zea mays* L.) seedlings. A prerequisite for colonization is conidial germination in the rhizosphere or on the rhizoplane. Jaronski et al. (2007) determined that conidial germination was almost nonexistent in root exudate of two-leaf sugar beets, but reached about 50% after 24 h in exudate from four-leaf sugar beets, cabbage and chard. In contrast, germination was >95% in oat, rye, or bean root exudate, as well as in 1% neopeptone, or in Sabouraud dextrose broth. The subject of fungal entomopathogens in the rhizosphere is treated further by Bruck (2009b).

Among the Protistan microfauna, soil amoebae have the potential to reduce fungal levels by direct mycophagy (Bryant et al. 1982). Several species of amoeba have the potential for direct ingestion of conidia but the fate of such conidia is not known. Members of the Vampyrellidae are known to perforate spores of plant pathogenic fungi (Anderson and Patrick 1985). They may also have the potential of attacking fungal entomopathogen conidia and mycelium. Soil mesofauna, such as Collembola, orbatid and prostigmatid mites, can also potentially affect fungal titer in the soil by feeding. Collembola have been proposed as biocontrol agents of plant pathogenic fungi (Curl 1988; Lartey et al. 1994). These insects seem to be somewhat refractory to infection by at least several isolates of fungal entomopathogens, and conidia were attractive to three species of Collembola (Dromph and Vestergaard 2002). Collembolan grazing can suppress *Rhizoctonia solani* Kühn and *R. cerealis* Hoeven, both in laboratory and field situations (Shiraishi et al. 2003 cited in Friberg et al. 2005). Effects on entomopathogenic titers in soil are not known. Much less is known about the impact of soil mites. I refer the interested reader to Friberg et al. (2005) for an introduction to the subject.

Summary and closing thoughts

Inundative use of fungal entomopathogens seeks to overcome by sheer numbers many of their disadvantages as classical biocontrol agents and the many environmental factors that mitigate their impact on a target insect population under natural conditions. Inundation can work to create transient epizootics to manage an insect population. To do so consistently, practicably, and within economic constraints is the challenge. We have learned a lot about how the key environmental variables of temperature, moisture and ultraviolet light in the foliar arena, and temperature, moisture and soil characteristics in the soil arena can affect the success or failure of these fungi. The bulk of studies have focused on one or a few variables at a time, for example as spore mortality factors, very often in the laboratory, not in real situations. That is a start. These variables interact, however, producing complex, dynamic effects on the spores. The existing body of knowledge is only a beginning to our understanding.

Development of models incorporating multiple variables is critically needed to better understand the many factors that operate together to affect efficacy of a fungal entomopathogen. Efforts have been made in this regard: Pinnock and Brand (1980) and Brand and Pinnock (1980) in a general sense; Galaini (1984) regarding Colorado potato beetle in potatoes; Yang et al. (1997) with the citrus rust mite; Feng et al. (1985) concerning European corn borer; Knudsen and Schotzko (1999) regarding modeling *B. bassiana* epizootics in Russian wheat aphid; Boulard et al. (2002) and Vidal et al. (2003) with regards to whiteflies in glasshouse tomatoes; Klass et al. (2007a, b) regarding climate suitability for locust control by *M. acridum*; and Polar et al. (2008) with ticks on livestock. Hesketh et al. (2009) further address models of fungi in natural populations of insects. The models should not be an end to themselves, but rather serve as tools to refine our understanding of environmental variables in a holistic perspective, comprehend how and when the fungi can work in an inundative use, and inspire methods to enhance efficacy. Further, the lessons from the models must be taken into operational situations.

We need to reconsider the best use of inundative release. These organisms are not chemicals. The chemical paradigm often involves use of insecticides, by themselves, once an outbreak has occurred and reached, or passed, the economic injury level. This "fire extinguishing" philosophy is not appropriate for fungal entomopathogens. The fungi work "too slowly", and repeated applications at short intervals must often be made. The best use of mycoinsecticides is pest outbreak prevention with applications targeting the first immigrants into the crop to prevent pest population establishment and increase to the economic injury level. This has become a guiding philosophy of at least one biopesticide company in advising growers how to use its fungus (Bioworks Inc 2008). A prime example of using fungal entomopathogens in a proactive, preventative, rather than reactive, strategy is deployment of *M. acridum* against African locust (Lecoq 2004; Showler 1997).

Lastly, we need to integrate these organisms with other tools, such as predators and parasites, and cultural practices, to create sustainable, biologically based systems, where possible, and not use the fungi by themselves. For example, predators and parasites can act efficiently on the survivors during mycopesticide use. In the glasshouse environment, combined use of a mycopesticide, parasites and predators has been a viable option (Labbe et al. 2009). The fungi can also be combined with other microbial pest control agents, such as *B. thuringiensis tenebrionis* to overcome the disadvantages of each, such as for Colorado potato beetle (Wraight and Ramos 2005). Only through intelligent use of fungal entomopathogens will they become significant tools for farmers.

References

Al-Mazra'Awi MS, Shipp L, Broadbent B, Kevan P (2006a) Biological control of *Lygus lineolaris* (Hemiptera: Miridae) and *Frankliniella occidentalis* (Thysanoptera: Thripidae) by *Bombus impatiens* (Hymenoptera: Apidae) vectored *Beauveria bassiana* in greenhouse sweet pepper. Biol Control 37(1):89–97

Al-Mazra'Awi MS, Shipp L, Broadbent B, Kevan P (2006b) Dissemination of *Beauveria bassiana* by honey bees (Hymenoptera: Apidae) for control of tarnished plant bug (Hemiptera: Miridae) on canola. Environ Entomol 35(6): 1569–1577

Altre JA, Vandenberg JD (2001) Factors influencing the infectivity of isolates of *Paecilomyces fumosoroseus* against diamondback moth, *Plutella xylostella*. J Invertebr Pathol 78(1):31–36

Alves RT, Bateman RP, Prior C, Leather SR (1998) Effects of simulated solar radiation on conidial germination of *Metarhizium anisopliae* in different formulations. Crop Prot 17(8):675–679

Anderson TR, Patrick ZA (1985) Mycophagous amoeboid organisms from soil that perforate spores of *Thielaviopsis basicola* and *Cochliobolus sativus*. Phytopathology 68: 1618–1626

Andrews JH, Buck JW (2002) Adhesion of yeasts to leaf surfaces. In: Lindow SE, Hect-Porter EI, Elliott VJ (eds) Phyllosphere microbiology. American Phytopathological Society, St Paul MN, pp 53–68

Arysta LifeScience (2009) Evito product information. Available via http://www.arystalifescience.com/default.asp?V_DOC_ID=1758 Cited 22 Sep 2009

Bateman RP, Matthews GA, Hall FA (2007) Ground-based application equipment. In: Lacey LA, Kaya HK (eds) Field manual of techniques in invertebrate pathology, 2nd edn. Springer, New York, pp 73–98

Baverstock J, Roy HE, Pell JK (2009) Entomopathogenic fungi and insect behavior: from unsuspecting hosts to targeted vectors. BioControl. doi 10.1007/s10526-009-9238-5 (this SI)

Beattie GA (2002) Leaf surface waxes and the process of leaf colonization by microorganisms. In: Lindow SE, Hect-Porter EI, Elliott VJ (eds) Phyllosphere microbiology. American Phytopathological Society, St Paul MN, pp 3–26

Behle RW (2006) Importance of direct spray and spray residue contact for infection of *Trichoplusia* in larvae by field applications of *Beauveria bassiana*. J Econ Entomol 99(4):1120–1128

Bidochka MJ, Kamp AM, Lavender TM, Dekoning J, deCroos JNA (2001) Habitat association in two genetic groups of the insect-pathogenic fungus *Metarhizium anisopliae*: uncovering cryptic species? Appl Environ Microbiol 67(3):1335–1342

Bioworks Inc (2008) Aphid management. BotaniGard ES Technical Bulletin B-Aphids/2-09. Available via http://www.bioworksbiocontrol.com/botanigard/pdf/BotaniGard%20AphidMgtTechBulletin.pdf. Cited 5 Sep 2009

Bonan GB (2008) Ecological climatology concepts and application, 2nd edn. Cambridge University Press, Cambridge

Boulard T, Mermier M, Fargues J, Smits N, Rougier M, Roy JC (2002) Tomato leaf boundary layer climate: implications for microbiological whitefly control in greenhouses. Agr Forest Metereol 110:159–176

Braga GUL, Flint SD, Miller CD, Anderson AJ, Roberts DW (2001a) Both solar UVA and UVB radiation impair conidial culturability and delay germination in the entomopathogenic fungus *Metarhizium anisopliae*. Photochem Photobiol 74(5):734–739

Braga GUL, Flint SD, Miller CD, Anderson AJ, Roberts DW (2001b) Variability in response to UV-B among species and strains of *Metarhizium* isolated from sites at latitudes from 61°N to 54°S. J Invertebr Pathol 78:98–108

Brand RJ, Pinnock DE (1980) Application of biostatistical modeling to forecasting the results of microbial control trials. In: Burges HD (ed) Microbial control of pests and plant diseases 1970–1980. Academic Press, London, pp 667–693

Brinkman MA, Gardner WA (2000) Enhanced activity of *Beauveria bassiana* to red imported fire ant workers (Hymenoptera: Formicidae) infected with *Thelohania solenopsae*. J Agric Urban Entomol 17(4):191–195

Bruck DJ (2005) Ecology of *Metarhizium anisopliae* in soilless potting media and the rhizosphere: implications for pest management. Biol Control 32:155–162

Bruck DJ (2009a) Impact of fungicides on *Metarhizium anisopliae* in the rhizosphere, bulk soil and *in vitro*. Biocontrol 54:597–606

Bruck D (2009b) Fungal entomopathogens in the rhizosphere. BioControl. doi 10.1007/s10526-009-9236-7 (this SI)
Bruck DJ, Lewis LC (2002) Rainfall and crop residue effects on soil dispersion and *Beauveria bassiana* spread to corn. Appl Soil Ecol 20(3):183–190
Bruck DJ, Snelling JE, Dreves AJ, Jaronski SJ (2005) Laboratory bioassays of entomopathogenic fungi for the control of *Delia radicum* (L.) larvae. J Invertebr Pathol 89:179–183
Bryant RJ, Woods LE, Coleman DC, Fairbanks BC, McClellan JF, Cole CV (1982) Interactions of bacterial and amebal populations in soil microcosms with fluctuating moisture content. Appl Environ Microbiol 43(4):747–752
Bugeme DM, Knapp M, Boga HI, Wanjoya AK, Maniania NK (2009) Influence of temperature on virulence of fungal isolates of *Metarhizium anisopliae* and *Beauveria bassiana* to the two-spotted spider mite *Tetranychus urticae*. Mycopathologia 167(4):221–227
Burges HD (1998) Formulation of mycoinsecticides. In: Burges HD (ed) Formulation of microbial biopesticides. Kluwer Academic Publishers, Dordrecht, pp 7–30
Burrage SW (1971) The micro-climate at the leaf surface. In: Preece TF, Dickinson CH (eds) Ecology of leaf surface micro-organisms. Academic Press, New York, pp 91–101
Carruthers RI, Larkin TS, Firstencel H (1992) Influence of thermal ecology on the mycosis of a rangeland grasshopper. Ecology 73(1):190–204
Chapple AC, Downer RA, Bateman RP (2007) Theory and practice of microbial insecticide application. In: Lacey LA, Kaya HK (eds) Field manual of techniques in invertebrate pathology, 2nd edn. Springer, Dordrecht, pp 9–34
Charnley AK (1989) Mechanisms of fungal pathogenesis in insects. In: Whipps JM, Lumsden RD (eds) Biotechnology of fungi for improving plant growth. Cambridge Univ Press, Cambridge, pp 85–125
Chu CC, Akey DH, Henneberry TJ (1994) Response of cotton plants to aldicarb soil application. J Agr Assoc China 167:75–90
Clark RA, Casagrande RA, Wallace DB (1982) Influence of pesticides on *Beauveria bassiana*, a pathogen of the Colorado potato beetle. Environ Entomol 11(1):67–70
Cohen E, Joseph T (2009) Photostabilization of *Beauveria bassiana* conidia using anionic dyes. Appl Clay Sci 42(3–4):569–574
Cory J, Ericsson JD (2009) Fungal entomopathogens in a tritrophic context. BioControl. doi 10.1007/s10526-009-9247-4 (this SI)
Cory JS, Hoover K (2006) Plant-mediated effects in insect-pathogen interactions Trends. Ecol Evol 21(5):278–286
Costa SD, Gaugler R (1989) Influence of *Solanum* host plants on Colorado potato beetle (Coleoptera: Chrysomelidae) susceptibility to the entomopathogen, *Beauveria bassiana*. Environ Entomol 18:531–536
Curl EA (1988) The role of soil macrofauna in plant disease suppression. CRC Crit Rev Pl Sci 7:175–198
da Silva RZ, Neves PMOJ (2005) Techniques and parameters used in compatibility tests between *Beauveria bassiana* (Bals) Vuill and *in vitro* phytosanitary products. Pest Manage Sci 61(7):667–674
De Croos JNA, Bidochka MJ (1999) Effects of low temperature on growth parameters in the entomopathogenic fungus *Metarhizium anisopliae*. Can J Microbiol 45(12):1055–1061
Devi KU, Sridevi V, Mohan CM, Padmavathi J (2005) Effect of high temperature and water stress on *in vitro* germination and growth in isolates of the entomopathogenic fungus *Beauveria bassiana* (Bals.) Vuillemin. J Invertebr Pathol 88(3):181–189
Down RE, Cuthbertson AGS, Mathers JJ, Walters KFA (2009) Dissemination of the entomopathogenic fungi, *Lecanicillium longisporum* and *L. muscarium*, by the predatory bug, *Orius laevigatus*, to provide concurrent control of Myzus persicae, *Frankliniella occidentalis* and *Bemisia tabaci*. Biol Control 50(2):172–178
Dromph KM (2003) Collembolans as vectors of entomopathogenic fungi. Pedobiologia 47(3):245–256
Dromph KM, Vestergaard S (2002) Pathogenicity and attractiveness of entomopathogenic hyphomycete fungi to collembolans. Appl Soil Ecol 21(3):197–210
Dubois T, Li Z, Jiafu H, Hajek AE (2004) Efficacy of fiber bands impregnated with *Beauveria brongniartii* cultures against the Asian longhorned beetle, *Anoplophora glabripennis* (Coleoptera: Cerambycidae). Biol Control 31(3):320–328
Edgington S, Segura H, Del a Rosa W, Williams T (2000) Photoprotection of *Beauveria bassiana*: testing simple formulations for control of the coffee berry borer. Intl J Pest Mgt 46(3):169–176
Eilenberg J, Hajek A, Lomer C (2001) Suggestions for unifying the terminology in biological control. BioControl 46(4):387–400
Ekesi S, Maniana NK, Lux SA (2002) Mortality in three African tephritid fruit fly puparia and adults caused by the entomopathogenic fungi, *Metarhizium anisopliae* and *Beauveria bassiana*. Biocont Sci Technol 12(1):7–17
Fargues J, Luz C (2000) Effects of fluctuating moisture and temperature regimes on the infection potential of *Beauveria bassiana* for *Rhodnius prolixus*. J Invertebr Pathol 75(3):202–211
Fargues J, Maniania NK (1992) Variabilité de la sensibilité de *Spodoptera littoralis* (Lep; Noctuidae) à l'hyphomycète entomopathogène *Nomuraea riley*. Entomophaga 37:545–554
Fargues J, Goettel MS, Smits N, Ouedraogo A, Vidal C, Lacey LA, Lomer CJ, Rougier M (1996) Variability in susceptibility to simulated sunlight of conidia among isolates of entomopathogenic hyphomycetes. Mycopathologia 135(3):171–181
Fargues J, Goettel MS, Smits N, Ouedraogo A, Rougier M (1997) Effect of temperature on vegetative growth of *Beauveria bassiana* isolates from different origins. Mycologia 89(3):383–392
Faria MR, Wraight SP (2007) Mycoinsecticides and mycoacaricides: a comprehensive list with worldwide coverage and international classification of formulation types. Biol Control 43:237–256
Feng Z, Carruthers RL, Roberts DW, Robson DS (1985) Age-specific dose-mortality effects of *Beauveria bassiana* on the European corn borer, *Ostrinia nubilalis*. J Invertebr Pathol 46:259–264
Feng MG, Poprawski TJ, Khachatourians GG (1994) Production, formulation, and application of the entomopathogenic

fungus *Beauveria bassiana* for insect control: current status. Biocont Sci Technol 4:3–34
Fernandes ÉKK, Rangel DEN, Moraes ÁML, Bittencourt VREP, Roberts DW (2007) Variability in tolerance to UV-B radiation among *Beauveria* spp. isolates. J Invertebr Pathol 96(3):237–243
Ferro DN, Chapman RB, Penman DR (1979) Observations on insect microclimate and insect pest management. Environ Entomol 8:1000–1003
Ferron P (1977) Influence of relative humidity on the development of fungal infection caused by *Beauveria bassiana* (Fungi Imperfecti, Moniliales) in imagines of *Acanthoscelides obtectus* (Col.: Bruchidae). Entomophaga 22(4):393–396
Ferron P (1981) Pest control by the fungi *Beauveria* and *Metarhizium*. In: Burges HD (ed) Microbial control of pests and plant diseases. Academic Press, London, pp 465–481
Friberg H, Lagerlöf J, Rämert B (2005) Influence of soil fauna on fungal plant pathogens in agricultural and horticultural systems. Biocont Sci Technol 15(7):641–658
Furlong MJ, Groden E (2003) Starvation induced stress and the susceptibility of the Colorado potato beetle, *Leptinotarsa decemlineata*, to infection by *Beauveria bassiana*. J Invertebr Pathol 83(2):127–138
Furlong MJ, Pell JK (2005) Interactions between entomopathogenic fungi and arthropod natural enemies. In: Vega FE, Blackwell M (eds) Insect-fungal associations: ecology and evolution. Oxford University Press, New York, pp 51–73
Fuxa JR (1987) Ecological considerations for the use of entomopathogens in IPM. Ann Rev Entomol 32:225–251
Galaini, S (1984) The efficacy of foliar applications of *Beauveria bassiana* conidia against *Leptinotarsa decemlineata.* MS thesis, Cornell University, Ithaca
Gardner WA, Sutton RM, Noblet R (1977) Persistence of *Beauveria bassiana*, *Nomuraea rileyi* and *Nosema necatrix* on soybean foliage. Environ Entomol 6:616–618
Gates DM (1968) Transpiration and leaf temperature. Rev Plant Physiol 19:211–238
Goettel MS, Eilenberg J, Glare T (2005) Entomopathogenic fungi and their role in regulation of insect populations. In: Gilbert LI, Iatrou K, Gill S (eds) Comprehensive molecular insect science. Elsevier, Amsterdam, pp 361–406
Golebiowski M, Malinski E, Bogus MI, Kumirska J, Stepnowski P (2008) The cuticular fatty acids of *Calliphora vicina*, *Dendrolimus pini* and *Galleria mellonella* larvae and their role in resistance to fungal infection. Insect Biochem Mol Biol 38(6):619–627
Hajek A, Delalibera I (2009) Fungal pathogens as classical biological control agents against arthropods. BioControl. doi 10.1007/s10526-009-9253-6 (this SI)
Hajek AE, McManus ML, Delalibera I (2005) Catalogue of introductions of pathogens and nematodes for the classical biological control of insects and mites. US Dept Agric For Serv, For Health Technol Enterprise Team 2005-05
Hare JD, Andreadis TG (1983) Variation in the susceptibility *of Leptinotarsa decemlineata* (Coleoptera: Chrysomelidae) when reared on different host plants to the fungal pathogen *Beauveria bassiana* in the field and in the laboratory. Environ Entomol 12:1892–1897
Hesketh H, Roy H, Eilenberg J, Pell J, Hails R (2009) Challenges in modelling complexity of fungal entomopathogens in semi-natural populations of insects. BioControl. doi 10.1007/s10526-009-9249-2 (this SI)
Ho WC, Ko WH (1986) Microbiostasis by nutrient deficiency shown in natural and synthetic soils. J Gen Microbiol 132(10):2807–2815
Hu G, St Leger RJ (2002) Field studies of a recombinant mycoinsecticide (*Metarhizium anisopliae*) reveal that it is rhizosphere competent. Appl Environ Microbiol 68: 6383–6387
Hunt TR, Moore D, Higgins PM, Prior C (1994) Effect of sunscreens, irradiance and resting periods on the germination of *Metarhizium flavoviride conidia.* Entomophaga 39(3/4):313–322
Hunter DM, Milner RJ, Spurgin PA (2001) Aerial treatment of the Australian plague locust, *Chortoicetes terminifera* (Orthoptera: Acrididae) with *Metarhizium anisopliae* (Deuteromycotina: Hyphomycetes). Bull Entomol Res 91:93–99
Ibrahim L, Butt TM, Beckett A, Clark SJ (1999) The germination of oil-formulated conidia of the insect pathogen, *Metarhizium anisopliae*. Mycol Res 103(7):901–907
Ignoffo C, Garcia C (1992) Influence of conidial color on inactivation of several entomogenous fungi (Hyphomycetes) by simulated sunlight. Environ Entomol 21(4): 913–917
Ignoffo CM, Garcia C, Hostetter DL, Pinnell RE (1977a) Vertical movement of *Nomuraea rileyi* through sand and loam soil. J Econ Entomol 70:163–164
Ignoffo CM, Hostetter DL, Sikorowski PP, Sutter G, Brooks WM (1977b) Inactivation of representative species of entomopathogenic viruses, a bacterium, fungus and protozoan by an ultraviolet light source. Environ Entomol 6:411–415
Ihara F, Toyama M, Mishiro K, Yaginuma K (2008) Laboratory studies on the infection of stink bugs with *Metarhizium anisopliae* strain FRM515. Appl Entomol Zool 43(4):503–509
Inglis GD, Goettel MS, Johnson DL (1993) Persistence of the entomopathogenic fungus, *Beauveria bassiana* on phylloplanes of crested wheatgrass and alfalfa. Biol Control 3:258–270
Inglis GD, Goettel MS, Johnson DL (1995a) Influence of ultraviolet light protectants on persistence of the entomopathogenic fungus, *Beauveria bassiana*. Biol Control 5:581–590
Inglis GD, Johnson DL, Goettel MS (1995b) Effects of simulated rain on the persistence of *Beauveria bassiana* on leaves of alfalfa and wheat. Biocont Sci Technol 5(3):365–369
Inglis GD, Johnson DL, Goettel MS (1996) Effects of temperature and thermoregulation on mycosis by *Beauveria bassiana* in grasshoppers. Biol Control 7(2):131–139
Inglis GD, Johnson DL, Goettel MS (1997a) Effects of temperature and sunlight on mycosis (*Beauveria bassiana*) of grasshoppers under field conditions. Environ Entomol 26:400–409
Inglis GD, Johnson DL, Goettel MS (1997b) Field and laboratory evaluation of two conidial batches of *Beauveria bassiana* (Balsamo) Vuillemin against grasshoppers. Can Entomol 129:171–186

Inglis GD, Duke GM, Kawchuk LM, Goettel MS (1999) Influence of oscillating temperatures on the competitive infection and colonization of the migratory grasshopper by *Beauveria bassiana* and *Metarhizium flavoviride*. Biol Control 14(1):111–120

Inglis GD, Ivie TJ, Duke GM, Goettel MS (2000) Influence of rain and conidial formulation on persistence of *Beauveria bassiana* on potato leaves and Colorado potato beetle larvae. Biol Control 18:55–64

Inglis GD, Goettel MS, Butt TM, Strasser H (2001) Use of hyphomycetous fungi for managing insect pests. In: Butt TM, Jackson C, Magan N (eds) Fungi as biocontrol agents: progress problems and potential. CABI Publishing, Wallingford, pp 23–70

Inyang EN, Butt TM, Beckett A, Archer S (1998) The effect of plant growth and topography on the acquisition of conidia of the insect pathogen *Metarhizium anisopliae* by larvae of *Phaedon cochleariae*. Mycol Res 102(11):1365–1374

Inyang EN, Butt TM, Beckett A, Archer S (1999) The effect of crucifer epicuticular waxes and leaf extracts on the germination and virulence of *Metarhizium anisopliae* conidia. Mycol Res 103(4):419–426

Inyang EN, McCartney HA, Oyejola B, Ibrahim L, Pye BJ, Archer SA, Butt TM (2000) Effect of formulation, application and rain on the persistence of the entomogenous fungus *Metarhizium anisopliae* on oilseed rape. Mycol Res 104(6):653–661

Jackson MA, Jaronski ST (2009) Production of microsclerotia of the fungal entomopathogen *Metarhizium anisopliae* and their use as a biocontrol agent for soil-inhabiting insects. Mycol Res 113(8):842–850

Jackson MA, Dunlap CA, Jaronski ST (2009) Ecological considerations in producing and formulating fungal entomopathogens for use in insect biocontrol. BioControl. doi 10.1007/s10526-009-9240-y (this SI)

James RR, Croft BA, Shaffer BT, Lighthart B (1998) Impact of temperature and humidity on host-pathogen interactions between *Beauveria bassiana* and a coccinellid. Environ Ent 27(6):1506–1513

Jaronski ST (1997) New paradigms in formulating mycoinsecticides. In: Goss GR, Hopkinson MJ, Collins HM (eds) Pesticide formulations and applications systems: 17th Volume, ASTM STP 1328. American Society for Testing and Materials, Philadelphia, pp 99–112

Jaronski ST (2007) Soil ecology of the entomopathogenic ascomycetes: a critical examination of what we (think) we know. In: Maniana K, Ekesi S (eds) Use of entomopathogenic fungi in biological pest management. Research SignPosts, Trivandrum India, pp 91–144

Jaronski ST, Jackson MA (2008) Efficacy of *Metarhizium anisopliae* microsclerotial granules. Biocont Sci Technol 18(8):849–863

Jaronski ST, Jackson MA (2009) Further Progress with *Metarhizium* microsclerotial production. IOBC/WPRS Bull 45:275–278

Jaronski ST, Rosinska J, Brown C, Osterlind R, Staten R, Craft R, Antilla L (1997) Impact of *Beauveria bassiana* Mycotech strain GHA, buprofezin, and pyriproxyfen on whitefly predators in Arizona cotton. In: Silverleaf whitefly 1997 supplement to the 5-Year National research and action plan: progress review, technology transfer and new research and action plan (1997–2001). p 49

Jaronski ST, Lord J, Rosinka J, Bradley C, Hoelmer K, Simmons G, Osterlind R, Brown C, Staten R, Antilla L (1998) Effect of *Beauveria bassiana*-based mycoinsecticide on beneficial insects under field conditions. In: The 1998 Brighton conference pests and diseases, British crop protection council Brighton, vol 2. pp 651–657

Jaronski ST, Grace JA, Schlothauer R (2005) *Metarhizium anisopliae* for biocontrol of sugarbeet root maggot: constraints and challenges. Proc 33rd Bienn Mtg Amer Soc Sugar Beet Technol, pp 185–187

Jaronski ST, Fuller-Schaeffer C, Jung K, Boetel M, Majumdar A (2007) Challenges in using *Metarhizium anisopliae* for biocontrol of sugarbeet root maggot, *Tetanops myopaeformis*. Bull IOBC/WPRS 30(7):119–124

Johnson DL, Goettel MS (1993) Reduction of grasshopper populations following field application of the fungus *Beauveria bassiana*. Biocont Sci Technol 3:165–175

Johnson DL, Goettel MS, Bradley C, van der Paaw H, Maiga B (1992) Field trials with the entomopathogenic fungus *Beauveria bassiana* against grasshoppers in Mali, West Africa, July 1990. In: Lomer CJ, Prior C (eds) Biological control of locusts and grasshoppers. CAB International, Wallingford, pp 296–310

Kabaluk JT, Ericsson JD (2007) Environmental and behavioral constraints on the infection of wireworms by *Metarhizium anisopliae*. Environ Entomol 36(6):1415–1420

Kabaluk T, Goettel M, Ericsson J, Erlandson M, Vernon B, Jaronski S, Mackenzie K, Cosgrove L (2007) Promise versus performance: Working toward the use of *Metarhizium anisopliae* as a biological control for wireworms. Bull IOBC/WPRS 30(7):69–76

Kalsbeek V, Mullens BA, Jespersen JB (2001) Field studies of Entomophthora (Zygomycetes: Entomophthorales)—induced behavioral fever in *Musca domestica* (Diptera: Muscidae) in Denmark. Biol Control 21(3):264–273

Klass JI, Blanford S, Thomas MB (2007a) Development of a model for evaluating the effects of environmental temperature and thermal behaviour on biological control of locusts and grasshoppers using pathogens. Agric For Entomol 9:189–199

Klass JI, Blanford S, Thomas MB (2007b) Use of a geographic information system to explore spatial variation in pathogen virulence and the implications for biological control of locusts and grasshoppers. Agric For Entomol 9:201–208

Klingen I, Haukeland S (2006) The soil as a reservoir for natural enemies of pest insects and mites with emphasis on fungi and nematodes. In: Eilenberg J, Hokkanen HMT (eds) An ecological and sociological approach to biological control. Springer, Dordrecht, pp 145–211

Knudsen GR, Schotzko DJ (1999) Spatial simulation of epizootics caused by *Beauveria bassiana* in Russian wheat aphid populations. Biol Control 16(3):318–326

Koppert (2009a). Mycotal product information. Available via http://www.koppert.com/pests/thrips/products-against-thrips/detail/mycotal-1/. Cited 26 Aug 2009

Koppert (2009b) Vertelac production information. Available via http://www.koppert.com/pests/aphid/products-against-aphids/detail/vertalec-2/. Cited 26 Aug 2009

Kouassi M, Coderre D, Todorova SI (2003) Effect of plant type on the persistence of *Beauveria bassiana*. Biocont Sci Technol 13(4):415–427

Labbe RM, Gillespie DR, Cloutier C, Brodeur J (2009) Compatibility of an entomopathogenic fungus with a predator and parasitoid in the biological control of greenhouse whitefly. Biocont Sci Technol 19(4):429–446

Lactin DJ, Johnson DL (1998) Environmental, physical, and behavioral determinants of body temperature in grasshopper nymphs (Orthoptera: Acrididae). Can Entomol 130(5):551–577

Lartey RT, Curl EA, Peterson CM (1994) Interactions of mycophagous Collembola and biological control fungi in the suppression of *Rhizoctonia solani*. Soil Biol Biochem 26(1):81–88

Laverlam International (2005) BotaniGard ES compatibility with fungicides. Tech Update Bull 98F1015A, Available via http://www.laverlamintl.com/pdf/BotaniCompiFungENG.pdf. Cited 1 Sep 2009

Lazzarini GMJ, Rocha LFN, Luz C (2006) Impact of moisture on *in vitro* germination of *Metarhizium anisopliae* and *Beauveria bassiana* and their activity on *Triatoma infestans*. Mycol Res 110(4):485–492

Lecoq M (2004) Vers une solution durable au problème du criquet pèlerin? Secheresse 15(3):217–224

Lecuona R, Clement JL, Riba G, Joulie C, Juarez P (1997) Spore germination and hyphal growth of *Beauveria* sp on insect lipids. J Econ Entomol 90(1):119–123

Lee A, Minho A, Yoon C-S, Yun T, Kim H-S, Yoo J-K (2002) Selection of a highly virulent *Verticillium lecanii* strain against *Trialeurodes vaporariorum* at various temperatures. J Microbiol Biotechnol 12(1):145–148

Lemaitre B, Hoffmann J (2007) The host defense of *Drosophila melanogaster*. Ann Rev Immunol 25:697–743

Li J, Feng M-G (2009) Intraspecific tolerance of *Metarhizium anisopliae* conidia to the upper thermal limits of summer with a description of a quantitative assay system. Mycol Res 113(1):93–99

Lobo Lima ML, Brito JM, Henry JE (1992) Biological control of grasshoppers in the Cape Verde Islands. In: Lomer CJ, Prior E (eds) Biological control of locusts and grasshoppers. CAB International, Wallingford, pp 287–295

Lockwood JA, Narisu, Schell SP, Lockwood DR (2001) Canola oil as a kairomonal attractant of rangeland grasshoppers: An economical liquid bait for insecticide formulation. Intl J Pest Mgt 47:185–194

Lord JC (2005) Low humidity, moderate temperature, and desiccant dust favor efficacy of *Beauveria bassiana* (Hyphomycetes: Moniliales) for the lesser grain borer, *Rhyzopertha dominica* (Coleoptera: Bruchidae). Biol Control 34(2):180–186

Luz C, Fargues J (1999) Dependence of the entomopathogenic fungus, *Beauveria bassiana*, on high humidity for infection of *Rhodnius prolixus*. Mycopathologia 146(1):33–41

Marcandier S, Khachatourians GG (1987) Susceptibility of the migratory grasshopper, *Melanoplus sanguinipes* (Fab.) (Orthoptera: Acrididae), to *Beauveria bassiana* (Bals.) Vuillemin (Hyphomycete): influence of relative humidity. Can Entomol 119:901–907

McCoy CW (1981) Pest control by the fungus *Hirsutella thompsonii*. In: Burges HD (ed) microbial control of pests and plant diseases 1970–1980. Academic Press, New York, pp 499–512

McCoy CW, Couch TL (1982) Microbial control of the citrus rust mite with the mycoacaricide, Mycar. Fla Entomol 65(1):116–126

McDowell JM, Funderburk JE, Boucias DG, Gilreath ME, Lynch RE (1990) Biological activity of *Beauveria bassiana* against lesser cornstalk borer, *Elasmopalpus lignosellus* (Lepidoptera; Pyralidiae) on both soil and leaf substrates. Environ Entomol 19(1):182–190

Mierzejewski K, Reardon RC, Thistle H, Dubois NR (2007) Conventional application equipment: aerial application. In: Lacey LA, Kaya HK (eds) Field manual of techniques in invertebrate pathology, 2nd ed. edn. Springer, Dordrecht, pp 99–126

Mietkiewski R, Gorski R (1995) Growth of selected entomopathogenic fungi species and isolates on media containing insecticides. Acta Mycol 30(1):27–33

Milner RJ (1997) *Metarhizium flavoviride* (FI985) as a promising mycoinsecticide for Australian acridids. Mem Ent Soc Can 171:287–300

Moore D, Bridge PD, Higgins PM, Bateman RP, Prior C (1993) Ultra-violet radiation damage to *Metarhizium flavoviride* conidia and the protection given by vegetable and mineral oils and chemical sunscreens. Ann Appl Environ Biol 122:605–616

Morris CE, Monier JM, Jacques MA (1997) Method for observing microbial biofilms directly on leaf surfaces and recovering them for isolation of culturable microorganisms. Appl Environ Microbiol 63:1570–1576

Muller C, Riederer M (2005) Plant surface properties in chemical ecology. J Chem Ecol 31(11):2621–2651

Müller U, Vogel P, Alber G, Schaub G (2008) The innate immune system of mammals and insects. Contrib Microbiol 15:21–44

Mullie W (2009) Does bird predation enhance the impact of Green Muscle for grasshopper control—experiences from cleared woodland in Central Senegal. Metaleptea special conference issue 10th international congress orthopterology. Available via http://140.247.119.138/OrthSoc/pdf/metaleptea_special_2009.pdf. cited 5 Sep 2009

Noma T, Strickler K (1999) Factors affecting *Beauveria bassiana* for control of lygus bug (Hemiptera: Miridae) in alfalfa seed fields. J Agricult Urban Ent 16(4):215–233

Noma T, Strickler K (2000) Effects of *Beauveria bassiana* on *Lygus hesperus* (Hemiptera: Miridae) feeding and oviposition. Environ Ent 29(2):394–402

Ouedraogo RM, Goettel MS, Brodeur J (2004) Behavioral thermoregulation in the migratory locust: a therapy to overcome fungal infection. Oecologia 138(2):312–319

Pell JD, Vandenberg JD (2002) Interactions Among the Aphid *Diuraphis noxia*, the Entomopathogenic Fungus *Paecilomyces fumosoroseus* and the Coccinellid *Hippodamia convergens*. Biocont Sci Technol 12(2):217–224

Pell JD, Hannam J, Steinkraus D (2009) Conservation biological control using fungal entomopathogens. BioControl. doi 10.1007/s10526-009-9245-6 (this SI)

Pereira RM, Stimac JL, Alves SB (1993) Soil antagonism affecting the dose response of workers of the red imported fire ant, *Solenopsis invicta*, to *Beauveria bassiana* conidia. J Invertebr Pathol 61:156–161

Pinnock DE, Brand RJ (1980) A quantitative approach to the ecology of the use of pathogens for insect control. In: Burges HD (ed) Microbial control of pests and plant diseases 1970–1980. Academic Press, London, pp 655–665

Polar P, Moore D, Kairo MTK, Ramsubhag A (2008) Topically applied myco-acaricides for the control of cattle ticks: overcoming the challenges. Exp Appl Acarol 46:119–148

Poprawski TJ, Carruthers RI, Speese J III, Vacek DC, Wendel LE (1997) Early-season applications of the fungus *Beauveria bassiana* and introduction of the hemipteran predator *Perillus bioculatus* for control of Colorado potato beetle. Biol Control 10(1):48–57

Poprawski TJ, Greenberg SM, Ciomperlik MA (2000) Effect of host plant on *Beauveria bassiana*- and *Paecilomyces fumosoroseus*-induced mortality of *Trialeurodes vaporariorum* (Homoptera: Aleyrodidae). Eviron Entomol 29(5):1048–1053

Ramoska WA (1984) The influence of relative humidity on *Beauveria bassiana* infectivity and replication in the Chinch bug, *Blissus leucopterus*. J Invertebr Pathol 43:389–394

Ramsay JA, Butler CG, Sang JH (1938) The humidity gradient at the surface of a transpiring leaf. J Exp Biol 15:255–265

Rath AC, Anderson GC, Worledge D, Koen TB (1995) The effect of low temperatures on the virulence of *Metarhizium anisopliae* (DAT F-001) to the subterranean scarab, *Adoryphorus couloni*. J Invertebr Pathol 65(2):186–192

Reddy NP, Khan PAA, Devi KU, Victor JS, Sharma HC (2008) Assessment of the suitability of Tinopal as an enhancing adjuvant in formulations of the insect pathogenic fungus *Beauveria bassiana* (Bals.) Vuillemin. Pest Manag Sci 64:909–915

Roberts DW, Campbell AS (1977) Stability of entomopathogenic fungi. Misc Publ Ent Soc Amer 10(3):19–76

Rostas M, Hilker M (2003) Indirect interactions between a phytopathogenic and an entomopathogenic fungus. Naturwissenschaften 90:63–67

Roy HE, Pell JK, Alderson PG (2001) Targeted dispersal of the aphid pathogenic fungus *Erynia neoaphidis* by the predator *Coccinella septempunctata*. Biocont Sci Technol 11(1):101–112

Sabbahi R, Merzouki A, Guertin C (2008) Efficacy of *Beauveria bassiana* (Bals.) Vuill. against the tarnished plant bug, *Lygus lineolaris* L., in strawberries. J Appl Entomol 132(2):124–134

Santiago-Alvarez C, Maranhao EA, Maranhao E, Quesada-Moraga E (2006) Host plant influences pathogenicity of *Beauveria bassiana* to *Bemisia tabaci* and its sporulation on cadavers. BioControl 51(4):519–532

Schuepp PH (1993) Tansley review no. 59, Leaf boundary layers. New Phytol 125:477–507

Scurlock JMO, Asner GP, Gower ST (2001) Worldwide historical estimates and bibliography of leaf area index, 1932–2000. ORNL Technical Memorandum TM-2001/268. Oak Ridge National Laboratory, Oak Ridge, Tennessee, USA

Shah PA, Pell JK (2003) Entomopathogenic fungi as biological control agents. Appl Microbiol Biotechnol 61(5/6):413–423

Shanley RP, Keena M, Wheeler MM, Leland J, Hajek AE (2009) Evaluating the virulence and longevity of non-woven fiber bands impregnated with *Metarhizium anisopliae* against the Asian longhorned beetle, *Anoplophora glabripennis* (Coleoptera: Cerambycidae). Biol Control 50:94–102

Shipp JL, Zhang Y, Hunt DWA, Ferguson G (2003) Influence of humidity and greenhouse microclimate on the efficacy of *Beauveria bassiana* (Balsamo) for control of greenhouse arthropod pests. Environ Entomol 32(5):1154–1163

Shiraishi H, Enami Y, Okano S (2003) *Folsomia hidakana* (Collembola) prevents damping-off disease in cabbage and Chinese cabbage by *Rhizoctonia solani*. Pedobiologia 47:33–38

Showler AT (1997) Proaction: strategic framework for today's reality. In: Krall S, Peveling R, Ba Diallo D (eds) New Strategies for Locust Control. Birkhauser, Basel, pp 461–465

Smith RJ, Grula EA (1982) Toxic components on the larval surface of the corn earworm (*Heliothis zea*) and their effects on the germination and growth of *Beauveria bassiana*. J Invertebr Pathol 39(1):15–22

Smits N, Fargues J, Rougier M, Goujet R, Itier B (1996) Effects of temperature and solar radiation interactions on the survival of quiescent conidia of the entomopathogenic hyphomycete *Paecilomyces fumosoroseus* (Wize) Brown and Smith. Mycopathologia 135(3):163–170

SosaGomez DR, Boucias DG, Nation JL (1997) Attachment of *Metarhizium anisopliae* to the southern green stink bug *Nezara viridula* cuticle and fungistatic effect of cuticular lipids and aldehydes. J Invertebr Pathol 69(1):31–39

Stewart WDP (1991) The importance of sustainable agriculture on biodiversity among invertebrates and microorganisms. In: Hawksworth DL (ed) The biodiversity of microorganisms and invertebrates, its role in sustainable agriculture. CAB International, Wallingford, pp 122–130

Storey GK, Gardner WA (1988) Movement of an aqueous spray of *Beauveria bassiana* into the profile of four Georgia soils. Environ Entomol 17:135–139

Storey GK, Gardner WA, Tollner EW (1989) Penetration and persistence of commercially formulated *Beauveria bassiana* conidia in soil of two tillage systems. Environ Entomol 18:835–839

Stotzky G (1972) Activity, ecology, and population dynamics of microorganisms in soil. CRC Crit Rev Microbiol 2(1):59–137

Thomas MB, Watson EL, Valverde-Garcia P (2003) Mixed infections and insect-pathogen interactions. Ecol Lett 6(3):183–188

Thompson SR, Brandeberg R, Arends JJ (2006) Impact of moisture and UV degradation on *Beauveria bassiana* (Balsamo) Vuillemin conidial viability in turfgrass. Biol Control 39:401–407

Thungrabeab M, Blaeser P, Sengonca C (2006) Effect of temperature and host plant on the efficacy of different entomopathogenic fungi from Thailand against *Frankliniella occidentalis* (Pergande) and *Thrips tabaci* Lindeman (Thysanoptera : Thripidae) in the laboratory. J Plant Dis Protect 113(4):181–187

Todorova S, Coderre D, Duchesne R-M, Cote J-C (1998) Compatibility of *Beauveria bassiana* with selected fungicides and herbicides. Biol Control 27(2):427–433

Tounou AK, Kooyman C, Douro-Kplndou OK, Poehling HM (2008) Interaction between *Paranosema locustae* and *Metarhizium anisopliae* var. *acridum*, two pathogens of the desert locust, *Schistocerca gregaria* under laboratory conditions. J Invertebr Pathol 97(3):203–210

Tugel A, Lewandowski A, Happe-von Arb D (eds) (2000) Soil Biology Primer. Soil and Water Conservation Society, Ankeny

Turnbow JH (1998) Temperature-sensitive *Beauveria bassiana* mycosis in the Mormon cricket, *Anabrus simplex*. MS Thesis. Montana State University, Bozeman

Ugine TA, Wraight SP, Sanderson JP (2007) A tritrophic effect of host plant on susceptibility of western flower thrips to the entomopathogenic fungus *Beauveria bassiana*. J Invertebr Pathol 96(2):162–172

Van der Valk H (2007) Review of the efficacy of *Metarhizium anisopliae* var. *acridum* against the desert locust plant production and protection division, locusts and other migratory pest group N. AGP/DL/TS/34. 81 pp. Available via http://www.fao.org/ag/locusts/common/ecg/1295/en/TS34e.pdf. Cited 10 Aug 2009

Vega FE, Dowd PF, Lacey LA, Pell JK, Jackson DM, Klein MG (2007) Dissemination of beneficial microbial agents by insects. In: Lacey LA, Kaya HK (eds) Field manual of techniques in invertebrate pathology, 2nd edn. Springer, New York, pp 127–146

Vesala T (1998) On the concept of leaf boundary layer resistance for forced convection. J Theor Biol 194:91–100

Vidal C, Fargues J, Rougier M, Smits N (2003) Effect of air humidity on the infection potential of hyphomycetous fungi as mycoinsecticides for *Trialeurodes vaporariorum*. Biocont Sci Technol 13(2):183–198

Watson DW, Mullens BA, Petersen JJ (1993) Behavioral fever response of *Musca domestica* (Diptera: Muscidae) to infection by *Entomophthora muscae* (Zygomycetes: Entomophthorales). J Invertebr Pathol 61(1):10–16

Willmer P (1986) Microclimate effects on insects at the plant surface. In: Juniper B, Southwood R (eds) Insects and the Plant Surface. E Arnold, New York, pp 65–80

Wojda I, Kowalski P, Jakubowicz T (2009) Humoral immune response of *Galleria mellonella* larvae after infection by *Beauveria bassiana* under optimal and heat-shock conditions. J Insect Physiol 55(6):525–531

Wraight SP, Carruthers RI (1999) Production, delivery and use of mycoinsecticides for control of insect pests of field crops. In: Hall FR, Menn JJ (eds) Methods in biotechnology, vol 5, Biopesticides: use and delivery. Humana Press, Totowa, New Jersey, pp 233–269

Wraight SP, Ramos ME (2002) Application parameters affecting field efficacy of *Beauveria bassiana* foliar treatments against Colorado potato beetle *Leptinotarsa decemlineata*. Biol Control 23:164–178

Wraight SP, Ramos ME (2005) Synergistic interaction between *Beauveria bassiana* and *Bacillus thuringiensis tenebrionis*-based biopesticides applied against field populations of Colorado potato beetle larvae. J Invertebr Pathol 90(3):139–150

Wraight SP, Carruthers RI, Bradley CA, Jaronski ST, Lacey LA, Wood P, Galaini-Wraight S (1998) Pathogenicity of the entomopathogenic fungi *Paecilomyces* spp. and *Beauveria bassiana* against the Silverleaf whitefly, *Bemisia argentifolii*. J Invertebr Pathol 71:217–226

Wraight SP, Carruthers RI, Jaronski ST, Bradley CA, Garza CJ, Galaini-Wright S (2000) Evaluations of the entomopathogenic fungi *Beauveria bassiana* and *Paecilomyces fumosoroseus* for the microbial control of the silverleaf whitefly, *Bemisia argentifolii*. Biol Control 17:203–217

Wraight SP, Ramos ME, Avery PB, Jaronski ST, Vandenberg JD (2009) Comparative virulence of *Beauveria bassiana* isolates against lepidopteran pests of vegetable crops. J Invertebr Pathol (in press)

Yang Y, Allen JC, Knapp JL, Stansly PA (1997) An age-structured population model of citrus rust mite: a fruit-mite-fungal pathogen system. Ecol Modeling 104(1):71–85

Yasuda K, Tokays K, Uehara K (1997) Effects of temperature, humidity and conidial density on infection by *Beauveria bassiana* of adult sweet potato weevil, *Cylas formicarius* (Fabricius) (Coleoptera: Curculionidae). Jap J Appl Entomol Zool 41(1):55–58

Zheng D, Hunt ER Jr, Running SW (1993) A daily soil temperature model based on air temperature and precipitation for continental applications. Clim Res 2:183–191

Zimmermann G (2007a) Review on safety of the entomopathogenic fungi *Beauveria bassiana* and *Beauveria brongniartii*. Biocont Sci Technol 17(5–6):553–596

Zimmermann G (2007b) Review on safety of the entomopathogenic fungus *Metarhizium anisopliae*. Biocont Sci Technol 17(9–10):879–920

Zimmermann G (2008) The entomopathogenic fungi *Isaria farinosa* (formerly *Paecilomyces farinosus*) and the *Isaria fumosorosea* species complex (formerly *Paecilomyces fumosoroseus*): biology, ecology and use in biological control. Biocont Sci Technol 18(9):865–901

Author Biography

Stefan Jaronski is a Research Entomologist with the Agricultural Research Service, US Department of Agriculture in Sidney Montana USA. His current research concerns development of microbial control of grasshoppers, and soil dwelling insect pests of sugar beets, part of bio-based integrated pest and plant pathogen management system. Before joining USDA in 2000, Dr. Jaronski spent 17 years in industry commercializing bacterial and fungal pest control agents, including *Beauveria bassiana* GHA. His research experience includes basic microbiology, mass production, formulation, bioassay systems, field trials, and regulatory aspects of entomopathogenic Hypocreales. He has a Ph.D. in Insect Pathology from Cornell University (1978).

BioControl (2010) 55:187–198
DOI 10.1007/s10526-009-9245-6

Conservation biological control using fungal entomopathogens

J. K. Pell · J. J. Hannam · D. C. Steinkraus

Received: 12 October 2009 / Accepted: 15 October 2009 / Published online: 17 November 2009

Abstract Conservation biological control relies on modification of the environment or management practices to protect and encourage natural enemies that are already present within the system, thereby enhancing and improving their ability to control pest populations in a reliable way. Such strategies are only possible when based on a strong understanding of the ecology of the species concerned at the individual, community and landscape scale. Conservation biological control with entomopathogenic fungi includes the manipulation of both the crop environment and also habitats outside the crop. Further investment in conservation biological control with entomopathogenic fungi could make a substantial contribution to sustainable crop production either as stand alone strategies or, more importantly, in support of other biological and integrated pest management strategies.

Handling Editor: Helen Roy.

J. K. Pell (✉)
Department of Plant and Invertebrate Ecology, Rothamsted Research, Harpenden, Hertfordshire AL5 2JQ, UK
e-mail: judith.pell@bbsrc.ac.uk

J. J. Hannam · D. C. Steinkraus
Department of Entomology, 319 AGRI, University of Arkansas, Fayetteville, AR 72701, USA

Keywords Ecology · Epizootiology · Entomophthorales · Hypocreales · Pest control · Conservation

Introduction

Unlike other biological control strategies, conservation biological control does not require the introduction or augmentation of natural enemies. Instead, it relies on modification of the environment or management practices to protect and encourage natural enemies that are already present within the system. This improves their ability to control pest populations in a reliable way and is only possible if the biology, behaviour and ecology of both the pests and their natural enemies are understood (Eilenberg et al. 2001; Hajek 2004; Pell 2007; Pimentel 2008). Unfortunately, for most entomopathogenic fungi, our understanding of their ecology and epizootiology is incomplete. The majority of examples of conservation biological control to date have been for arthropod natural enemies (e.g. Barbosa 1998; Gurr et al. 2004; Fiedler et al. 2008; Griffiths et al. 2008; Jonsson et al. 2008; Wade et al. 2008). However, similar approaches are relevant to entomopathogenic fungi where fungi are principal enemies of the target pest and where their ecology and epizootiology are understood (Fuxa 1998; Pell et al. 2001; Pell 2007; Tscharntke et al. 2008).

The entomopathogenic fungi are a diverse assemblage of fungi with one thing in common: they infect and cause disease in insects and other arthropods. Most are found within two main groups: the order Hypocreales within the phylum Ascomycota (subkingdom Dikarya) and the order Entomophthorales (Hibbett et al. 2007; Blackwell 2009).

Natural control by entomopathogenic fungi

Entomopathogenic fungi play major roles in the natural regulation of many insect and mite species. It is well known that they can develop dramatic epizootics that lead to rapid declines in host populations. In these systems the regularity and intensity of epizootics could be enhanced through conservation biological control and should be a primary target. However, in systems where fungi currently appear to have little regulatory impact on pest populations, there remains the possibility that this is as a result of management practices and could still be improved through conservation biological control.

Without doubt, the monetary value of un-manipulated, natural control of pests exerted by fungi worldwide is already substantial. Examples from the Entomophthorales include *Entomophaga grylli* that is capable of reducing destructive grasshopper outbreaks to negligible proportions in some years (MacLeod 1963). Gypsy moth (*Lymantria dispar*) populations in North America are regularly controlled by outbreaks of *Entomophaga maimaiga* (Hajek 1999), cotton aphid (*Aphis gossypii*) populations by *Neozygites fresenii* (Steinkraus 2007; Abney et al. 2008), various aphid populations by *Pandora neoaphidis* (Pell et al. 2001) and spider mite (*Tetranychus* spp.) populations on soybean by *Neozygites floridana* (Klubertanz et al. 1991). Examples from the Hypocreales are more commonly associated with hosts that spend some or all of their time in the soil, where hypocrealean fungi are ubiquitous. Epizootics of *Beauveria bassiana* in the scarab beetle *Costelytra zealandica* can reduce the host population by 99% (Townsend et al. 1995). *Nomuraea rileyi* has been shown to greatly reduce populations of *Pseudoplusia gemmatalis* overwintering in soil (Carruthers and Soper 1987) and *Tolypocladium cylindrosporum* severely reduces populations of *Agrotis segetum* dormant in soil (Steenberg and Ogaard 2000).

Although our understanding of the ecology and epizootiology of entomopathogenic fungi is often incomplete (Vega et al. 2009), a conservation biological control approach could have significant potential if we identified and filled the gaps in our ecological knowledge. By understanding the factors that promote or inhibit epizootic development, strategies can be identified that ensure favourable conditions for entomopathogenic fungi, and consequently reliable epizootics (Pell et al. 2001; Pell 2007). This is a significant challenge requiring an understanding of the persistence, transmission, dispersal and host range of fungi in cropped and semi-natural areas within managed ecosystems. Furthermore, these factors will vary considerably depending on the species of fungus and its life history strategy. As described by Hesketh et al. (2009) entomophthoralean and the anamorphic stages of hypocrealean fungi have contrasting life history attributes (with few exceptions). Entomophthoralean fungi are generally associated with foliar insect hosts, they are biotrophic and have limited host ranges. The soil is purely a reservoir environment in which their conidia and resting spores must persist, often when hosts are absent in the foliar environment above. In contrast, many hypocrealean fungi are hemibiotrophic, have broad host ranges and are associated with hosts that spend at least some of their life cycle in or on the soil. The soil is not just a reservoir environment in which they persist but also the habitat in which much of their lifecycle occurs, including multiplication within hosts and also, potentially, saprophytic growth. These differences have important implications for the conservation biological control approaches used.

Conservation biological control strategies applied within the crop; factors for consideration

Abiotic environment

Without doubt, high relative humidity is the most essential criterion for fungal activity. Ambient humidities in excess of 90% are usually required for germination, sporulation and infection (Tanada and Kaya 1993). Increasing the relative humidity through crop irrigation can, therefore, significantly enhance the activity of many entomopathogenic fungi. Clear examples of this come from species in the Entomophthorales, but the principle also applies to species in the Hypocreales where increasing humidity has been used widely to improve their efficacy in inundative control. Irrigation increased the prevalence of *P. neoaphidis* in aphid populations in alfalfa, field beans, pecan and spinach (Hall and Dunn 1957; Wilding et al. 1986; Pickering et al. 1989; McLeod and Steinkraus 1997). Prevalence of *Erynia ithacensis* in mushroom gnats was greatly increased by spraying water in the mushroom houses (Huang et al. 1992). Increasing relative humidity by water mists, irrigation and sprinkler systems is a relatively simple method that could be applicable for a wide range of crops, but can prove too costly if the value of the crop is low.

There are also methods to increase humidity without application of water. In crops that are harvested more than once, such as alfalfa, appropriate timing of the first cut allows manipulation of humidity and associated enhanced pest control by fungi. An early first cut concentrated weevil pests in the humid windrows where they were more likely to become infected by the fungal pathogen *Zoophthora phytonomi*. Although the profit from the first cut was reduced using this method, because the harvest was early, the resulting weevil control improved the yield from the second cut significantly (Nordin 1984; Brown and Nordin 1986).

All entomopathogenic fungi spend some part of their life cycle outside of their hosts, as conidia or resting structures (resting spores, chlamydospores, hyphal bodies), on leaf surfaces, bark and soil. In the phylloplane, conidia of all fungi are particularly susceptible to UV degradation (e.g. Furlong and Pell 1997; Fargues et al. 1996). Reducing the row spacing of crops, thereby increasing canopy cover, is one relatively simple method for improving protection from UV while simultaneously elevating ambient relative humidity. In the soybean system, Sprenkel et al. (1979) found higher prevalence rates of *N. rileyi* in lepidopteran larvae from plots that had been planted early, in narrow rows and at a high seed density compared to conventionally planted plots. If higher density plantings are compatible with other agronomic/economic aspects of soybean production, this represents a simple way to increase control by entomopathogenic fungi in soybean and may be applicable in other crops. Intercropping with plant species that increase canopy cover may also prove useful by raising ambient humidity.

Soil composition and disturbance

Soil structure, temperature, pH and water availability will all influence the species diversity and abundance of entomopathogenic fungi in soil and indeed which species may predominate in a given field or region (Klingen and Haukeland 2006; Meyling and Eilenberg 2007). Although it is important to recognize these influences, they are not open to easy manipulation for conservation biological control. However, physical and chemical perturbations due to tillage practices are open to manipulation within conservation biological control (Pell et al. 2001; Meyling and Eilenberg 2007; Pell 2007).

Any fungal propagule could be affected by tillage practices. This could have negative effects if fungal structures are buried deep within the soil where they would no longer be able to contact potential hosts or positive if they are moved closer to hosts at the surface, or if dispersal to new host populations is facilitated. Such factors can only be understood by detailed evaluation of each system and examples of this are scarce. However, Bing and Lewis (1993) found that *B. bassiana* infected more *Ostrinia nubilalis* in no-till corn crops than in corn from conventionally ploughed fields. In a different study, *B. bassiana*, *Metarhizium anisopliae*, and *Isaria* spp. were all more abundant in pest populations in no-till compared to tilled plots (Sosa-Gomez and Moscardi 1994). Furthermore, in a study in vegetable crops, the number of *G. mellonella* that became infected by *B. bassiana* and *M. anisopliae*, when used as bait on the soil, was significantly greater when they were exposed to soil from no-till rotations compared to soil that had been tilled (Hummel et al. 2002). In soybean crops in Brazil, selective media were used to compare the abundance of entomopathogenic fungi from tilled and no-till soils. This study found significantly more colony forming units (a measure of fungal abundance) in soil that had not been tilled compared to tilled soil. However, the number of colony forming units of the same pathogens in the canopy was not significantly different in the two systems. The reduced quantity of inoculum in the soil did not, therefore, translate into reduced exposure of insects inhabiting plants in the same field (Sosa-Gomez et al. 2001). Conservation tillage practices are now widely available and could enhance the level of control provided by entomopathogenic fungi above and below ground in a number of cropping systems, although this requires further evaluation. Within-crop strategies already employed to encourage arthropod natural enemies, such as mulching and beetle banks, may also have the potential to enhance the efficacy of entomopathogenic fungi, although this is as yet unproven (Meyling and Eilenberg 2007).

Pesticide applications

Applications of insecticides, fungicides and herbicides are a common component of crop management and these could impact entomopathogenic fungi in both the soil and foliar environment directly (by killing or inhibiting fungal propagules) and indirectly (by removing hosts) (Wekesa et al. 2008; Klingen and Haukeland 2006; Mochi et al. 2005; Morjan et al. 2002; Chandler et al. 1998; Lagnaoui and Radcliffe 1998; McLeod and Steinkraus 1997; Mietkiewski et al. 1997). Some of the effects are not easy to interpret, particularly when many studies have been done *in vitro* (Meyling and Eilenberg 2007). However, overall, Klingen and Haukeland (2006) suggested that insecticides and herbicides were less harmful than fungicides, although this was dependent on particular circumstances; insecticides may not be damaging directly but can remove hosts for subsequent transmission. Interestingly, the importance of entomopathogenic fungi for the control of pests is sometimes revealed in studies with fungicides. In a study on the green peach aphid (*Myzus persicae*), on potato, Ruano-Rossil et al. (2002) found that when fungicides were applied, extremely high aphid populations developed. They found that the fungicides were disrupting the natural control provided by *P. neoaphidis*, *Entomophthora planchoniana*, and *Conidiobolus obscurus*.

Reducing or targeting pesticide applications is the simplest way to mitigate any potential negative impacts and can be achieved by identifying and monitoring the activity of beneficial fungi in the crop, predicting their efficacy and thereby recommending when insecticides need not be applied (Pell et al. 2001; Pell 2007). The best example of this is for the entomophthoralean fungus *Neozygites fresenii* and cotton aphid control in the southern states of the USA. Studies in the USA showed that *N. fresenii* epizootics in *Aphis gossypii* occurred annually between June and August over wide areas of cotton production and that their occurrence could be predicted by diagnosis of aphid samples (Hollingsworth et al. 1995; Steinkraus et al. 1995). When fungus prevalence reached 15% in the aphid samples tested, declines caused by epizootics were certain within the week (within days if prevalence reached 50%) and recommendations could be made to farmers not to spray insecticides. Because this approach not only conserved fungal and insect natural enemies but also saved farmers money it has been widely adopted. An extension-based service to determine fungal prevalence and provide advice was established in 1993 in Arkansas and eventually covered Alabama, Florida, Georgia, Louisiana, Mississippi, Missouri, North Carolina, South Carolina and Tennessee (Steinkraus et al. 1998; Steinkraus and Zawislak 2005). Aphid samples submitted by farmers, extension agents, crop consultants and others were diagnosed and reports on prevalence provided to the senders. The diagnosis service provided detailed information via a website (http://www.uark.edu/misc/aphid) so that farmers could follow the spread of the fungus in their area and rationalise their pesticide use in response (Steinkraus et al. 1996, 1998; Steinkraus and Boys 1997).

Burning of crop residues

Very little research exists on the effects of fire on entomopathogenic fungi. However, it is likely that they could be important in areas where burning crop residues is still used as a management tool. Unfortunately, with such a limited body of literature on the topic, we can only make speculative hypotheses. Fire is known to reduce the activity of many plant pathogenic fungi (Hardison 1976). Some Entomophthorales, like *N. fresenii*, produce resting structures that persist on plant material, including crop stubble (Byford and Ward 1968). Presumably, in areas where fire is used to clear stubble, these resting structures would be destroyed. In Australia, aphid populations increased following controlled burns, suggesting either a direct positive effect of fire on aphid population growth or negative effects on their natural enemies, such as entomopathogenic fungi (Briese 1996). Fire could also have impacts on the soil environment, thereby indirectly affecting entomopathogenic fungi. After fire, soil can become hydrophobic (MacDonald and Huffman 2004), its pH rises (Hennig-Sever et al. 2001), and the soil nutrient composition changes (Kaufmann et al. 1994). There is some evidence that *M. anisopliae* may become more abundant in the soil from forests which have been burned (Bastias et al. 2008).

Extending conservation biological control strategies beyond the crop; the importance of reservoirs and complexity

Biological control must be effective in the crop environment and so conservation strategies that can be applied within the crop are an obvious first target. However, while some aspects of crop management can be modified to improve the efficacy of entomopathogenic fungi, many are not easy to modify. The soil structure and profile on a farm are controlled principally by the geology of the site, some level of tillage is essential and pesticides will need to be applied, even when integrated pest management strategies are practiced. Entomopathogenic fungi also need populations of hosts for their multiplication and, when these hosts are pests, a delicate balance between host and pathogen populations on the crop must be achieved. For these reasons, conservation biological control strategies that manage areas outside of the crop to encourage natural enemies have advantages. These semi-natural habitats can provide alternative hosts for multiplication of enemies and will not receive pesticide applications. Ensuring appropriate humidity and UV protection through canopy management could be easier and furthermore, they are semi-permanent and not tilled. However, the entomopathogenic fungi utilizing these resources must have the capacity to disperse from the reservoirs into adjacent crops. The value of dispersal potential has been identified for *F. virescens* infecting *Pseudaletia unipuncta* on undisturbed fescue. The same insect in surrounding wheat crops never became infected suggesting that the fungus may have been unable to disperse between habitats (Steinkraus et al. 1993). Although research in this area has been led by studies on arthropod natural enemies, as described previously, they are also extremely relevant for entomopathogenic fungi.

Alternative hosts as inoculum sources

Using a modelling approach to understand the parameters influencing epizootic development of *Entomophthora muscae* in the onion fly, *Delia antiqua*, the presence of a secondary host (the seed corn maggot, *D. platura*) in field border plants significantly increased the prevalence of *E. muscae* in *D. antiqua* on onions (Carruthers et al. 1985; Carruthers and Soper 1987). Elegant observational studies have also shown that hedgerows are important for the persistence and spread of *E. muscae* and *E. schizophorae* in other dipteran populations, e.g. carrot root fly, *Chamaepsila rosae* (Eilenberg 1985, 1988). The prevalence of fungus was always greater in carrot flies from hedges than from carrot fields. Hedges were the preferred sites for flies to rest and where infected flies died. This made the hedges important sites for transfer of conidia from one host to the next (Eilenberg 1987). Similar observations have been made in aphid populations in Switzerland (Keller and Suter 1980). Large populations of economically unimportant aphid species developing in meadows (lucerne and alfalfa) in the spring, were correlated with *P. neoaphidis* and *C. obscurus* rapidly achieving levels sufficient to regulate aphid populations in adjacent fields of annual crops. When aphids were scarce in the spring this did not happen suggesting that the presence of alternative aphid hosts in nearby meadows was critical. Grass and legume rich field margins and woodlands are also thought to have great potential as reservoirs for the aphid pathogen *P. neoaphidis* in South Africa (Hatting et al. 1999a, b). Other aphid pathogenic species, such as *Zoophthora aphidis, Z. phalloides* and *E. planchoniana*, are also known to overwinter in hosts in hedges and forest borders (Keller 1987a, b; Nielsen et al. 2001).

All the studies described above demonstrate the potential that managed habitats outside crops could have for pest control within the crop. However, the underpinning ecological data that would allow optimization in these systems is often incomplete and, where it is available, demonstrates the levels of ecological complexity that must be considered. An interesting case study on the potential utility of managed field margins to encourage *P. neoaphidis* is currently receiving significant attention and practical and ecological data sets in support of this are being collected and integrated. In Europe farmers receive subsidies for planting a diversity of field margins to encourage biodiversity. Some of these schemes have demonstrable benefits for particular arthropod natural enemy abundance, and in some cases relationships between arthropod natural enemy abundance in margins and pest suppression in adjacent crops has also been demonstrated (Collins et al. 2002; Powell et al. 2003; Holland 2007; Pell 2007). These margins could also be useful reservoirs of *P. neoaphidis* (and potentially other entomopathogens) if they contain plants that support alternative, susceptible aphid hosts throughout the season and if virulent isolates of the fungus could disperse from the margin into adjacent crops and initiate infection.

Pandora neoaphidis is an aphid specialist and has been recorded from numerous aphid species on crops, weeds and wildflowers (Pell et al. 2001). Laboratory bioassays against a range of pest aphid species identified considerable variability in susceptibility (e.g. Shah et al. 2004a). In these studies the pea aphid, *Acyrthosiphon pisum*, was the most susceptible pest aphid evaluated. Many non-crop legume plants are common in existing non-crop habitat mixes and also support *A. pisum* suggesting that the pea aphid could be a useful source of fungal inoculum when feeding in non-crop habitats and also a relatively easy target when on the crop. As a large species it would also produce more inoculum when dead than smaller species (Baverstock et al. 2005). In contrast *Rhopalosiphum padi*, an aphid pest on cereals, was far less susceptible than *A. pisum* and may therefore be less useful as a reservoir for *P. neoaphidis* in non-crop habitats and a harder target in the crop. It should be noted that these results were for a limited number of isolates and single biotypes of each aphid species. Biotypes of *A. pisum* can vary significantly in their susceptibility to *P. neoaphidis* (Ferrari et al. 2001), and infected *R. padi* have been recorded in the field (Pell et al. 2001) highlighting the complexity of the interactions (Pell 2007).

To avoid encouraging pest aphid species at field boundaries, non-pest aphids as sources of *P. neoaphidis* infection would be valuable. Ekesi et al. (2005) demonstrated that some non-pest aphids were also susceptible to infection: *Microlophium carnosum* a specialist on the perennial stinging nettle (*Urtica dioica*) was very susceptible. Furthermore, isolates of *P. neoaphidis* from field collected *M. carnosum* were virulent against a number of pest aphid species, indicating the potential for transmission from non-pest aphid reservoirs to pest aphids on crops (Shah et al. 2004a). This was confirmed by molecular studies that found no relationship between the aphid host from which an isolate originated and its host range (Tymon et al. 2004; Tymon and Pell 2005). *Microlophium carnosum* populations peak very early in the season (Perrin 1975) providing a source of *P. neoaphidis* for infection of adjacent crop aphids before their populations reach damaging levels. In food web studies of aphids and their fungal enemies in a natural meadow, aphids on nettles were identified as an important source of *P. neoaphidis* for infection of other aphid species (van Veen et al. 2008). Other hedgerow plants that support non-pest aphids throughout the season include hogweed (*Heraclium sphondylium*), teasel (*Dipsacus fullonum*) and bramble (*Rubus fruticosus*) and, therefore, also have potential as reservoirs for *P. neoaphidis* (Shah et al. 2004b). Such habitats could also be important for overwintering of *P. neoaphidis* as they are undisturbed and protected. *Pandora neoaphidis* is likely to persist in overwintering anholocyclic aphids via continuous cycles of infection and as conidia on the soil (Nielsen et al. 2007). *Pandora neoaphidis* remains able to infect aphids under simulated winter conditions and preliminary studies suggest that managed non-crop habitats with dense canopies also improve inoculum survival (Baverstock et al. 2008a).

Dispersal into crops from reservoirs

The studies above have identified plants that could be useful in supporting alternative hosts for *P. neoaphidis* and circumstantial evidence for the ability of *P. neoaphidis* to transmit between aphids in semi-natural habitats and crops. However, concrete evidence is required to confirm that this actually happens—a challenge in any conservation biological control approach. As with other entomophthoralean fungi, *P. neoaphidis* produces conidia that are actively discharged, leaving the boundary layer and entering the airstream (Hemmati et al. 2001a, b). Field studies have shown that they travel at least 20 m in the air, and probably considerably further, giving them the potential to move between distant habitats (Hemmati 1999). Sentinel aphids placed downwind from sources in the field and in polytunnel experiments became infected, demonstrating that conidia remained viable in the airstream, at least over short distances (Shah et al. 2004b; Ekesi et al. 2005). However, this mechanism of dispersal is entirely passive, diminishing chances of landing on a suitable host. More directed and long distance dispersal can occur through the movement of infected winged (alate) aphids between plants in response to overcrowding or during dispersal between primary and secondary host plants (e.g. Feng and Chen 2002; Feng et al. 2004). Of course, the movement of pest aphids from margins into crops, even if they are infected, is a significant trade off and would require careful consideration. Furthermore, there are other mechanisms of targeted dispersal that do not rely on the movement of aphids. In both laboratory and field studies, predators such as the ladybird *C. septempunctata*, become contaminated with conidia of *P. neoaphidis* while foraging on aphids on both crop and non-crop plants and are able to carry sufficient conidia to healthy aphid populations to initiate infection (Pell et al. 1997; Roy et al. 2001; Ekesi et al. 2005). As ladybirds also use non-crop habitats as reservoirs, particularly nettles, early in the year before moving into the crop, this represents a very important targeted mechanism of dispersal for *P. neoaphidis* both within and between non-crop and crop habitats as they will be carried with the predator that is actively seeking out aphid prey.

Considering the wider natural enemy community

Entomopathogenic fungi do not occur in isolation but within diverse guilds of natural enemies. The importance of considering the entire guild when developing conservation biological control is important but has not always been considered. For example, the predatory ladybird *C. septempunctata* will consume *P. neoaphidis*—infected aphids, inhibiting transmission (Pell et al. 1997; Roy et al. 1998, 2003). However, they can simultaneously significantly increase local transmission from sporulating cadavers which greatly outweighs the detrimental effect of feeding (Roy et al. 1998; Ekesi et al. 2005). This enhanced transmission in conjunction with passive vectoring of inoculum, as described previously, is likely to benefit *P. neoaphidis* significantly (Roy et al. 2001). Parasitoid wasps also enhance local transmission of *P. neoaphidis* although they do not contribute significantly to passive vectoring of inoculum (Fuentes-Contreras et al. 1998; Baverstock et al. 2008b, 2009a). However, in contrast to the fungus/ predator interaction, *P. neoaphidis* and parasitoids compete within individual aphid hosts: parasitoids take longer to develop than the fungus and so are often outcompeted in aphids that are already infected by fungi (Powell et al. 1986; Fuentes-Contreras et al. 1998; Furlong and Pell 2005). Parasitoids are also detrimentally affected by intraguild predation by predators such as ladybirds although some parasitoid species can recognise chemical trails produced by the predator and so avoid oviposition in aphid populations in which predators are foraging (Nakashima et al. 2004).

These outcomes are context specific, continually co-evolving and can be variable both for the enemies themselves and for overall aphid population control (Sunderland et al. 1998; Brodeur and Boivin 2006). From the point of view of aphid management, field studies demonstrate that different natural enemy groups are responsible for aphid control in different years (Sunderland et al. 1998) and, in the laboratory, that a combination of predators, parasitoids and *P. neoaphidis* has the greatest impact on aphid population suppression, although it can also lead to exclusion of some natural enemy species in the short term (Baverstock et al. 2009a). Furthermore, we know that susceptibility to *P. neoaphidis* varies amongst aphid species and biotypes (Shah et al. 2004a; Ferrari et al. 2001) but that the pathogen—resistant forms are attacked by predators and parasitoids. For these reasons it is a widely accepted belief that, for long term and resilient pest management, a diversity of natural enemies with contrasting requirements is required to deliver pest management in a constantly changing environment (Tscharntke et al. 2005, 2008; Pell 2007).

Conclusions and considerations for the future

Development of entomopathogenic fungi within conservation biological control strategies has received far less attention than their development for augmentation (Pell 2007). In the cases where conservation approaches have been considered, the focus has often been with the Entomophthorales because their epizootiology is generally better understood than the Hypocreales. Understanding the ecology of hypocrealean fungi in their favoured habitats and their relationships with above and below ground hosts would be a major step forward in untapping their potential (Bruck 2009; Cory and Ericsson 2009).

For example, *B. bassiana* is ubiquitous in soil but has recently been shown also to be common in the plant canopy, active against plant pathogens and even systemically active within plants, providing further opportunities for exploitation (Meyling and Eilenberg 2007; Pell 2007; Vega et al. 2009; Ownley et al. 2009). These aspects of their ecology would certainly improve the opportunities for exploitation in conservation biological control but should also underpin their use in other strategies (Roy et al. 2009; Jackson et al. 2009; Jaronski 2009; Hajek and Delalibera 2009; Baverstock et al. 2009b). It is certainly likely that if conservation approaches were used in conjunction with augmentation that the effectiveness of the augmentation strategy would be improved. There remain significant gaps in ecological understanding and examples of 'proof of concept' for conservation biological control with entomopathogenic fungi are rare. It is clear that for us to advance, greater investment in long-term, in depth studies, aimed at understanding the most important factors governing survival and spread of entomopathogenic fungi are essential. These studies should be coupled with replicated experimentation at the field and landscape scale to evaluate the strategies robustly.

A key factor for further study is the dispersal capability of entomopathogenic fungi between host populations and has particular relevance for conservation strategies in which the fungus multiplies outside of the crop and moves into the crop to be effective. Such studies would benefit from insights from theory, particularly metapopulation theory that considers populations linked by dispersal (Meyling and Hajek 2009) and modeling studies that seek to understand the role of reservoirs in pathogen population dynamics (Hesketh et al. 2009).

The implications of interactions between fungal and arthropod enemies and the requirement for diverse enemy guilds for resilient pest control should also receive more attention. When enemy interactions are complex and can have both positive and negative impacts, it is a challenge for conservation biological control but one that could be achieved by manipulating habitat diversity at a landscape scale. The particular requirements of each enemy group must be considered alongside the aspect and location of managed non-crop habitats and farm practices. The quantity and distribution of the various alternative habitats could have a significant effect on natural enemy diversity and pest management function and all these aspects require further study to link function at the individual level through to populations and communities at the field, farm and landscape scale. Such studies will become increasingly important as crop ecosystems respond to changes in climate and as new crops are introduced for other purposes (e.g. bioenergy).

The value of biodiversity in delivering a particular ecosystem function applies more generally than described above. There are undoubtedly many isolates and/or species of fungi that play as yet unknown roles in the regulation of pest populations. This may be because the systems in which they are active are understudied or because they are considered unimportant based on our existing knowledge of their ecology—but this could change as our climate and cropping landscapes change.

Conservation biological control with entomopathogenic fungi could make a substantial contribution to sustainable crop production, either as a stand alone strategy or, more importantly, in support of other biological and integrated management strategies. Its development and implementation must be underpinned by fundamental ecological understanding of the fungi concerned and their complex interactions with their hosts and the wider community at the individual to landscape scale. Although, like any control strategy, uptake will rely on economics and incentives (Gelernter 2005; Pell 2007; Cullen et al. 2008; Griffiths et al. 2008) the potential is there and warrants further investment.

Acknowledgements JKP was funded by the Department for Environment, Food and Rural Affairs of the United Kingdom (Defra) and the Biotechnology and Biological Sciences Research Council (BBSRC) of the United Kingdom. Rothamsted Research is an Institute of the BBSRC.

References

Abney MR, Ruberson JR, Herzog GA, Kring TJ, Steinkraus DC, Roberts PM (2008) Rise and fall of cotton aphid (Hemiptera: Aphididae) populations in southeastern cotton production systems. J Econ Entomol 101:23–35

Barbosa P (ed) (1998) Conservation biological control. Academic Press, San Diego, 396 pp

Bastias BA, Anderson IC, Rangel-Castro JI, Parkin PI, Prosser JI, Cairney JWG (2008) Influence of repeated prescribed burning on incorporation of ^{13}C from cellulose by soil

forest fungi as determined by RNA stable Isotope probing. Soil Biol Biochem 41:467–472
Baverstock J, Elliot SL, Alderson PG, Pell JK (2005) Response of the aphid pathogenic fungus *Pandora neoaphidis* to aphid-induced plant volatiles. J Invertebr Pathol 89: 157–164
Baverstock J, Clark SJ, Pell JK (2008a) Effect of seasonal abiotic conditions and field margin habitat on the infectivity and survival of *Pandora neoaphidis*. J Invertebr Pathol 97:282–290
Baverstock J, Baverstock KE, Clark SJ, Pell JK (2008b) Transmission of *Pandora neoaphidis* in the presence of co-occurring arthropods. J Invertebr Pathol 98:356–359
Baverstock J, Clark SJ, Alderson PG, Pell JK (2009a) Intraguild interactions between the entomopathogenic fungus *Pandora neoaphidis* and an aphid predator and parasitoid at the population scale. J Invertebr Pathol 102:167–172
Baverstock J, Roy HE, Pell JK (2009b) Entomopathogenic fungi and insect behaviour: from unsuspecting hosts to targeted vectors. BioControl. doi:10.1007/s10526-009-9238-5 (this SI)
Bing LA, Lewis LC (1993) Occurrence of the entomopathogen *Beauveria bassiana* (Balsamo) Vuillemin in different tillage regimes and in *Zea mays* L. and virulence towards *Ostrinia nubilalis* (Hübner). Agric Ecosys Environ 45: 147–156
Blackwell M (2009) Fungal evolution and taxonomy. BioControl. doi:10.1007/s10526-009-9243-8 (this SI)
Briese DT (1996) Biological control of weeds and fire management in protected natural areas: are they compatible strategies? Biol Conserv 77:135–141
Brodeur J, Boivin G (2006) Trophic and guild interactions in biological control. Springer, Berlin, 249 pp
Brown GC, Nordin GL (1986) Evaluation of an early harvest approach for induction of *Erynia* epizootics in alfalfa weevil populations. J Kansas Ent Soc 59:46–453
Bruck D (2009) Fungal entomopathogens in the rhizosphere. BioControl doi:10.1007/s10526-009-9236-7 (this SI)
Byford WJ, Ward LK (1968) Effect of the situation of the aphid host at death on the type of spore produced by *Entomphthora* spp. Trans Brit Mycol Soc 51:598–600
Carruthers RI, Soper RS (1987) Fungal diseases. In: Fuxa JR, Tanada Y (eds) Epizootiology of insect diseases. Wiley, New York, pp 357–416
Carruthers RI, Haynes DL, MacLeod DM (1985) *Entomophthora muscae* (Entomophthorales: Entomophthoraceae) in the onion fly, *Delia antiqua* (Diptera: Anthomyiidae). J Invertebr Pathol 45:81–93
Chandler D, Mietkiewski RT, Davidson G, Pell JK, Smits PH (1998) Impact of habitat type and pesticide application on the natural occurrence of entomopathogenic fungi in UK soils. IOBC-WPRS Bull 21:81–84
Collins KL, Boatman ND, Wilcox A, Holland JM, Chaney K (2002) Influence of beetle banks on cereal aphid population predation in winter wheat. Agric Ecosyst Environ 93:337–350
Cory J, Ericsson JD (2009) Fungal entomopathogens in a tritrophic context. BioControl. doi:10.1007/s10526-009-9247-4 (this SI)
Cullen R, Warner KD, Jonsson M, Wratten SD (2008) Economics and adoption of conservation biological control. Biol Contr 45:272–280
Eilenberg J (1985) Relationship between the carrot fly (*Psila rosae* F.) and its fungal pathogens from Entomophthorales, particularly *Entomophthora muscae* (C.) Fres. Ph.D. thesis, Department of Zoology, Royal Veterinary and Agricultural University, Copenhagen, Denmark, 109 pp
Eilenberg J (1987) Abnormal egg-laying behaviour of female carrot flies (*Psila rosae*) induced by the fungus *Entomophthora muscae*. Entomol Exp et Appl 52:17–24
Eilenberg J (1988) Occurrence of fungi from Entomophthorales in a population of carrot flies (*Psila rosae* F.). Results 1985 and 1986. IOBC Bull 11:53–59
Eilenberg J, Hajek AE, Lomer C (2001) Suggestions for unifying the terminology in biological control. BioControl 46:387–400
Ekesi S, Shah PA, Clark SJ, Pell JK (2005) Conservation biological control with the fungal pathogen *Pandora neoaphidis*; implications of aphid species, host plant and predator foraging. Agric For Entomol 7:21–30
Fargues J, Goettel MS, Smits N, Ouedraogo A, Vidal C, Lacey LA, Lomer CJ, Rougier M (1996) Variability in susceptibility to simulated sunlight of conidia among isolates of entomopathogenic hyphomycetes. Mycopath 135: 171–181
Feng M-G, Chen C (2002) Incidences of infected *Myzus persicae* alatae in flight imply place to place dissemination of entomophthoralean fungi in aphid populations through migration. J Invertebr Pathol 81:53–56
Feng M-G, Chen B, Chen C (2004) Wide dispersal of aphid-pathogenic Entomophthorales among aphid relies on migratory alates. Enviro Microbiol 6:510–516
Ferrari J, Müller CB, Kraaijeveld AR, Godfray HCJ (2001) Clonal variation and covariation in aphid resistance to parasitoids and a pathogen. Evolution 55:1805–1814
Fiedler AF, Landis DA, Wratten SD (2008) Maximizing ecosystem services from conservation biological control: the role of habitat management. Biol Contr 45:254–271
Fuentes-Contreras E, Pell JK, Niemeyer HM (1998) Influence of plant resistance at the third trophic level: interactions between parasitoids and entomopathogenic fungi of cereal aphids. Oecologia 117:426–432
Furlong MJ, Pell JK (1997) The influence of environmental factors on the persistence of *Zoophthora radicans* conidia. J Invertebr Pathol 69:223–233
Furlong MJ, Pell JK (2005) Interactions between fungal entomopathogens and other beneficial organisms. In: Vega FE, Blackwell M (eds) Insect–fungal associations ecology and evolution pp 51–73
Fuxa JR (1998) Environmental manipulation for microbial control of insects. In: Barbosa P (ed) Conservation biological control. Academic Press, San Diego, pp 255–268
Gelernter WD (2005) Biological control products in a changing landscape. In: Proceedings of the BCPC international congress—crop science and technology. Glasgow, pp 293–300
Griffiths GJK, Holland JM, Bailey A, Thomas MB (2008) Efficacy and economics of shelter habitats for conservation biological control. Biol Contr 45:200–209
Gurr GM, Wratten SD, Altieri MA (eds) (2004) Ecological engineering for pest management: advances in habitat manipulation for arthropods. CSIRO Publishing, Australia, 232 pp

Hajek AE (1999) Pathology and epizootiology of *Entomophaga maimaiga* infections in forest lepidoptera. Microbiol Mol Biol Rev 63:814–935

Hajek AE (2004) Natural enemies. An introduction to biological control. Cambridge Univ. Press, Cambridge

Hajek AE, Delalibera I (2009) Fungal pathogens as classical biological control agents against arthropods. BioControl. doi:10.1007/s10526-009-9253-6 (this SI)

Hall IM, Dunn PH (1957) Entomogenous fungi on the spotted alfalfa aphid. Hilgardia 27:159–181

Hardison JR (1976) Fire and flame for plant disease control. Ann Rev Phytopathol 14:359–379

Hatting JL, Humber RA, Poprawski TJ, Miller RM (1999a) A survey of fungal pathogens of aphids from South Africa, with special reference to cereal aphids. Biol Contr 16:1–12

Hatting JL, Poprawski TJ, Miller RM (1999b) Managing the entomopathogenic fungus *Conidiobolus thromboides* in Russian wheat aphid cultures. Southwest Entomol 24: 99–106

Hemmati F (1999) Aerial dispersal of the entomopathogenic fungus *Erynia neoaphidis*. PhD Thesis, University of Reading, 186 pp

Hemmati F, Pell JK, McCartney HA, Deadman ML (2001a) Airborne concentrations of conidia of *Erynia neoaphidis* above cereal fields. Mycol Res 105:485–489

Hemmati F, Pell JK, McCartney HA, Deadman ML (2001b) Aerodynamic diameter of conidia of *Erynia neoaphidis* and other entomophthoralean fungi. Mycol Res 106:233–238

Hennig-Sever N, Poliakov D, Broza M (2001) A novel method for estimation of wild fire intensity based on ash pH and soil microarthropod community. Pedobiologia 45:98–106

Hesketh H, Roy HE, Eilenberg J, Pell, JK, Hails RS (2009) Challenges in modelling complexity of fungal entomopathogens in semi-natural populations of insects. BioControl. doi:10.1007/s10526-009-9249-2 (this SI)

Hibbett DS, Binder M, Bischoff JF, Blackwell M, Cannon PF, Eriksson OE, Huhndorf S, James T, Kirk PM, Lücking R, Thorsten Lumbsch H, Lutzoni F, Brandon Matheny P, McLaughlin DJ, Powell MJ, Redhead S, Schoch CL, Spatafora JW, Stalpers JA, Vilgalys R, Aime MC, Aptroot A, Bauer R, Begerow D, Benny GL, Castlebury LA, Crous PW, Dai YC, Gams W, Geiser DM, Griffith GW, Gueidan C, Hawksworth DL, Hestmark G, Hosaka K, Humber RA, Hyde KD, Ironside JE, Kõljalg U, Kurtzman CP, Larsson K-H, Lichtwardt R, Longcore J, Miądlikowska J, Miller A, Moncalvo JM, Mozley-Standridge S, Oberwinkler F, Parmasto E, Reeb V, Rogers JD, Roux C, Ryvarden L, Sampaio JP, Schüßler A, Sugiyama J, Thorn RG, Tibell L, Untereiner WA, Walker C, Wang Z, Weir A, Weiss M, White MM, Winka K, Yao YJ, Zhang N (2007) A higher-level phylogenetic classification of the Fungi. Mycol Res 111:509–547

Holland JM (2007) The potential of agri-environment schemes to enhance biocontrol in arable crops. Asp Appl Biol 81:127–134

Hollingsworth RG, Steinkraus DC, McNew RW (1995) Sampling to predict fungal epizootics on cotton aphids (Homoptera: Aphididae). Environ Entomol 24:1414–1421

Huang Y, Zhen B, Li Z (1992) Natural and induced epizootics of *Erynia ithacensis* in mushroom hothouse populations of yellow-legged fungus gnats. J Invertebr Pathol 60:254–258

Hummel RL, Walgenbach JF, Barbercheck ME, Kennedy GG, Hoyt GD, Arellano C (2002) Effects of production practices on soil-borne entomopathogens in western North Carolina vegetable systems. Environ Entomol 31:84–91

Jackson M, Dunlap CA, Jaronski S (2009) Ecological considerations in producing and formulating fungal entomopathogens for use in insect biocontrol. BioControl. doi: 10.1007/s10526-009-9240-y (this SI)

Jaronski S (2009) Ecological factors in the inundative use of fungal entomopathogens. BioControl. doi:10.1007/s10526-009-9248-3 (this SI)

Jonsson M, Wratten SD, Landis DA, Gurr GM (2008) Recent advances in conservation biological control of arthropods by arthropods. Biol Contr 45:172–175

Kaufmann JB, Cummings DL, Ward DE (1994) Relationships of fire, biomass and nutrient dynamics along a vegetation gradient in the Brazilian cerrado. J Ecol 82:519–531

Keller S (1987a) Die Bedeutung ökologischer ausgleichsflächen für den Pflanzenschutz. Mitt Schweiz Land 35:56–65

Keller S (1987b) Observations on the overwintering of *Entomophthora planchoniana*. J Invertebr Pathol 50:333–335

Keller S, Suter H (1980) Epizootiologische Untersuchungen über das Entomophthora-Auftreten bei feldbaulich wichtigen Blattlausarten. Oecologica Applicata 1:63–81

Klingen I, Haukeland S (2006) The soil as a reservoir for natural enemies of pest insects and mites with emphasis on fungi and nematodes. In: Eilenberg J, Hokkanen HMT (eds) An ecological and societal approach to biological control. Springer, Berlin, pp 145–212

Klubertanz TH, Pedigo LP, Carlson RE (1991) Impact of fungal epizootics on the biology and management of the twospotted spider mite (Acari: Tetranychidae) in soybean. Environ Entomol 20:731–735

Lagnaoui A, Radcliffe EB (1998) Potato fungicides interfere with entomopathogenic fungi impacting population dynamics of green peach aphid. Amer J Pot Res 75:19–25

MacDonald LH, Huffman EL (2004) Post-fire soil water repellency. Soil Sci Soc Amer J 68:1729–1734

MacLeod DM (1963) Entomophthorales infections. In: Steinhaus EA (ed) Insect pathology, an advanced treatise, vol 2. Academic Press, London, pp 189–231

McLeod PJ, Steinkraus DS (1997) Influence of irrigation and fungicide sprays on prevalence of *Erynia neoaphidis* (Entomophthorales: Entomophthoraceae) infections of green peach aphid (Homoptera: Aphididae) on spinach. J Agric Urban Entomol 16:279–284

Meyling NV, Eilenberg J (2007) Ecology of the entomopathogenic fungi *Beauveria bassiana* and *Metarhizium anisopliae* in temperate agroecosystems: potential for conservation biological control. Biol Contr 43:145–155

Meyling NV, Hajek AE (2009) Principles from community and metapopulation ecology: application to fungal entomopathogens. BioControl. doi:10.1007/s10526-009-9246-5 (this SI)

Mietkiewski RT, Pell JK, Clark SJ (1997) Influence of pesticide use on the natural occurrence of entomopathogenic fungi in arable soils in the UK: field and laboratory comparisons. Biocontr Sci Technol 7:565–575

Mochi DA, Monteiro AC, Barbosa JC (2005) Actions of pesticides to *Metarhizium anisopliae* in soil. Neotrop Entomol 34:961–971

Morjan WE, Pedigo LP, Lewis LC (2002) Fungicidal effects of glyphosate and glyphosate formulations on four species of entomopathogenic fungi. Environ Entomol 31:1206–1212

Nakashima Y, Birkett MA, Pye BJ, Pickett JA, Powell W (2004) The role of semiochemicals in the avoidance of the seven-spot ladybird, *Coccinella septempunctata* by the aphid parasitoid, *Aphidius ervi*. J Chem Ecol 30:1097–1110

Nielsen C, Eilenberg J, Dromph KM (2001) Entomophthorales on cereal aphids: characterisation, growth, virulence, epizootiology and potential for microbial control. Ministry of the Environment, Copenhagen, 58 pp

Nielsen C, Jensen AB, Eilenberg J (2007) Survival of entomophthoralean fungi infecting aphids and higher flies during unfavourable conditions and implications for conservation biological control. In: Ekesi S, Maniania N (eds) Use of entomopathogenic fungi in biological pest management. Research Signpost, Kerala, India, pp 13–38

Nordin GL (1984) Enhancement strategies for entomogenous fungi in integrated pest management. In: Proceedings of the China national academy of science/United States national academy of science joint symposium on biological control of insects, Beijing, China (pp 122–141). Science Press, Beijing, 445 pp

Ownley B, Gwinn KD, Vega FE (2009) Endophytic fungal entomopathogens with activity against plant pathogens: ecology and evolution. BioControl. doi:10.1007/s10526-009-9241-x (this SI)

Pell JK (2007) Ecological approaches to pest management using entomopathogenic fungi; concepts, theory, practice and opportunities. In: Ekesi S, Maniania N (eds) Use of entomopathogenic fungi in biological pest management. Research Signpost, Kerala, India, pp 145–177

Pell JK, Pluke R, Clark SJ, Kenward MG, Alderson PG (1997) Interactions between two aphid natural enemies, the entomopathogenic fungus, *Erynia neoaphidis* and the predatory beetle, *Coccinella septempunctata*. J Invertebr Pathol 69:261–268

Pell JK, Eilenberg J, Hajek AE, Steinkraus DC (2001) Biology, ecology and pest management potential of Entomophthorales. In: Butt TM, Jackson C, Magan N (eds) Fungi as biocontrol agents, progress, problems and potential. CABI Publishing, Wallingford, pp 71–153

Perrin RM (1975) The role of perennial stinging nettle, *Urtica dioica*, as a reservoir of beneficial natural enemies. Ann Appl Biol 81:289–297

Pickering J, Dutcher JD, Ekbom BS (1989) An epizootic caused by *Erynia neoaphidis* and *E. radicans* (Zygomycetes: Entomophthoraceae) on *Acrythosiphon pisum* (Homoptera: Aphidae) on legumes under overhead irrigation. J App Entomol 107:331–333

Pimentel D (2008) Preface special issue: conservation biological control. Biol Contr 45:171

Powell W, Wilding N, Brobyn PJ, Clark SJ (1986) Interference between parasitoids (Hym, Aphidiidae) and fungi (Entomophthorales) attacking cereal aphids. Entomophaga 31:193–199

Powell W, Walters K, A'Hara S, Ashby J, Stevenson H, Northing P (2003) Using field margin diversification in agri-environment schemes to enhance aphid natural enemies. In: Rossing WAH, Poehling HM, Burgio G (eds) Landscape management for functional biodiversity. IOBC/WPRS Bulletin 26/4:123–128

Roy HE, Pell JK, Clark SJ, Alderson PG (1998) Implications of predator foraging on aphid pathogen dynamics. J Invertebr Pathol 71:236–247

Roy HE, Pell JK, Alderson PG (2001) Targeted dispersal of the aphid pathogenic fungus *Erynia neoaphidis* by the aphid predator *Coccinella septempunctata*. Biocon Sci Technol 11:99–110

Roy HE, Alderson PG, Pell JK (2003) Effect of spatial heterogeneity on the role of *Coccinella septempunctata* as an intra-guild predator of the aphid pathogen *Erynia neoaphidis*. J Invertebr Pathol 82:85–95

Roy HE, Brodie EL, Chandler D, Goettel MS, Pell JK, Wajnberg E, Vega FE (2009) Deep space and hidden depths: understanding the evolution and ecology of fungal entomopathogens. BioControl. doi:10.1007/s10526-009-9244-7 (this SI)

Ruano-Rossil JM, Radcliffe EB, Ragsdale DW (2002) Disruption of entomopathogenic fungi of green peach aphid, *Myzus persicae* (Sulzer), by fungicides used to control potato late blight. In: Simon JC, Dedryver CA, Rispe C, Hullé M (eds) Aphids in a New Millenium. Institut National de la Recherche Agronomique, Paris, pp 365–370

Shah PA, Clark SJ, Pell JK (2004a) Assessment of aphid host susceptibility and isolate variability in *Pandora neoaphidis* (Zygomycota: Entomophthorales). Biol Contr 29:90–99

Shah PA, Tymon AM, Pell JK (2004b) Novel strategies for aphid control using entomopathogenic fungi HGCA. Project Report Sustainable Arable LINK Project 1159. London, HGCA, 92 pp

Sosa-Gomez DR, Moscardi F (1994) Effect of till and no-till soybean cultivation on dynamics of entomopathogenic fungi in the soil. Fla Entomol 77:284–287

Sosa-Gomez DR, Delpin KE, Moscardi F, Farias JRB (2001) Natural occurrence of the entomopathogenic fungi *Metarhizium, Beauveria,* and *Paecilomyces* in soybean under till and no-till cultivation systems. Neotrop Entomol 30:407–410

Sprenkel RK, Brooks WM, Van Duyn JW, Deitz LL (1979) The effects of three cultural variables on the incidence of *Nomuraea rileyi*, phytophagous Lepidoptera, and their predators on soybeans. Environ Entomol 8:334–339

Steenberg T, Ogaard L (2000) Mortality in hibernating turnip moth larvae, *Agrotis segetum*, caused by *Tolypocladium cylindrosporum*. Mycol Res 104:87–91

Steinkraus DC (2007) Documentation of naturally-occurring pathogens and their impact in agroecosystems. In: Lacey LA, Kaya HK (eds) Field manual of techniques in invertebrate pathology. Springer, Dordrecht, pp 267–281

Steinkraus DC, Boys GO (1997) Update on prediction of epizootics with extension-based sampling service (pp 1047–1048). In: Proceedings of Beltwide cotton conferences, Jan 6–10 1997, New Orleans. Nat Cotton Council Am, Memphis, TN

Steinkraus DC, Zawislak J (2005) Utilizing natural biological control for cotton aphids: cotton aphid fungus sampling service twelth year. Summ Ark Cotton Res 220–223

Steinkraus DC, Mueller AJ, Humber RA (1993) *Furia virescens* (Thaxter) Humber (Zygomycetes: Entomophthoraceae) infections in the armyworm, *Pseudaletia unipuncta*

(Haworth) (Lepidoptera: Noctuidae) in Arkansas with notes on other natural enemies. J Entomol Sci 28:376–386

Steinkraus DC, Hollingsworth RG, Slaymaker PH (1995) Prevalence of *Neozygites fresenii* (Entomophthorales: Neozygitaceae) on cotton aphids (Homoptera: Aphididae) in Arkansas cotton. Environ Entomol 24:465–474

Steinkraus DC, Boys GO, Hollingsworth RG, Bacheler JS, Durant JA, Freeman BL, Gaylor MJ, Harris FA, Knutson A, Lentz GL, Leonard BR, Luttrell R, Parker D, Powell JD, Ruberson JR, Sorenson C (1996) Multistate sampling for *Neozygites fresenii* in cotton. In: Proceedings 1996 Beltwide cotton research conferences, Nashville, TN, pp 735–738

Steinkraus DC, Boys GO, Bagwell RD, Johnson DR, Lorenz GM, Meyers H, Layton MB, O'Leary PF (1998) Expansion of extension-based aphid fungus sampling service to Louisiana and Mississippi. 1998 Beltwide Cotton Conferences, San Diego, CA, pp 1239–1242

Sunderland KD, Axelsen JA, Dromph K, Freier B, Hemptinne J-L, Holst NH, Mols PJM, Petersen MK, Powell W, Ruggle P, Triltsch H, Winder L (1998) Pest control by a community of natural enemies. Acta Jutlandica 72:271–326

Tanada Y, Kaya HK (1993) Insect pathology. Academic Press, London, 666 pp

Townsend RJ, Glare TR, Willoughby BE (1995) The fungi *Beauveria* spp. cause epizootics in grass grub populations in Waikato. In: Proceedings of the 48th New Zealand plant protection conference, pp 237–241

Tscharntke T, Klein AM, Kruess A, Steffan-Dewenter I, Thies C (2005) Landscape perspectives on agricultural intensification and biodiversity—ecosystem service management. Ecol Let 8:857–874

Tscharntke T, Bommarco R, Clough Y, Crist TO, Kleijn D, Rand TA, Tylianakis JM, van Nouhuys S, Vidal S (2008) Conservation biological control and enemy diversity on a landscape scale. Biol Contr 45:238–253

Tymon AM, Pell JK (2005) Use of ISSR, ERIC and RAPD techniques to detect genetic diversity in the aphid pathogen *Pandora neoaphidis*. Mycol Res 109:285–290

Tymon AM, Shah PA, Pell JK (2004) PCR-based molecular discrimination of *Pandora neoaphidis* isolates from related entomopathogenic fungi and development of species-specific diagnostic primers. Mycol Res 108:1–15

van Veen FJF, Muller CB, Pell JK, Godfray HCJ (2008) Food web structure of three guilds of natural enemies: predators, parasitoids and pathogens of aphids. J Anim Ecol 77:191–200

Vega FE, Goettel MS, Blackwell M, Jackson MA, Keller S, Koike M, Maniania NK, Monzón A, Ownley B, Pell JK, Rangel D, Roy HE (2009) Fungal entomopathogens: new insights on their ecology. Fungal Ecol 2:149–159

Wade MR, Zalucki MP, Wratten SD, Robinson KA (2008) Conservation biological control of arthropods using artificial food sprays: current status and future challenges. Biol Contr 45:185–199

Wekesa VW, Knapp M, Delalibera I (2008) Side-effects of pesticides on the life cycle of the mite pathogenic fungus *Neozygites floridana*. Exp Appl Acarol 46:287–297

Wilding N, Mardell SK, Brobyn PJ (1986) Introducing *Erynia neoaphidis* into a field population of *Aphis fabae*: form of the inoculum and effect of irrigation. Ann Appl Biol 108:373–385

GPSR Compliance

The European Union's (EU) General Product Safety Regulation (GPSR) is a set of rules that requires consumer products to be safe and our obligations to ensure this.

If you have any concerns about our products, you can contact us on ProductSafety@springernature.com

In case Publisher is established outside the EU, the EU authorized representative is:

Springer Nature Customer Service Center GmbH
Europaplatz 3
69115 Heidelberg, Germany

Batch number: 10762370

Printed by Printforce, the Netherlands